Ergebnisse der Mathematik und ihrer Grenzgebiete

3. Folge · Band 14

A Series of Modern Surveys in Mathematics

Mark Goresky Robert MacPherson

Stratified Morse Theory

Springer-Verlag
Berlin Heidelberg NewYork
London Paris Tokyo

Mark Goresky
Department of Mathematics
Northeastern University
Boston, MA 02115, USA

Robert MacPherson
Department of Mathematics
Massachusetts Institute of Technology
Cambridge, MA 02139, USA

Mathematics Subject Classification (1980): 14F35, 57D70

ISBN 3-540-17300-5 Springer-Verlag Berlin Heidelberg New York
ISBN 0-387-17300-5 Springer-Verlag New York Berlin Heidelberg

Library of Congress Cataloging in Publication Data
Goresky, Mark, 1950–. Stratified Morse theory.
(Ergebnisse der Mathematik und ihrer Grenzgebiete; 3. Folge, Bd. 14)
Bibliography: p. Includes index.
1. Morse theory. 2. Analytic spaces. 3. Topology.
I. MacPherson, Robert, 1944–. II. Title. III. Series.
QA331.G655 1988 515.7′3 87-26418
ISBN 0-387-17300-5

Printed in Germany
Typesetting, printing and binding: Universitätsdruckerei H. Stürtz AG, 8700 Würzburg
2141/3140-543210

To René Thom

who contributed three essential steps to the story presented here

- the idea that a Morse function leads to a cell decomposition (1949)
- the idea of studying complex varieties using Morse theory (1957)
- the isotopy lemmas of stratification theory (1969)

Preface

This book explores the natural generalization of Morse theory to stratified spaces. Applications are given, primarily to the topology of complex analytic varieties. The main theorems are proven here for the first time, although they are heirs to a long line of historical development (see Sect. 1.7 and Sect. 2.8 of the introduction and Sect. 1.0 and Sect. 2.0 of Part I).

The work presented here was first announced in 1980 [GM1]. The original proofs were discouragingly complicated and technical. During the intervening years we have developed methods which have greatly simplified the arguments, while at the same time making them more geometric (see Chap. 4 and Chap. 5 of Part I).

Conversations with P. Deligne, R. Lazarsfeld, and especially W. Fulton about potential applications to complex varieties were instrumental in persuading us to take up the study of stratified Morse theory. We have also profited from valuable conversations with G. Bartel, L. Kaup, P. Orlik, P. Schapira, B. Teissier, R. Thom, R. Thomason, K. Vilonen, H. Hamm, Lê D.T., D. Massey, W. Pardon, and D. Trotman. We would like to thank these last five, and particularly D. Trotman, for their very careful reading of portions of several preliminary versions of this manuscript.

We received support from several research institutions while this book was being written, including the Eidgenössische Technische Hochschule (Zürich), the Ecole Normale Supérieure (Paris) and the National Science Foundation (grants # DMS 850-2422 and # DMS 820-1680). We would particularly like to thank the Consiglio Nazionale delle Ricerche of Italy, the University of Rome (La Sapienza) and the University of Rome II (Tor Vergata) for their support and hospitality during 1985 and the Institut des Hautes Etudes Scientifiques (Paris) for their support and hospitality during 1981 and 1986.

October 1987 M. Goresky R. MacPherson

Table of Contents

Part II. Morse Theory of Complex Analytic Varieties

Part III. Complements of Affine Subspaces

Introduction

This book contains mathematical results from three distinct subject areas. These correspond to the three parts of the book. Part I contains a systematic exploration of the natural extension of Morse theory to include singular spaces. Part II gives a large collection of theorems on the topology of complex analytic varieties. Part III presents the calculation of the homology of the complement of a collection of flat subspaces of Euclidean space.

The reason for including these three disparate subject areas in one volume is that the results of the second and the third are proved by applying the Morse theory of the first. However, the statements of the results themselves are independent from one part to another. Also the three subject areas may be of interest to different sets of readers. For these reasons, this introduction is written in completely independent chapters. Anyone interested mainly in the topology of complex analytic varieties can skip now to Chap. 2 of the introduction, p. 23. Readers interested in flat subspaces of Euclidean space may skip to Chap. 1 of Part III of this book, p. 237.

Chapter 1. Stratified Morse Theory

Suppose that X is a topological space, f is a real valued function on X, and c is a real number. Then we will denote by $X_{\leq c}$ the subspace of points x in X such that $f(x) \leq c$. The fundamental problem of Morse theory is to study the topological changes in the space $X_{\leq c}$ as the number c varies.

1.1. Morse-Smale Theory

In classical Morse theory, the space X is taken to be a compact differentiable manifold. This is best illustrated by the following standard diagram: Consider a two-dimensional torus $\mathscr{T}$ embedded in three-dimensional Euclidean space.

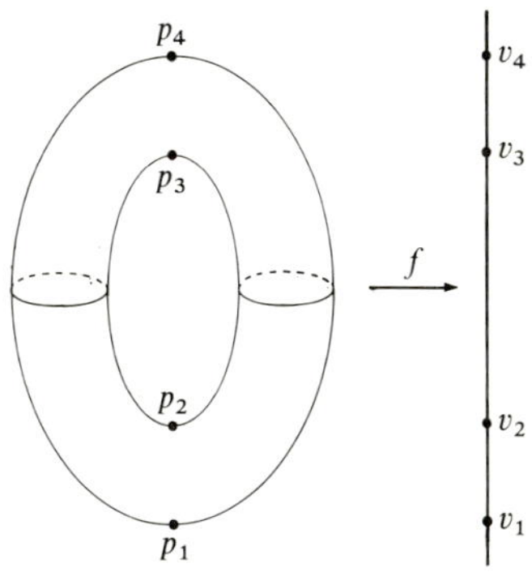

Let f be the projection onto the vertical coordinate axis. So, $f(x)$ measures the height of the point x. For any real number c, the subspace $\mathscr{T}_{\leq c}$ is the wet part after the torus has been filled with water to height c.

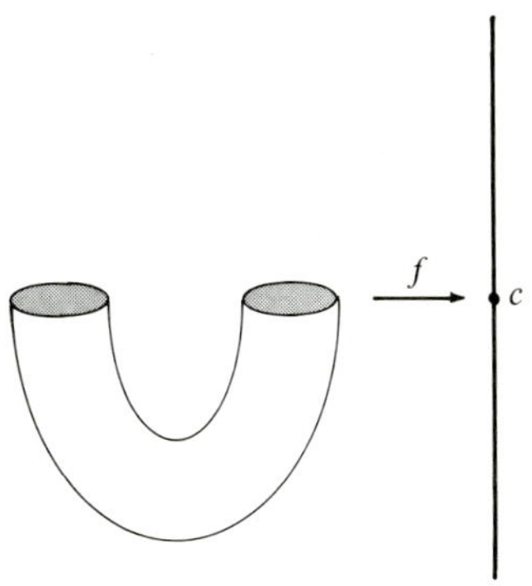

We imagine slowly increasing c and we watch how the topology of $\mathcal{T}_{\leq c}$ changes. We observe that it changes only when c crosses one of the four critical values $v_1, \ldots, v_4$ corresponding to the critical points $p_1, \ldots, p_4$. (The *critical points* of a differentiable function on a smooth manifold X are the points where the differential df of f vanishes. The *critical values* are the values f takes at the critical points.) This observation about $\mathcal{T}$ illustrates Part A of the fundamental result of classical Morse theory:

Theorem (CMT Part A). *Let f be a differentiable function on a compact smooth manifold X. As c varies within the open interval between two adjacent critical values, the topological type of $X_{\leq c}$ remains constant.*

Next, we want to examine the way in which the topological type of $\mathcal{T}_{\leq c}$ changes as c crosses one of the critical values v_i. If c is less than v_1, then $\mathcal{T}_{\leq c}$ is empty. As c crosses v_1, the space $\mathcal{T}_{\leq c}$ changes by adding a two-disk (shaped like a bowl). As c crosses v_2, the space $\mathcal{T}_{\leq c}$ is changed by gluing in a rectangle along two opposite edges.

Crossing the critical value v_2

As c crosses v_3, another rectangle is glued in along two opposite edges.

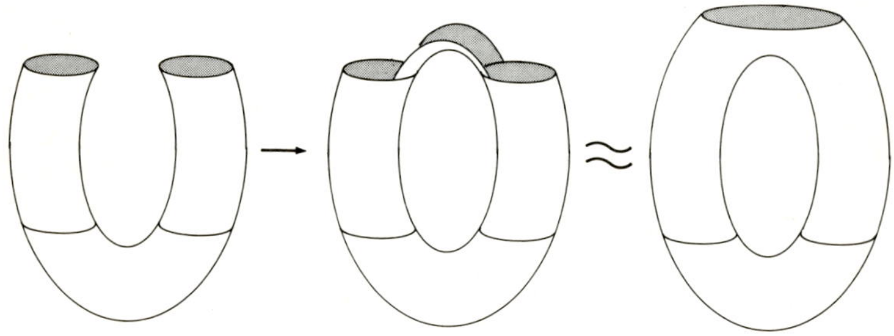

Crossing the critical value v_3

Finally, as c crosses v_4, a two-disk (shaped like a cap) is glued in along its boundary, thus completing $\mathcal{T}$.

We define *Morse data* for a function f at a critical point p in a space X to be a pair of topological spaces (A, B) where $B \subset A$ with the property that as c crosses the critical value $v = f(p)$, the change in $X_{\leq c}$ can be described by gluing in A along B. The descriptions above of the changes in $\mathcal{T}_{\leq c}$ may be summarized by the following table of Morse data for $\mathcal{T}$:

Critical point	Morse data (A, B)
p_1	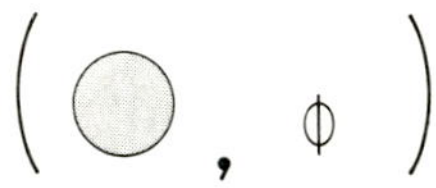 $=(D^0 \times D^2, \partial D^0 \times D^2)$
p_2 or p_3	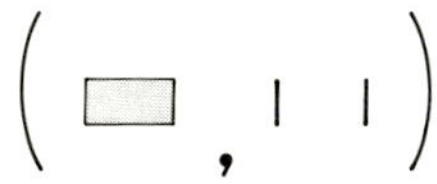 $=(D^1 \times D^1, \partial D^1 \times D^1)$
p_4	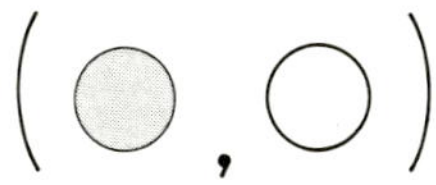 $=(D^2 \times D^0, \partial D^2 \times D^0)$

Here, D^i denotes the closed i-dimensional disk and ∂^i denotes its boundary $i-1$ sphere. (Note that 0-disk is a point and that its boundary is empty.)

This table of Morse data for $\mathscr{T}$ illustrates Part B of the fundamental result of classical Morse theory:

Theorem (CMT Part B). *Let f be a Morse function (see Sect. 1.3 of the introduction) on a smooth manifold X. Morse data measuring the topological change in $X_{\leq c}$ as c crosses the critical value v of the critical point p is given by the "handle" $(D^\lambda \times D^{n-\lambda}, (\partial D^\lambda) \times D^{n-\lambda})$, where λ is the Morse index of f at p, i.e., the number of negative eigenvalues of the Hessian matrix of second derivatives at p, and n is the dimension of X.*

In the case of $\mathscr{T}$, the Morse index λ is 0 for p_1, 1 for p_2 and p_3, and 2 for p_4.

1.2. Morse Theory on Singular Spaces

In this book, we generalize Morse theory by extending the class of spaces to which it applies. This increase in generality allows us to apply Morse Theory to several new questions. The most easily understood of these is to the study of singular spaces.

Consider the following singular space $\mathscr{R}$ embedded in Euclidean three space. (Topologically, $\mathscr{R}$ may be obtained from the torus $\mathscr{T}$ by shrinking the circle going around the left side to a point and stretching a taut disk across the circle around the hole.)

As before, let the function f measure the height. It is clear by inspection that the topological type of $\mathscr{R}_{\leq c}$ changes only when c passes one of the values $v'_1, \ldots, v'_5$, and that the cause of the exceptional nature of these values is that they are the images of the points $p'_1, \ldots, p'_5$. So to generalize Morse theory to singular spaces, we need a general definition of critical points which singles out the five points $p'_1, \ldots, p'_5$ in this case.

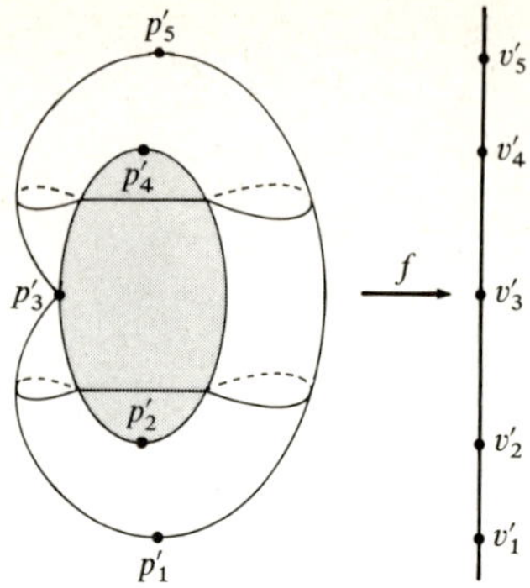

A Whitney stratification of a space X is a decomposition of X into submanifolds called *strata* satisfying the Whitney condition given in Part I, Sect. 1.2. The intuitive meaning of the Whitney condition is that the topological nature of the singularities of the space (including the singularities of the stratification itself) should be locally constant along each stratum. For the space $\mathscr{R}$, the singular set consists of the circle which bounds the disk. The largest stratum is the complement of this circle. Although the circle is itself nonsingular, the point p'_3 is distinguished by the fact that $\mathscr{R}$ has a different kind of singularity there. This point is the smallest stratum, and the rest of the singular circle is the middle stratum.

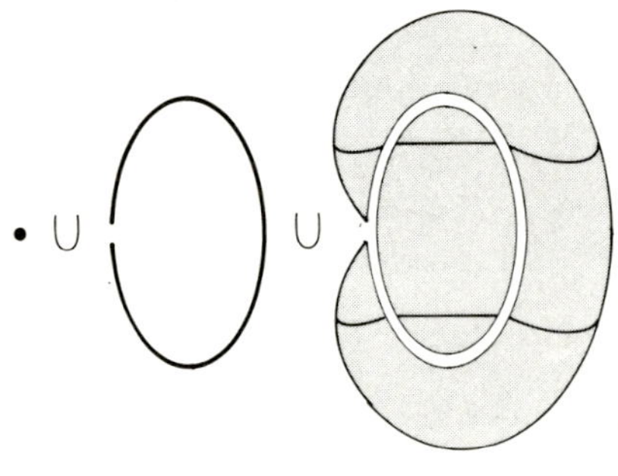

Stratification of $\mathscr{R}$

Now suppose that X is a compact Whitney stratified subspace of a manifold M and that f is the restriction to X of a smooth function on M. We define a *critical point* of f to be a critical point of the restriction of f to any stratum. (In particular, all zero-dimensional strata are critical points.) A *critical value* is, as before, the value of f at a critical point. With these definitions, Theorem CMT Part A now generalizes to give the first fundamental result of stratified Morse theory, which has the same statement:

Theorem (SMT Part A). *As c varies within the open interval between two adjacent critical values, the topological type of* $X_{\leq c}$ *remains constant.*

Now we wish to investigate how the topological type of $\mathscr{R}_{\leq c}$ changes as c crosses a critical value v'_i. As before if c is less than v'_1, then $\mathscr{R}_{\leq c}$ is empty, and as c crosses v'_1, the space $\mathscr{R}_{\leq c}$ changes by adding a two-disk. Also, as c crosses v'_5, the space $\mathscr{R}_{\leq c}$ changes by gluing in a two-disk along its boundary.

As c crosses the critical values v'_2, v'_3, and v'_4, the change in $\mathscr{R}_{\leq c}$ is described by Morse data, as shown by the following sequence of pictures. It is not immediately obvious what the pattern is, except that the Morse data is determined by the local picture of $\mathscr{R}$ and f near the critical points.

Crossing the critical value v'_2

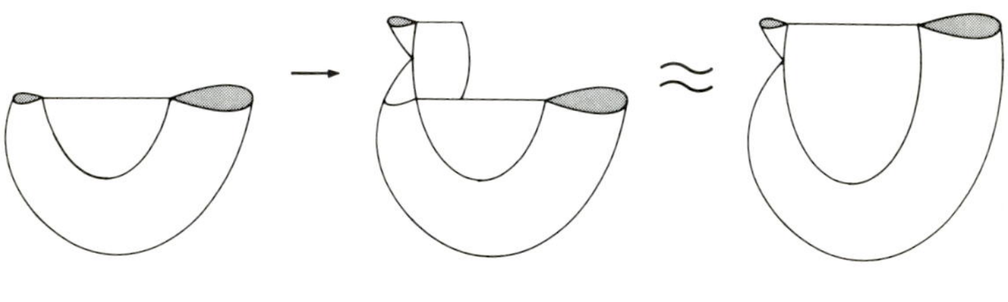

Crossing the critical value v'_3

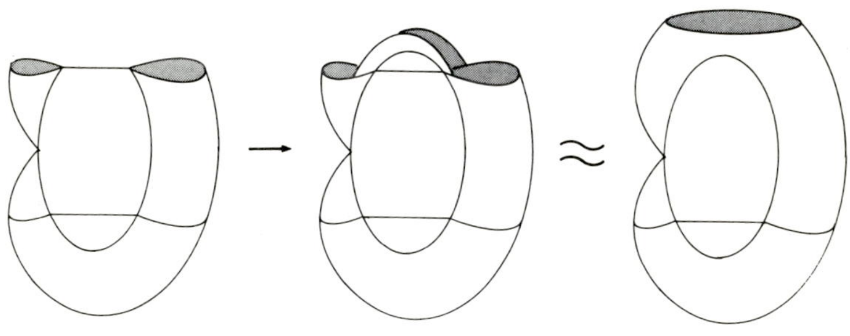

Crossing the critical value v'_4

If X is a Whitney stratified subspace of a manifold M, then we denote by $D(p)$ a small disk in M transverse to the stratum S containing p such that $D(p) \cap S = p$. (The dimension of $D(p)$ will necessarily be the dimension of the manifold M minus the dimension of the stratum S.) The intersection of $D(p)$ with X is called the normal slice at p and is denoted $N(p)$. The normal slice $N(p)$ is a key construction for a singular space. It has a boundary $L(p) = \partial D(p) \cap X$ which is called the *link* of the stratum S. Topologically, $N(p)$ is the cone over the link of S with its vertex at p. The topological type of the link may be thought of as measuring the singularity type of X along the stratum S. If X is nonsingular along S, then the link $L(p)$ is a sphere. The Whitney conditions guarantee that the connected component of S containing p has a neighborhood which is a fibre bundle over S and whose fibre is $N(p)$.

Consider $D(p)$ and $N(p)$ for the points $p'_1, \ldots, p'_5$ in our example $\mathscr{R}$. The disk $D(p'_3)$ is a three-ball around p'_3 since p'_3 lies in a zero-dimensional stratum, so the normal slice $N(p'_3)$ is a regular neighborhood of p'_3 in $\mathscr{R}$.

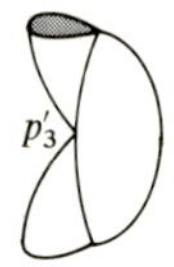

The point p'_1 is equal to its normal slice since it lies in a top dimensional stratum; likewise for p'_5. The following picture shows the disks $D(p_2)$ and $D(p_4)$ for $\mathscr{R}$, along with the normal slices at p'_2 and p'_4.

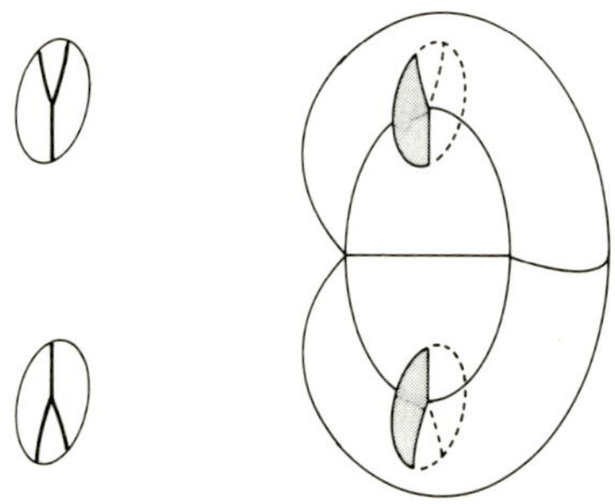

For any critical point p in X with critical value v, we define *normal Morse data* at p to be the pair of spaces (A, B) where A is the set of points x in the normal slice $N(p)$ such that $v-\varepsilon \leq f(x) \leq v+\varepsilon$ and B is the set of points x in $N(p)$ such that $f(x)=v-\varepsilon$, for very small ε. We may think of normal Morse data at p as Morse data for the restriction of f to the normal slice at p. We define *tangential Morse data* at p to be Morse data for the restriction of f to the stratum S of X containing p. Tangential Morse data may be computed using Theorem CMT Part B of the last section. Now we are in a position to state part two of the fundamental theorem of stratified Morse theory.

Theorem (SMT Part B). *Let f be a Morse function (see Sect. 1.4 of the introduction) on a compact Whitney stratified space X. Then, Morse data measuring the change in the topological type of $X_{\leq c}$ as c crosses the critical value v of the critical point p is the product of the normal Morse data at p and the tangential Morse data at p.*

The notion of product of pairs used in this theorem is the standard one in topology, namely $(A, B) \times (A', B') = (A \times A', A \times B' \cup B \times A')$. This theorem is illustrated by the following table of Morse data for our example $\mathscr{R}$.

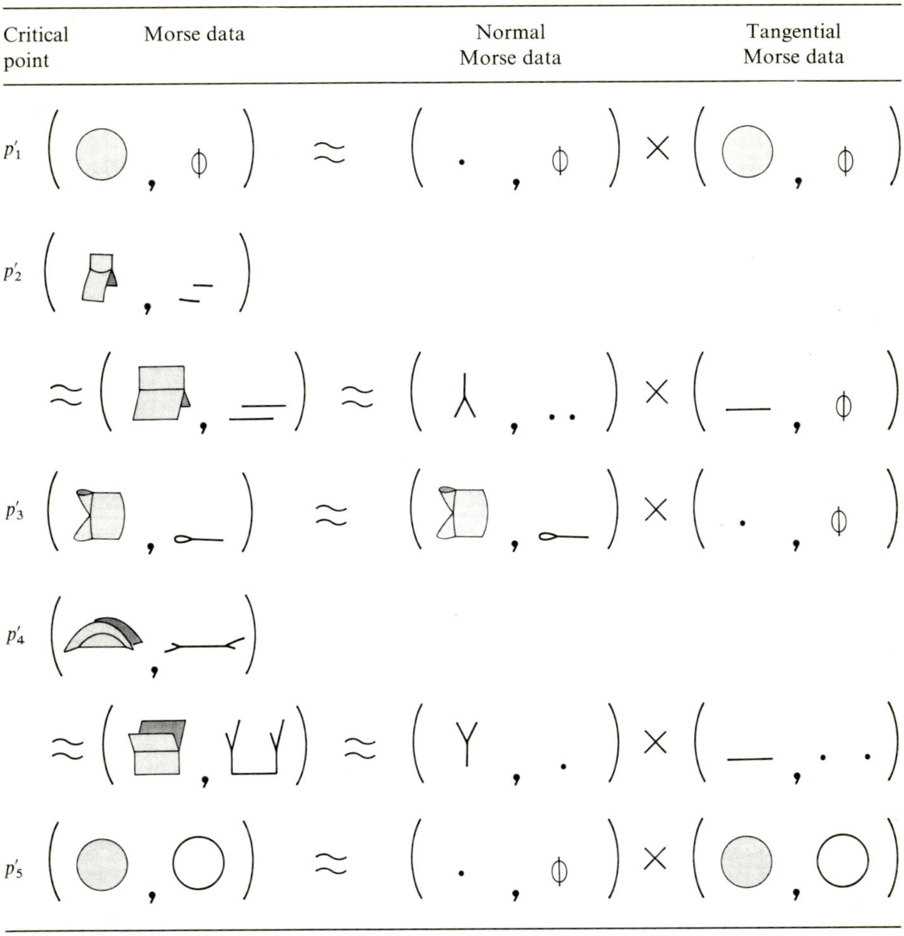

Theorem SMT Part B, although very natural and geometrically evident in examples, takes 100 pages to prove rigorously in this book. We are interested in applying it to establish results about the topology of X. This is possible since X is built up in a series of steps, one for each critical point of f, and the change brought about by each step is given by Theorem SMT Part II. However, in order to use it we must have information about *both* the normal Morse data and about the tangential Morse data for each critical point. The quest for this information is complicated by the fact that the normal Morse data can differ for various critical points in a connected stratum, as observed above.

In this book, we describe two classes of spaces X for which miraculous accidents give us a priori information on the normal and the tangential Morse data. One is complex varieties, described in Sect. 1.5 of the introduction. The other is complements of collections of flat subspaces of Euclidean space, described in the introduction to Part III of this book.

1.3. Two Generalizations of Stratified Morse Theory

So far, we have only considered stratified Morse theory for a function defined on a compact Whitney stratified space. The extension of this to the case of a proper function on a noncompact space requires no further modification of the results of the last section. (A function is proper if the inverse image of each closed interval is compact.) We now wish to consider two extensions of stratified Morse theory. The first is to certain nonproper functions. The second, which we call relative Morse theory, is to composed functions.

These two extensions broaden the range of questions to which stratified Morse theory may be applied, beyond the study of singular spaces. In fact some of the most important applications in this book (for example in proving Deligne's conjecture (Part II, Sect. 1.1)) are about nonsingular spaces.

(a) *Morse theory for nonproper functions.* Consider the example of the open unit disk with the origin removed. We call this space $\mathscr{D}$.

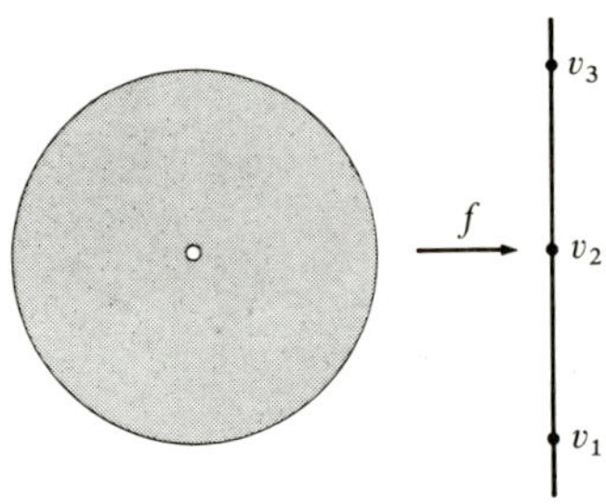

As usual, we study the height function f. There are three values v_1, v_2, and v_3 with the property that the topological type of $\mathscr{D}_{\leq c}$ changes as c crosses them. (At v_3, although there is no change in homotopy type, it changes from a manifold with boundary to a manifold without boundary.) So, by the general philosophy of Morse theory, there should be three critical points. However $\mathscr{D}$ is nonsingular, and f has no critical points in $\mathscr{D}$ at all. So the philosophy of Morse theory would not appear to apply to this example. (Even if we try to apply Morse theory to the closure $\bar{\mathscr{D}}$ of $\mathscr{D}$ in the plane, it will not work since the function f on the closure has only two critical points.)

The trick is to consider the closure $\bar{\mathscr{D}}$ *with an appropriate stratification:* a stratification such that the original space $\mathscr{D}$ is one of the strata. The simplest such stratification has three strata: a two-dimensional stratum – the open punctured disk $\mathscr{D}$ itself; a one-dimensional stratum – the circle at its edge; and a zero-dimensional stratum – the origin. The origin is forced to be considered as a separate stratum, even though the space $\bar{\mathscr{D}}$ is nonsingular there, by the requirement that $\mathscr{D}$ should be a stratum. The function f on $\bar{\mathscr{D}}$ with this stratification has three critical points as we wanted. (Even though $\bar{\mathscr{D}}$ is nonsingular at the origin, it has a critical point there in the sense of stratified Morse theory, since any zero-dimensional stratum is a critical point.)

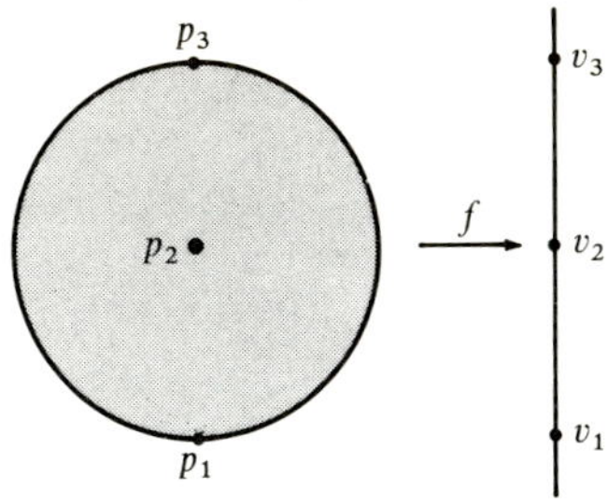

Note that these critical points for the height function on $\mathscr{D}$ lie *outside* $\mathscr{D}$.

In general, to study a nonproper Morse function f on a stratified space X, we require that X is a dense union of strata in some other stratified space Z, and that the function f extends to a *proper* Morse function on Z. We call the resulting diagram

$$X \subset Z \xrightarrow{f} \mathbb{R}$$

the *setup* for Morse theory of nonproper functions. Now f has two types of critical points: those that lie in X and those that lie in the complement $Z \setminus X$. In either case, the topological type of $X_{\leq c}$ can change as c crosses the corresponding critical value. However, between two adjacent critical values, the topological type of $X_{\leq c}$ remains the same, so we still have the first fundamental result of stratified Morse theory.

We will complete the fundamental theorem for nonproper Morse functions by giving a result which calculates the change in $X_{\leq c}$ as c passes a critical value. Before doing this, we make our second generalization:

(b) *Relative Morse theory.* We replace the setup for Morse theory of nonproper functions by a more general diagram:

$$X \xrightarrow{\pi} Z \xrightarrow{f} \mathbb{R}.$$

This diagram is called the *relative stratified Morse theory set-up* if f is a proper Morse function (see Sect. 1.4 of the introduction) and π satisfies the following technical condition: π has a factorization $X \subset \bar{X} \to Z$ such that X is a union of strata of $\bar{X}$, and $\bar{X} \to Z$ is a proper stratified mapping. (A stratified mapping is defined in Part I, Sect. 1.6. The idea behind the definition is that over each stratum of Z, the map should be a fibration in a stratum-preserving way.)

Any algebraic map $\pi: X \to Z$ admits stratifications of X and Z such that this technical condition is satisfied. One use of the additional generality of relative stratified Morse theory is to study the topology of a complex algebraic variety X mapping to a complex projective space Z through Morse functions on Z (see Part II, Sects. 2.6 and 3.4). Also, the quotient map for a compact group action satisfies this condition, so relative Morse theory could be used to study equivariant Morse functions.

For the relative stratified Morse theory setup, we define a critical point p of f in Z, and tangential Morse data for f at p just as before: it is the Morse data at p of the restriction $f|S$, where S is the stratum of Z which contains the critical point p. For this setup, however, normal Morse data at

p is defined differently. For any critical point p in Z with critical value v, we define *normal Morse data* at p to be the pair of spaces (A, B) where A is the set of points x in the inverse image of the normal slice $\pi^{-1}(N(p))$ such that $v-\varepsilon \leq \pi(f(x)) \leq v+\varepsilon$ and B is the set of points x in $\pi^{-1}(f(p))$ such that $\pi(f(x)) = v-\varepsilon$, for very small ε. Here the normal slice at p, $N(p) \subset Z$ is defined as before. Note that normal Morse data is defined as a pair of subspaces of X, whereas tangential Morse data is constructed using only the behavior of f on Z.

We can now state most general version of the fundamental theorem of stratified Morse theory. Here $X_{\leq c}$ refers to the composed function $f \circ \pi$, i.e., $X_{\leq c}$ is $\{x \in X \mid f\pi(x) \leq c\}$.

Theorem (SMT for the relative and nonproper cases). *Assume that the composition*

$$X \xrightarrow{\pi} Z \xrightarrow{f} \mathbb{R}$$

is a relative stratified Morse theory setup.

Part A. As c varies between two adjacent critical values, the topological type of $X_{\leq c}$ remains constant.

Part B. Morse data measuring the change in the topological type of $X_{\leq c}$ as c crosses the critical value v of the critical point p is the product of the normal Morse data at p and the tangential Morse data at p.

In the case that $\pi: X \to Z$ is the identity, this theorem specializes to Theorem SMT of Sect. 1.2 of the introduction.

The reader can easily check (in the case $\pi: X \to Z$ is the inclusion of the example $\mathscr{D}$ into its closure $\bar{\mathscr{D}}$) that this theorem correctly describes the changes in $\mathscr{D}_{\leq c}$ which occur as c crosses the critical values v_1, v_2, and v_3.

1.4. What is a Morse Function?

The object of this book is to give the natural generalization of classical Morse theory on a manifold X to stratified spaces. We have shown how the fundamental theorem relating the singularities of a function to the topology of X generalizes to the stratified context. Now we examine the class of functions to which this analysis applies. These are called Morse functions. They are the natural generalization of classical Morse functions on a manifold. We recall the classical case first.

In classical Morse theory, Morse functions are singled out from all proper smooth functions on a differentiable manifold X by two requirements:

0. The critical values of f must be distinct.

1. Each critical point of f is nondegenerate, i.e., the Hessian matrix of second derivatives has nonvanishing determinant.

It follows from this definition that the set of critical points is discrete in X and the set of critical values is discrete in $\mathbb{R}$.

In addition to leading to the beautiful fundamental theorem of Morse theory described in Sect. 1.1 of the introduction, Morse functions have two further desirable properties. The first is that they are plentiful. There are several theorems of this type. For example, Morse functions form an open dense set in the space of all proper smooth functions with the appropriate (Whitney) topology, so any proper smooth function on X may be approximated by a Morse function. Another example is that if X is embedded in a Euclidean space $\mathbb{R}^k$ as a closed proper subspace, then for almost all points e in $\mathbb{R}^k$, the function measuring the distance to e is a Morse function. The second desirable property of Morse functions is that they are C^∞ structurally stable. In other words, if f is Morse and if f' is close enough to f in the Whitney topology, then there exist C^∞ diffeomeorphisms h and h' such that the following diagram commutes:

$$\begin{array}{ccc} X & \xrightarrow{f} & \mathbb{R} \\ {\scriptstyle h}\downarrow & & \downarrow{\scriptstyle h'} \\ X & \xrightarrow{f'} & \mathbb{R} \end{array}$$

The existence of such a commutative diagram means that f and f' have the same C^∞ topological type.

In stratified Morse theory we consider Whitney stratified spaces X embedded in some smooth manifold M. In order to find the analogue of the definition of Morse functions in this context, we first need an analogue of the class of smooth functions. A function on X is called *smooth* if it is the restriction to X of a smooth function on M. If X is an algebraic variety, then this notion of smoothness is intrinsic to X, since it can be seen to be independent of the choice of an algebraic embedding of X in M. By definition (due originally to Lazzeri and Pignoni), *Morse functions* are singled out from all proper smooth functions on a Whitney stratified space X by three requirements:

(0) The critical values of f must be distinct.

(1) At each critical point p of f, the restriction of f to the stratum S containing p is nondegenerate.

(2) The differential of f at any critical point p does not annihilate any limit of tangent spaces to any stratum S' other than the stratum S containing p.

It follows from this definition that the set of critical points is discrete in X and the set of critical values is discrete in $\mathbb{R}$.

Conditions (0) and (1) together imply that the restriction of f to each stratum is Morse in the classical sense. Condition (1) is a nondegeneracy requirement in the tangential directions to S, while Condition (2) is a nondegeneracy requirement in the directions normal to S.

The geometric significance of Condition (2) is illustrated by the following example. On the left and on the right are cusps stratified with a zero-dimensional stratum at the cusp point. In each case, we consider the height function:

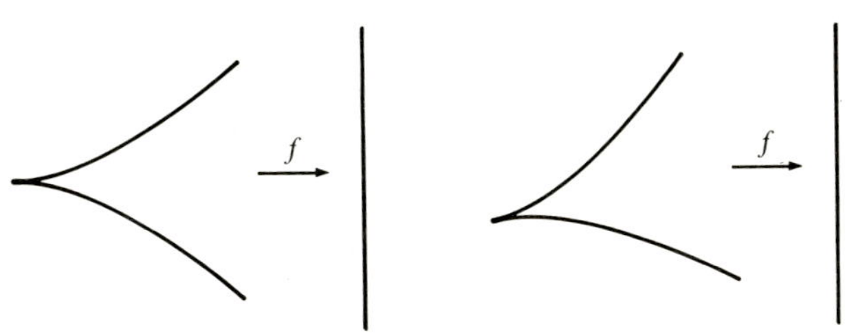

The height function is not Morse on the left, but is Morse on the right. Consider a sequence of tangent spaces to the one-dimensional stratum at a sequence of points approaching the zero-dimensional statum.

The limit lines to these sequences of tangent are shown in the following diagram.

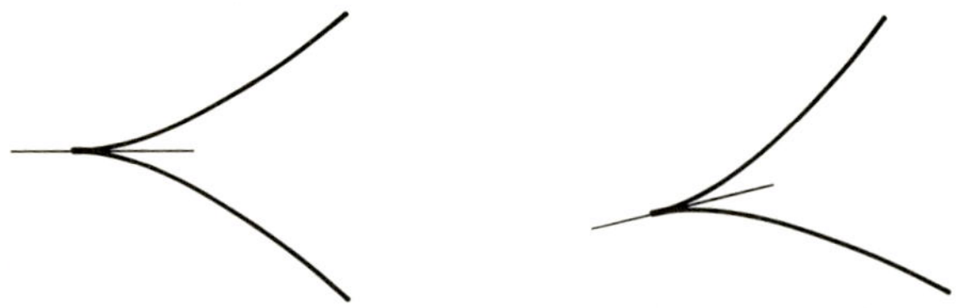

The function on the left is not Morse, because the limit on the left is horizontal and is annihilated by the differential of the height function. The limit on the right is not.

The fundamental result of stratified Morse theory of Sects. 1.2 and 1.3 of the introduction relating the topology of X to the critical points of f holds for Morse functions as just defined. In addition, these Morse functions satisfy the two further desirable properties of classical Morse functions described above. They form an open dense set in the space of all proper smooth functions with the appropriate (Whitney) topology. So as before, any proper smooth function on X may be approximated by a Morse function. As with classical Morse functions, if X is embedded in a Euclidean space $\mathbb{R}^k$ as a closed proper subspace, then for almost all points e in $\mathbb{R}^k$, the function measuring the distance to e is a Morse function. Also, Morse functions are C^0 structurally stable [P1]. In other words, if f is Morse and if f' is close enough to f in the Whitney topology, then there exist (C^0) homeomorphisms h and h' such that the above diagram commutes, i.e., f and f' have the same topological type. (In general, there are no C^∞ (or even C^1) structurally stable functions on a Whitney statified space.)

The fundamental theorems of stratified Morse theory (Sects. 1.2 and 1.3 of the introduction) remain valid for a wider class of functions than Morse functions. Condition (1) of the definition of Morse functions can be replaced by a condition that we call *nondepraved* (see Part I, Sect. 2.3). This is a Whitney-like condition on a critical point of a function on a smooth manifold, which may prove useful in other contexts.

1.5. Complex Stratified Morse Theory

In Sect. 1.2 of this introduction, we saw that in order to use stratified Morse theory to study the topology of a space X, we need to know properties of both the tangential Morse data and the normal Morse data for the critical points of a Morse function on X. In general, this knowledge may be as difficult to obtain as the knowledge of the topology of X itself. However, if X is a complex analytic space, then two miracles of complex geometry allow a partial calculation of the Morse data.

For purposes of exposition, in this introduction we will consider only the case that the space X has an embedding in $\mathbb{C}^k$ as a closed subspace, and the function f that we are considering is the distance function to a point e in $\mathbb{C}^k$. The first miracle is this:

Lemma. *Suppose that $S \subset \mathbb{C}^k$ is any complex submanifold of complex dimension s and that f is the distance function to a point $e \in \mathbb{C}^k$. Then, any nondegenerate critical point p of f on S has Morse index λ at most equal to s.*

Viewed as a real submanifold, S has dimension $2s$, so we would expect all Morse indices from 0 up to $2s$ to be possible, but this statement says that half of these possibilities are ruled out for reasons of complex geometry. This lemma may be deduced from the fact that if the Hessian quadratic form is negative on a tangent vector v, then it is positive on $\sqrt{-1} \cdot v$ (but not conversely).

This lemma was first applied to classical Morse theory by Thom, who exploited it to prove results about complex varieties. For example, suppose that X itself is nonsingular and that it has complex dimension s. Then for a generic center point e, the distance function f is Morse and all of its critical points have Morse index at most s. This gave the first proof that a Stein manifold of complex dimension s has *homotopy dimension* at most s, i.e., has the homotopy type of a cell complex of dimension at most s [AF1] (since any Stein space is homeomorphic to a closed subspace of some $\mathbb{C}^k$).

This lemma enables us to find bounds on the homotopy dimension of tangential Morse data in certain cases, because tangential Morse data are precisely the classical Morse data of the stratum.

The second miracle of complex geometry is that the normal Morse data at a critical point p depend only on the stratum S in which p lies, not on the function f or the point p. In fact there is a geometric construction of the normal Morse data in terms of an auxilliary complex variety associated to S which we call the *complex link* of S.

The complex link $\mathscr{L}(S)$ of a stratum S of X is constructed as follows: Let N be a complex analytic manifold in $\mathbb{C}^k$ transverse to the stratum S at some point $p \in S$ such that $N \cap S = p$. (The dimension of N will necessarily be $k-s$ where s is the dimension of the stratum S.) Let $D(p)$ be a small disk in N around p. Let H be a generic codimension one hyperplane in N that passes very close to p but not through p. Then, $\mathscr{L}(S)$ is the intersection $X \cap D(p) \cap H$. This is illustrated in the following diagram (which gives a real analogue of the situation).

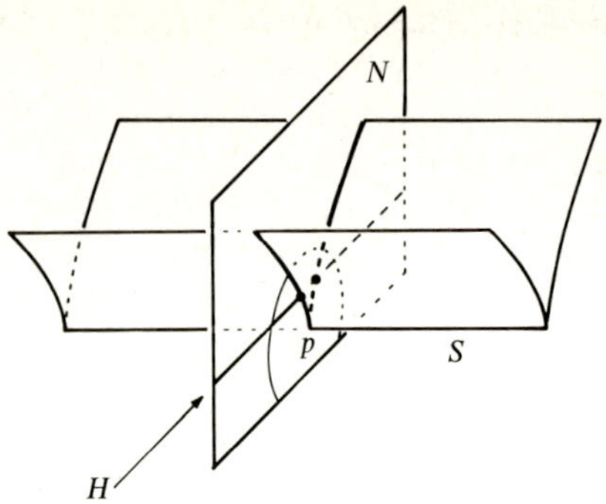

We believe that the complex link $\mathscr{L}(S)$ is a very important construction in its own right. It is a complex analytic space with a boundary $\partial\mathscr{L}(S)=X\cap\partial D(p)\cap H$. Its interior $\mathscr{L}(S)\smallsetminus\partial\mathscr{L}(S)$ is a Stein space. Up to homeomorphism, both the complex link and its boundary depend only on the stratum S; they are independent of all the other choices in its construction. Just as the ordinary link of a stratum in a real stratified space measure the singularity at that stratum, the complex link $\mathscr{L}(S)$ measures the singularity at S. For example, if X is nonsingular at S, then the complex link will be a complex disk. If X is the complex cone over $Y\subset\mathbb{CP}^{k-1}$ and S is the vertex, then the complex link is Y minus a neighborhood of the hyperplane section. If X is a curve and S is a singular point, then the complex link is a set of points of cardinality the multiplicity of X at S.

We can now state the fundamental theorem of *c*omplex *s*tratified *M*orse *t*heory:

Theorem (CSMT Part A). *Suppose that p is a critical point for a proper Morse function f on a complex analytic variety and $S\subset X$ is the stratum containing p. Then, the normal Morse data for f at p is homotopy equivalent to the pair (Cone $\mathscr{L}(S)$, $\mathscr{L}(S)$) consisting of the cone on the complex link of S and the base of the cone.*

As an illustration of this theorem, consider the curve singularity given by three lines meeting at a point. An embedded real picture is given on the left, and a topologically correct but not embedded complex picture is given on the right.

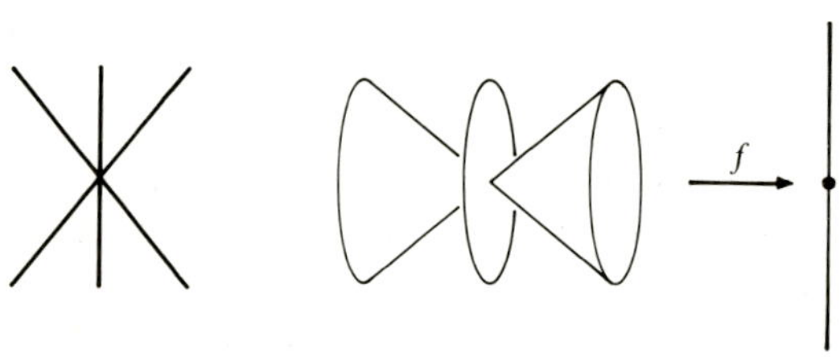

The complex link of the singularity consists of three points. The theorem shows that homotopy Morse data for the singular point is given by the pair

There is a calculation of the normal Morse data at p in terms of the complex link of S which is precise up to homeomorphism (Part II, Sect. 2.5). This is more complicated and involves a "monodromy map".

Theorem CSMT Part A is particularly useful in inductive proofs. As an example, we give a sketch of the proof of the theorem of Hamm and Karchyauskas that any Stein space X has homotopy dimension at most equal to its complex dimension, (Part II, Sect. 1.1*). Embed X topologically as a closed subspace of some $\mathbb{C}^k$, and use as a Morse function f the distance function to an appropriate point e. We need a bound on the homotopy dimension of the Morse data for all of the critical points. The lemma gives bounds on the homotopy dimension of the tangential Morse data. Theorem CSMT Part A bounds the homotopy dimension of the normal Morse data in terms of the homotopy dimension of the complex link $\mathscr{L}(S)$. However, the complex link is homotopy equivalent to its interior, which is a Stein space of smaller dimension. So, by induction on the dimension, we are done. The detailed argument is carried out in Part II, Sect. 5.1*.

We wish to make a philosophical point about this sort of induction, which is prototypical for most of the applications of Morse theory in this book. The study of the topology of X by Morse theory always involves passage from local information (Morse data at a critical point $p \in X$) to global information about X. In complex stratified Morse theory, the Morse data at p is calculated from global information about the complex link $\mathscr{L}(S)$ of the stratum S containing p. In this induction, the required global information about $\mathscr{L}(S)$ is itself calculated by Morse theory, using a naturally defined Morse function on $\mathscr{L}(S)$. Thus, $\mathscr{L}(S)$ is described in terms of local information (i.e., Morse data at a critical point $p_1 \in \mathscr{L}(S)$). Intuitively, points in the complex link represent complex directions away from S, so local information in the complex link is "local in the space of directions from p". In the language of Hörmander [Hö], "local" in the complex link is called *microlocal* in X. Morse data at p_1 is in turn calculated using global information about $\mathscr{L}_1$, the complex link in $\mathscr{L}(S)$ of the stratum containing p_1. The induction proceeds further to calculate this by Morse theory, reducing it to local information in $\mathscr{L}_1$ (Morse data at a point $p_2 \in \mathscr{L}_1$). This is micro-micro-local or (micro)^2-local information. This information is obtained from (micro)^3-local information, and so on. This accumulation of micro's seems essential to stratified Morse theory, and indeed to the study of nonisolated singularities in general.

We may also use stratified Morse theory to study nonproper Morse functions in the complex case. The setup for the complex version of this is a diagram

$$X \overset{\pi}{\subset} Z \xrightarrow{f} \mathbb{R}.$$

Here $\pi: X \subset Z$ is an inclusion of complex analytic varieties stratified by complex analytic strata such that X is a union of strata of Z, and $f: Z \to \mathbb{R}$ is a proper Morse function on Z. The critical points of f are of two types: those in X and those in the complement $Z \setminus X$. Normal Morse data for the first type is given up to homotopy by Theorem CSMT Part 1 above. For the second, the corresponding result is the following:

Theorem (CSMT Part B). *Suppose that p is a critical point for a Morse function f on a complex analytic variety and the stratum S containing p does not lie in X. Then, the normal Morse data for f at p is homotopy equivalent to the pair* $(\mathcal{L}(S), \partial\mathcal{L}(S)) \times (D^1, \partial D^1)$.

This theorem may be illustrated by considering the square of the distance function to a point e in the variety $\mathscr{C}$ consisting of the complex line with the origin removed. The origin is now a critical point, and the space $\mathscr{C}_{\leq c}$ looks like this before and after c crosses the corresponding critical value:

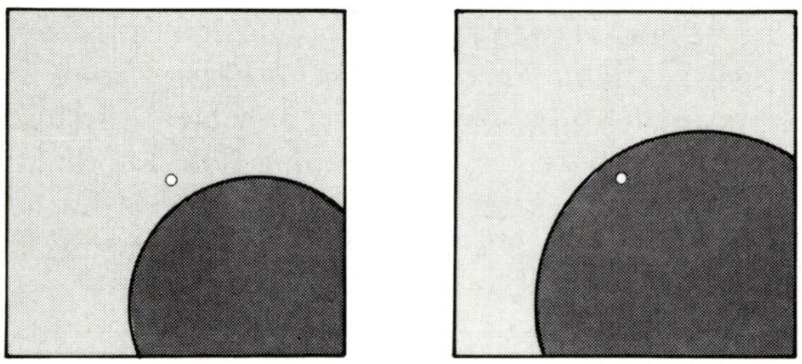

The complex link of the origin is a point with no boundary, so homotopy Morse data predicted by the theorem is the pair $(D^1, \partial D^1)$.

1.6. Morse Theory and Intersection Homology

At this point, the reader may be feeling nostalgia for classical Morse theory, where all the information about the Morse data (A, B) was contained in one number: the Morse index. The Morse index may be homologically characterized as the unique integer i for which $H_i(A, B)$ is nonzero. This characterization of the Morse index as the unique degree in which the homology of the Morse data does not vanish is the only fact Morse used to prove Morse inequalities.

Such a simple situation is impossible for singular spaces, as is shown by taking X to be the suspension of an arbitrary Y:

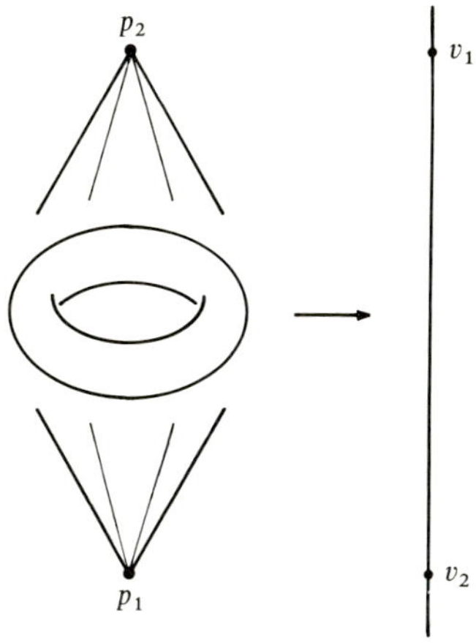

The homology of the Morse data at p_2 is, up to a degree shift by one, the reduced homology of Y. This can be nonzero in many degrees. Such a simple situation is even impossible for complex varieties, as shown by the complex cone over a complex algebraic variety Y embedded in projective space. For these, the homology of the Morse data at the vertex is, up to a degree shift by one, the reduced homology of the complement of the hyperplane section of Y, which can have arbitrarily great homological complexity.

As is often the case, however, the essential simplicity of the nonsingular case may be restored by considering the intersection homology of a complex variety.

Theorem (see Part II, Sect. 6.4). *Let f be a proper Morse function on a purely n-dimensional complex analytic variety X. If (A, B) denotes the Morse data for a critical point p in a stratum S of dimension s, then the intersection homology group $IH_i(A, B)$ vanishes for all i except for $i=\lambda_S+n-s$, where λ_S is the Morse index of the restriction of f to S.*

So for intersection homology, critical points have a true analogue to the classical Morse index, namely $\lambda=\lambda_S+n-s$. It is no longer true, however, that the group $IH_\lambda(A, B)$ is one-dimensional. Instead, it is an important and poorly understood invariant of the singularity of X along the stratum S.

In order to prove this result, we need the full calculation of the Morse data up to homeomorphism, since intersection homology is not a homotopy invariant.

1.7. Historical Remarks

The paper in which Morse introduced Morse theory to the world [Mo4] was submitted in 1923 and published in 1926. By an interesting coincidence, this was exactly when Lefschetz published his work on the topology of algebraic varieties [Lef1] (1924), which was the starting point of the other main theme of this book. This original version of Morse theory was the homological version. It related the critical points of a proper differentiable function on a smooth manifold X to the homology of X.

At this time, algebraic topology was, in the words of Lefschetz, "hardly further along" than "its infancy" ([Lef2], p. 15). Homology theory had long since been created by Riemann, Betti, and finally Poincaré [Po1], [Po2] (1895, 1899). The first book on algebraic topology (at that time called analysis situs) had just been written by Veblen [Veb] (1922). However, the theory did not yet have rigorous foundations.

The theory of Riemann, which was never published, was based on an intuitive notion of a k-cycle in a space X, as an oriented k-surface with singularities contained in X, and an intuitive notion of a homology between two cycles, a $(k+1)$-surface with singularities bounding their union which establishes their homological equivalence. These notions of a cycle and a homology were not defined precisely, but their properties were established by pictures and an appeal to geometric intuition. Poincaré attempted to rigorize them by defining cycles and homologies as semianalytic subsets, an idea that was carried to completion in 1975 [Ha1]. Morse refers to Veblen's rigorous book, which concerns cellular homology of regular cell complexes. However, to establish that his version of homology is a topological invariant rather than a combinatorial one, Veblen refers to [Ax] of Alexander. This contains an attempt at defining what is now called singular homology (finally achieved by Lefschetz [Lef3] and Eilenberg [E]). But, Alexander implicitly assumes that space filling curves do not exist, as was noticed by Lefschetz [Lef3]. Furthermore, even to apply Veblen's cellular homology, Morse must use the fact that a differentiable manifold with boundary can be cell-decomposed, which he asserts without proof. Morse must have been aware that there was a difficulty here, since the year after the paper was published he suggested it as a thesis problem to Cairns ([Bo2], p. 913). The century-long story of the taming of homology theory is one of the greatest in mathematical history, and has not yet been adequately recorded by historians. In any case, it was far from over when Morse and Lefschetz began their pioneering work.

If mathematical journals in 1924 had the same standards of rigor that they have today, neither Morse theory nor Lefschetz theory could have been published. Morse and Lefschetz both attributed their success to their use of intuitive homology theory without insisting on adequate foundations. In 1951, as the taming of homology theory was reaching its completion, Morse wrote "Mathematicians of today are perhaps too exuberant in their desire to build new logical foundations for everything. Forever the foundation and never the cathedral" ([Mo6], p. 58). (We feel a kinship with this sentiment. We developed intersection homology through free use of intuitive cycles. It took us four years to find a rigorous version for public presentation.) In conversations, Morse and Lefschetz were both often critical of the highly algebraic turn that topology took after World War II.

Like most mathematical advances, Morse theory had its precursors. Poincaré had a Morse inequality for vector fields in two dimensions ([Po3], p. 129, 1885), i.e., half of the "Hopf Index Theorem". The first indication that there is a connection between critical points of a function and the topology of its domain of definition was G.D. Birkhoff's "minimax principle" ([Bir], p. 240). This gives a lower bound on the number of saddle points of a function defined on a 2-manifold in terms of the number of relative minima and the homoloy of the manifold. Morse's work was inspired by Birkhoff's, but it is far enough beyond its predecessors to call it qualitatively new.

Morse's original work inspired a long history of later developments. Smale has termed Morse theory the most significant single contribution to mathematics by an American mathematician. It has been extended many times, always maintaining its original flavor. These extensions have usually consisted of generalizing the setup or of finding new techniques to calculate the Morse data. The extensions have usually been made with a view of giving new applications. Since Morse theory relates the singularities of the function f to the topology of the space X, applications consist of knowing something about one of these two so as to deduce something about the other. What follows is only a sketch of some highlights. More complete versions have been recorded in several places [Bo4], [Sma3], [Maz], [Bo2].

The first extension was by Morse himself, almost immediately after his original work. This was to spaces X which are infinite-dimensional, such as the path space of a manifold, and to functions f which are functionals in the sense of calculus of variations, like the length [Mo5]. He also found a technique to calculate the Morse index in terms of Jacobi vector fields, the Morse Index Theorem. He was able to prove, for example, that two points on a sphere with any metric are joined by infinitely many geodesics. Bott extended Morse theory by allowing certain nonisolated critical points of the function f, called nondegenerate critical submanifolds. He also found group theoretical methods for calculating Morse indices on Lie groups and their path spaces. This led to the first proof of the periodicity theorem [Bo5], [Bo6]. Thom first exploited the fact that complex geometry can be used to bound Morse indices [T9]. This led to results on the topology of complex analytic spaces, of which this book contains many more.

In the original version of Morse theory, only the homology of X entered. Lysternik and Schnirelmann extended this to an invariant which can be finer, the category of X [LS]. Thom ([T8], 1949) showed the existence of a cell complex structure on X, with one cell for each singular point of f. This gave results on the homotopy type of X. Then Smale introduced his "Handlebody decomposition" of X, with one handle for each critical point of f. This gives results on the diffeomorphism type of X. (It is the version that we presented in the beginning of the introduction.) This led to many developments in differential topology such as the proof of the Poincaré conjecture in dimension five or more. This is summarized in [Sma2], [Sma3], and [Maz]. Smale and Conley have developed the idea of extending Morse theory by replacing the function f by a dynamical system ([Sma1], [Co]) and used it, for example, to show the existence of fixed points and closed orbits. More recently, Atiyah and Bott have developed equivariant Morse theory and applied it to equivariant cohomology [Bo4].

Morse theory on manifolds with boundary, originally due to Baiada and Morse [BaM], has been applied by Thom to give bounds on the Betti numbers of a real algebraic variety [T10]. Morse theory extended to manifolds with boundaries and corners was further developed by Hamm, Karchyauskas, Lê and Siersma ([Kr1], [Kr2], [H3], [H4], [HL2], [H13], [S]) and was applied to the study of the homotopy type of Stein spaces and to Lefschetz theorems for quasiprojective varieties.

Finally, the extension of Part I of this book to stratified spaces X, and the applications in Parts II and III, can be considered to be part of this line of development.

1.8. Remarks on Geometry and Rigor

As is shown by the above history of Morse theory or by the history of stratification theory (Part I, Sect. 1.0), there is often a creative tension between geometry and rigor. Rigor follows the initial conception with a much greater time delay in geometry than it does in algebra. Also, when it comes, true geometers often feel its language misses the essential geometric ideas. Language is not well adapted to describing geometry, as the facilities for language and geometry live on opposite sides of the human brain. This perhaps accounts for the presence in the current literature on singularities of expressions like "using the isotopy lemma, it can be shown" without the forty pages of geometric constructions and estimates needed to apply the isotopy lemma.

Nevertheless, a geometrically apt rigorization of a geometric idea can actually add to its ease of visualization. Major examples of this are the final versions of singular homology and of stratified spaces.

We have tried to alleviate the incredible complexity of the arguments in this book with two technical innovations that we hope are geometrically apt. The first is *moving the wall* (Part I, Sect. 4), a technique for rigorously constructing isotopies of stratified spaces by examining pictures of "characteristic vectors" in an auxiliary space. The second is *fringed sets* (Part I, Sect. 5), a method of handling estimates geometrically. We hope that the combination of these allows us to approach the geometer's ideal of giving proofs that are both rigorous and visual.

Chapter 2. The Topology of Complex Analytic Varieties and the Lefschetz Hyperplane Theorem

One of the main sets of mathematical results proved in this book is a collection of theorems on the topology of complex analytic varieties. There are generalizations of the Lefschetz hyperplane theorem for complex projective varieties, and generalizations of the theorem that the homotopy dimension of a Stein manifold is bounded by its complex dimension. In this section of the introduction, we give a sketch of the statements of the theorems with motivation and some history. Technically precise statements of the theorems in their most general form are grouped together in Chapter 1 of Part II of the book.

The proofs of these theorems are applications of stratified Morse theory. However, both this section of the introduction and Chapter 1 of Part II may be read without any knowledge of stratified Morse theory.

2.1. The Original Lefschetz Hyperplane Theorem

The idea of the *L*efschetz *h*yperplane *t*heorem is that a complex projective variety resembles its hyperplane section:

Theorem (the LHT) [Lef1]. *Let X be a closed nonsingular purely n-dimensional algebraic subvariety of complex projective space, and let H be a generic hyperplane. Then, $H_i(X, X \cap H)=0$ for $i<n$.*

Combined with the long exact sequence of the pair $(X, X \cap H)$ and Poincaré duality, this theorem implies that all of the Betti numbers b_i of X are determined by those of $X \cap H$ except for three: b_{n-1}, b_n, and b_{n+1}; and that b_{n-1} and b_{n+1} are bounded by the $n-1^{\text{st}}$ Betti number of $X \cap H$. The primary use of this theorem is in studying projective varieties by induction on their dimension.

The Lefschetz hyperplane theorem has a dual (in a sense explained in Sect. 2.7 of the introduction). This states that a nonsingular complex affine variety has the homology of a space half its real dimension:

Theorem (the LHT*). *Let X be a closed nonsingular purely n-dimensional algebraic subvariety of complex affine space. Then $H_i(X)=0$ for $i>n$.*

A tremendous amount of effort has gone into generalizing these two fundamental theorems. Significant contributions have been made by many mathematicians: [AF1], [AF2], [Art], [Bar1], [BL], [Ber], [Bo1], [C1], [Ch2], [D1], [Fa], [FK1 to FK5], [Fr1], [FH], [FL1], [FL2], [GK], [GM2], [GM3], [Gr1],

[Gro], [H1 to H6], [HL1], [HL2], [HL3], [Kr1], [Kr2], [Kr3], [Kp1 to Kp5], [KW], [L], [La], [Mi1], [Mi2], [N], [Og], [Oka], [Oko1 to Oko3], [R], [Sm1 to Sm6], [SV], [SoV], [T9], [Wa], [Z1]. These authors have used widely differing techniques.

This book contributes further generalizations to the list. But, the main advantage of stratified Morse theory is that, at least for complex varieties, it provides a unified approach through which a wide variety of generalizations can be proved and understood. What follows is a nonhistorical account. Original references to the literature for specific results are given immediately after their statements in the main portion of the text.

2.2. Generalizations Involving Varieties which May be Singular or May Fail to be Closed

One of the most dramatic generalizations is that the LHT holds for quasiprojective varieties and the LHT* holds for singular varieties, both without modifying the statements:

Theorem. *The hypothesis "closed" may be omitted from the statement of the LHT and the hypothesis "nonsingular" may be omitted from the statement of the LHT*.*

The reverse is not true. Easy examples show that it is not possible to omit the hypothesis "nonsingular" from the LHT or the hypothesis "closed" from the LHT* (see Part II, Chap. 8).

So, singularities of X can cause failure of the LHT. We want to measure quantitatively the effect of singularities on the validity of the theorem. Local complete intersection singularities have no effect on its validity. We define a measure $S(p)$ of the degree of singularity of X at a point p to be (the number of equations needed to define X near p) minus (the codimension of X in projective space). The number $S(p)$ is zero when X is a local complete intersection at p.

Theorem (the LHT for singular spaces). *Let X be a purely n-dimensional algebraic subvariety of complex projective space, and let H be a generic hyperplane. Then $H_i(X, X\cap H)=0$ for $i<n-\sup_{p\in X} S(p)$.*

Similarly, removing a subvariety V from X can cause failure of the LHT*. Again, we want a quantitative measure of this. Removing a Cartier divisor from X has no effect. We define a measure $S^*(p)$ of the degree to which V fails to be a Cartier divisor near p to be one less than the number of equations needed to define V as a subvariety of X near p.

Theorem (the LHT* for non-closed subspaces). *Let X be a closed purely n-dimensional algebraic subvariety of complex affine space and let V be a subvariety of X. Then $H_i(X-V)=0$ for $i>n+\sup_{p\in V} S^*(p)$.*

2.3. Generalizations Involving Large Fibres

Another direction of generalization is to consider varieties X that are mapped to complex projective space or affine space, rather than subvarieties. Recall that any algebraic map $\pi: X \to Y$ has the property that X can be finitely decomposed into subvarieties V_i of varying dimension, each of which maps to Y with constant fibre dimension. Suppose that X has pure dimension so that codimensions make sense. We call π *semismall* if for each i, its fibre dimension in V_i is at most the codimension of V_i in X. If the map of X to complex projective space or affine space is semismall, then the LHT and the LHT* remain true. We define a measure $D(\pi)$ of deviation of π from semismallness to be the supremum over i of (the fibre dimension of π in V_i) minus (the codimension of V_i in X).

Theorem (the LHT with large fibres). *Let $\pi: X \to \mathbb{CP}^N$ be a (not necessarily proper) map of a nonsingular purely n-dimensional algebraic variety into complex projective space, and let H be a generic hyperplane. Then $H_i(X, \pi^{-1}(H))=0$ for $i<n-D(\pi)$.*

The above theorem (or rather the homotopy refinement of it; see Part II, Sect. 1.1) was conjectured by Deligne [D1]. It contains as a special case the classical Bertini theorem (Part II, Sect. 1.1).

Theorem (the LHT* with large fibres). *Let $\pi: X \to \mathbb{C}^N$ be a proper map of a purely n-dimensional (possibly singular) algebraic variety into complex affine space. Then $H_i(X)=0$ for $i>n+D(\pi)$.*

2.4. Further Generalizations

Many refinements of the above statements can be made, and are incorporated into the sharper statements of Part II, Chap. 1 of this book. First, the above homology statements all have homotopy analogues. The statements that relative homology groups vanish can be strengthened to statements that the corresponding relative homotopy groups vanish. The vanishing of homology groups of X of degree greater than n can be strengthened to the assertion that X has the homotopy type of a CW complex of dimension n. (We shall shorten this by saying that X has *homotopy dimension n*.) In the LHT*, X need only be analytic. This leads to the same statement for a complex Stein space.

The assumptions on the singularities or the size of the fibres in the LHT statements need only be imposed on the singularities of fibres away from the hyperplane. The hyperplane need not be taken to be generic, provided that it is replaced in the statement by a small tubular neighborhood. The hyperplane may be replaced by an arbitrary linear subspace, provided that the range of vanishing in the conclusion is appropriately modified. More generally, the projective space and the linear subspace may be replaced by any pair such that the subspace is the zeros of a nonnegative function with a suitable Levi form. Dually, in the LHT* statements, the complex affine space may be replaced by complex projective space minus a linear space, with a similar modification of the conclusion. Or, it may be replaced by any analytic manifold that admits a real valued function with a suitable Levi form.

Finally, the LHT may be strengthened to a local version. To give a projective variety X is the same as to give a conical variety K of one more dimension, namely the cone on X. Philosophically, any statement about the projective variety or its embedding really comes from a statement about the singularity at the point of the cone. Theorems about projective varieties should be consequences of more general theorems about singularities which are no longer required to be conical. This is the case for the LHT, which is a consequence of the following:

Theorem (Local LHT). *Suppose that K is a purely $(n+1)$-dimensional analytic subvariety of complex affine space with an isolated singularity at p, H is a generic hyperplane through p, and ∂B_ε is the boundary of a small enough ball around p. Then, $\pi_i(X \cap \partial B_\varepsilon, X \cap H \cap \partial B_\varepsilon)=0$ for $i<n$.*

The local LHT comes with generalizations for the case that K has singularities aside from the one at p, or that it is no longer embedded but is mapped in with large fibres just as in the generalizations of the LHT above. It also has a dual version, the local LHT*.

2.5. Lefschetz Theorems for Intersection Homology

The (middle perversity) intersection homology $IH_i(X)$ of a singular complex algebraic variety X behaves in many ways like the ordinary homology of a nonsingular one. The LHT and the LHT* are examples of this phenomenon.

Theorem (the LHT for intersection homology). *Let X be a possibly singular purely n-dimensional quasiprojective algebraic subvariety of complex projective space, and let H be a generic hyperplane. Then $IH_i(X, X \cap H)=0$ for $i<n$.*

Theorem (the LHT* for intersection homology). *Let X be a possibly singular n-dimensional complex Stein space. Then $IH_i(X)=0$ for $i>n$.*

Note that as a result of the refinements of the LHT* above, we already knew that an n-dimensional Stein space has the homotopy type of an n-dimensional CW-complex. However, this does not imply the LHT* for intersection homology, since intersection homology is not a homotopy invariant.

2.6. Other Connectivity Theorems

So far, we have described results that we prove directly by Morse theory. Many other very interesting results on the connectivity of algebraic varieties can be proved by using one of the above theorems together with an auxiliary construction. The subject of connectivity theorems was pioneered in recent times by W. Fulton and several other mathematicians, especially P. Deligne, G. Faltings, T. Gaffney, J. Hansen, K. Johnson, J.P. Jouanolou, R. Lazarsfeld, J. Roberts, and F.L. Zak. This body of work was one of our primary motivations for undertaking this book. We particularly recommend the survey article by Fulton and Lazarsfeld [FL1] to readers interested in this subject. We will only give a few examples of results here.

The first example is a further generalization of the LHT. It addresses the question, "what happens to the LHT if the linear space H is replaced by a more general variety?" Of course if the more general variety is a complete intersection, then it is a linear section of some embedding of the projective space in some larger projective space, so the LHT applies directly. If it is only a local complete intersection, then we have the following statement of Fulton, proved in [FL1]:

Theorem. *Suppose that X and H are closed local complete intersections in complex projective space $\mathbb{CP}^m$, that X has dimension n and H has codimension d. Then the map*

$$\pi_i(X, X \cap H) \to \pi_i(\mathbb{CP}^m, H)$$

is an isomorphism for $i \leq n-d$ and is surjective for $i = n-d+1$.

This theorem simultaneously generalizes the LHT and the strong homotopy version of the Barth theorem: for a local complete intersection X, $\pi_i(\mathbb{CP}^m, X)=0$ for $i \leq 2n-m+1$. (The latter follows by taking $X=H$.) It generalizes to the case that X is mapped in by a finite map.

The second example is the connectedness theorem, a generalization by Deligne of a theorem of Fulton and Hansen:

Theorem. *Let X be a closed connected purely n-dimensional local complete intersection in $\mathbb{CP}^m \times \mathbb{CP}^m$, and let Δ be the diagonal.*

(a) *If $n-m \geq 1$, then $\pi_1(X, X \cap \Delta)=0$ and $X \cap \Delta$ is connected.*

(b) *If $n-m \geq 2$, then there is an exact sequence*

$$\pi_2(X \cap \Delta) \to \pi_2(X) \to \mathbb{Z} \to \pi_1(X \cap \Delta) \to \pi_1(X) \to 0.$$

(c) *If $2 < i \leq n-m$, then $\pi_i(X \cap \Delta)=0$.*

A similar result holds if the subvariety X is replaced by a finite morphism from X to $\mathbb{CP}^m \times \mathbb{CP}^m$. Even the statements for π_0 and π_1 (which were proved without using any results of this book) have spectacular geometric applications. For example, every immersion into projective space $\mathbb{CP}^m$ of a variety of dimension more than $m/2$ is an embedding (Fulton and Hansen). For every branched covering of projective space $\mathbb{CP}^m$ with at most $m+1$ sheets, there is a point over which all the sheets come together (Gaffney and Lazarsfeld). The fundamental group of the complement of a plane curve with only node singularities is abelian (Fulton and Deligne).

2.7. The Duality

There is a duality which pervades the whole of complex stratified Morse theory and Lefschetz hyperplane theory. We have emphasized this by the notation * in the numbering of statements in the introduction, and of the sections in Part II. This duality is in some sense a form of Poincaré duality.

The simplest case of Theorem LHT* of Sect. 2.1 in the introduction actually follows from theorem LHT by duality. The affine space $\mathbb{C}^k$ of LHT* is compactified by projective space $\mathbb{CP}^k$. Consider first the case that the subvariety $X \subset \mathbb{C}^k$ of theorem LHT* has a nonsingular closure $\bar{X}$ in $\mathbb{CP}^k$ and the hyperplane H at infinity is in general position with respect to $\bar{X}$. Then, Poincaré duality (or more properly Lefschetz duality) says that $H^i(X) \cong H_{2n-i}(\bar{X}, \bar{X} \cap H)$. Theorem LHT says that this vanishes for $i > n$, and by the universal coefficient theorem, this implies that $H_i(X)$ vanishes for $i > n$. The unwanted hypotheses on $\bar{X}$ in this proof can be dispensed with if appropriately stronger versions of LHT and Lefschetz duality are used.

In general, the duality is not a consequence of Poincaré duality, although it seems related to both this and the duality between nonsingularity and compactness in mixed Hodge theory. The more complicated dual pairs of statements do not imply each other, although they often have dual proofs. It is hard to formulate the duality precisely, but one can give a rough dictionary:

Vanishing of low homology groups or low homotopy groups	Vanishing of high homology groups
Vanishing of low degree intersection homology	Vanishing of high degree intersection homology
Nonsingular, or local complete intersection	Closed in Affine space, or Stein
The singularity defect $S(p)$	The local noncompactness defect $S^*(p)$
Large fibres	Large fibres
The Morse function f	The Morse function $-f$
Distance from a codimension c subspace	Distance from a subspace of dimension c

We have used this duality as a guide to discover several of the theorems of this book, as well as their proofs.

2.8. Historical Remarks

The Lefschetz hyperplane theorem first appeared in Lefschetz's book *L'Analysis Situs et la Géométrie Algébrique* published in 1924, simultaneously with Morse's creation of Morse theory. The main technique of proof in Lefschetz's book was the local topological study of generic singularities of a pencil of hyperplane sections. These generic singularities are locally equivalent to quadratic singularities of a complex function. So their study, which is commonly called Picard-Lefschetz theory, is the complex analogue of Morse theory, viewed as the local topological study of quadratic singularities of a real function.

Lefschetz's book initiated the topological study of nonsingular projective varieties. It was known since Riemann that any oriented 2-manifold is homeomorphic to a projective curve, so nonsingular projective curves have no special topological properties. Lefschetz asks the question whether the analogous statement could be true in higher dimensions, and finds that already nonsingular projective surfaces have homological properties not shared by general oriented 4-manifolds ([Lef1], p. 306). In addition to the hyperplane theorem, Lefschetz's book originated two other staples of modern mathematics – the intersection product in homology, and the hard Lefschetz theorem (which states that on a nonsingular projective n-fold, intersecting with a generic i-plane induces an isomorphism between $H_{n+i}(X, \mathbb{Q})$ and $H_{n-i}(X, \mathbb{Q})$). One must agree with Lefschetz's own asessment that this book "planted the harpoon of algebraic topology into the body of the whale of algebraic geometry" ([Lef2], p. 13).

As was already described in Sect. 1.7 of this introduction, algebraic topology was in a very primitive state at this time. Lefschetz credits the fortunate fact that he "made use most uncritically of early topology a la Poincaré" to his relative mathematical isolation in Nebraska and Kansas. Lefschetz spent the next eighteen years of his life largely devoted to the program of rigorizing algebraic topology, culminating in his 1942 Colloquium lectures [Lef4].

It should be noted that noone has ever filled a gap in the original proof of the hard Lefschetz theorem ([Lef2], p. 316 in the middle of the proof of Theorem 13 of Chapter II). In fact, in spite of two other attempts at geometric proofs ([AF2] and [Wa]) it is still unknown whether or not there exists a proof which does not use analysis. (See [Lam].) Of the two known proofs, the one of Hodge uses harmonic analysis and the one of Deligne [D4] uses p-adic analysis.

The dual train of theorems LHT* is much more recent. The theorem that the integral homology of an affine variety vanishes above the complex dimension must have been evident to Lefschetz, since it is a direct combination of Lefschetz duality with the Lefschetz hyperplane theorem. The first published result that we have found was a theorem of Serre in 1953 that the homology with complex coefficients for a nonsingular Stein space vanishes above its complex dimension (together with the above observation for affine varieties). Serre's proof was an application of Cartan's Theorems A and B [C1], [C2]. Serre poses the problem of whether the torsion similarly vanishes, which was shortly answered by Morse theory throught the work of Thom [T9] and Andreotti and Frankel [AF1].

The idea of applying topology to algebraic geometry was not due to Lefschetz. It was, in fact, one of the prime motivations of Riemann and Poincaré for developing homology theory (as is made clear in [Po1]). Picard and Simart in 1897 had studied the homology of affine algebraic surfaces using Picard-Lefschetz theory [PS] (so their book should be viewed also as a precursor of Morse theory). The realization that the homology of nonsingular projective varieties has such incredibly beautiful properties is, however, due to Lefschetz.

The history of developments after Lefschetz could be seen as the whole history of algebraic geometry, and is too vast for treatment here. Since one of the themes of this book is singular spaces, however, we include one observation. The beautiful picture of the homology of a nonsingular projective variety, as completed by Hodge with his (p, q) decomposition, has now been completely reproduced for the intersection homology of a singular projective variety (see [MP2]). The Lefschetz hyperplane theorem is in this book, Poincaré duality is in [GM4], the hard Lefschetz theorem is in [BBD], and the Hodge decomposition is in [Sai].

Part I. Morse Theory of Whitney Stratified Spaces

Chapter 1. Whitney Stratifications and Subanalytic Sets

1.0. Introduction and Historical Remarks

In this chapter we develop the tools from stratification theory which are needed in the proof of the main theorems of Part I (Sects. 3.7, 3.10, 9.5, 10.5, 11.5, 12.5). Sections 1.2 through 1.8 constitute a short course on the main results and techniques of stratification theory, and they summarize the work of many people: [Ab], [AR], [BW], [Ch], [G1], [Ha1], [Ha2], [Hi1], [Hi2], [Hi3], [J], [La], [LT2], [Lo1], [Lo2], [Ma1], [Ma2], [O], [P1], [P2], [T1], [T2], [T3], [T4], [T5], [T6], [Tr1], [Tr2], [V], [Ve1], [Ve2], [W1], [W2], [W3]. Sections 1.8 through 1.10 contain the results on characteristic covectors which are used along with "moving the wall" (Chap. 4) in creating the deformations which are the heart of our main theorems. Section 1.11 is a basic result in stratification theory that is often quoted and whose proof turns out to be surprisingly tricky: the transversal intersection of a manifold M and a Whitney stratified space W admits a "tubular neighborhood" in W, i.e., a neighborhood which is homeomorphic to the total space of a vectorbundle $E \to W \cap M$.

Historical Remarks on Stratification Theory. Singularities were observed by the algebraic geometers of the nineteenth century, and Poincaré in his first paper on homology [Po1], exhibits a singular space and indicates that the singular point destroys Poincaré duality. The natural idea of dividing a singular space into manifolds was at least partially realized in the study of simplicial complexes and regular cell complexes ([Veb]), even before the notion of a manifold was well defined. In fact, there were a number of early attempts to triangulate algebraic sets, including Poincaré [Po2], Lefschetz [Lef3] (1930), Koopman and Brown [KB] (1932), and Lefschetz and J.H.C. Whitehead [LW] (1933). (It is now known that Whitney stratified sets can be triangulated ([G1], [J]), and so up to homeomorphism, the class of Whitney stratified sets coincides with the class of simplicial complexes.)

Perhaps the first attempt at an abstract theory of stratifications appears in Whitney's concept of a "complifold", or complex of manifolds [W4] (1947). However, we will concentrate on the history of stratification theory during the period between 1950 and 1970, when complete proofs of the isotopy lemmas appeared. Although stratification theory developed together with the theory of singularities of smooth mappings, it quickly became an important tool with a broad range of applications which extends well beyond the study of singularities of mappings (see, for example, [Lo3] (1959), [Sc] (1965), [W2] (1965), [Fe1] (1965), [Fe2] (1966), [Z3] (1971), [MP1] (1974)).

Whitney discussed the program of trying to understand the singularities of a generic smooth map during a lecture in Strasbourg in 1950. He announced that he could show that folds and cusps were the only generic singularities for maps from the plane to the plane. His results were not published until [W1] (1955).

As part of this program to understand the topology of a generic smooth map, R. Thom [T1] (1955) defined the iterated singularity sets $S_i(S_j(f))$ of a smooth map $f: \mathbb{R}^n \to \mathbb{R}^m$, and showed that for certain generic functions these were manifolds. Thus, the domain of a smooth function was decomposed into a collection of naturally defined manifolds. These were called "manifold collections" in Whitney [W3] (1955), and this name remained for the next ten years. (See Thom [T1] (1955–56), [T7] (1956), Haefliger [Hae1] (1956/57), Fox [Fox] (1957), Whitney [W5] (1958), Feldman [Fe1] (1964).) It is shown in [W6] (1957) that every real algebraic variety can be decomposed into a finite collection of smooth manifolds. None of these papers suggest any conditions on how the strata of a stratification should fit together, other than that they satisfy the axiom of the frontier. In [Lo3] (1959), Lojasiewicz (following ideas of Osgood [Os] (1929)), decomposed real analytic sets into a locally finite collection of manifolds, and obtained inequalities on the distance between points in these "strata". He used this geometric result to prove the division conjecture of L. Schwartz: the result of dividing a distribution by a real analytic function is again a distribution. See also [Lo4] (1961).

Until at least 1958 it was hoped that the smooth maps which were transverse to a natural stratification of the jet space would be smoothly stable. Thus, the transversality lemma of Thom [T7] (1956) would imply that the smoothly stable maps were dense in the space of all smooth functions. This was hinted at in Thom [T7] (1956) and was explicitly conjectured by Whitney in [W5] (1958). Whitney even carried out this program in special cases: for Morse functions, for maps from the plane to the plane, and for maps to large dimensional spaces.

In his 1959 book on singularities of maps (which was never formally published), H. Levine [Lev] recorded Thom's example showing that smoothly stable maps did not form a dense subset of the space of all maps. This example must have come as a shock for all concerned, for it meant that completely new techniques would be needed in order to study the topology of generic smooth maps.

In the paper [T6] of 1962, R. Thom outlines his enormous program for showing that generic smooth maps are topologically (rather than smoothly) stable, thus resolving the questions raised by this counterexample. He introduces the word "stratification" and proposes a "regularity" condition on how the strata of a stratification should fit together: each stratum should have a neighborhood which fibres over that stratum, which should be controlled by a "fonction tapis", and these various fibrations should be compatible with each other. He states that algebraic sets admit regular stratifications. He then goes on to give an early version of mappings "sans eclatement" (or *Thom mappings*, as they are now known): that the dimension of the fibre over a point p should not increase as p specializes to a smaller stratum. He conjectures that such maps can be triangulated (this conjecture is still open, but believed to be true), and states the first and second isotopy lemmas, applying these to show that functions which are transverse to the natural stratification of the jet space would be topologically stable! The amazing vision which is indicated in this paper was not fully realized until almost eight years later.

These ideas were reiterated with a little more detail in [T3] (1964), in which he proposes a condition – let us call it condition T – on a stratification of an algebraic set. His condition is that a smooth function which is transverse to one stratum of a stratification should consequently be transverse to nearby strata. (This is now known [Tr2] (1979) to be almost equivalent to Whitney's condition A, which had not yet appeared.) Thom also indicates that such a stratification of a real algebraic variety should be locally trivial (or "regular") in the above sense.

It is not surprising that such a visionary paper might contain gaps. In [W2] (1964) Whitney replaced Thom's condition T with his own condition A, and showed that A implies T. (See also Feldman [Fe1] (1965).) Whitney also gave an example to show that condition A alone is not sufficient for local triviality of a stratification, and then proposed his condition B as a candidate sufficient condition. Besides showing that complex analytic varieties admit stratifications satisfying A and B, Whitney showed that these conditions implied *semianalytic* local triviality in low codimensions, and conjectured that any complex analytic variety admits a stratification which is semianalytically locally trivial. (This conjecture has only recently been verified by R. Hardt.) The details of Whitney's theorems appear in [W1] (1965), which was the last paper he wrote on the subject.

Triangulations and Whitney stratifications of semianalytic sets were constructed in [Lo2] (1964) and [Lo1] (1965) respectively.

During the period between 1965 and 1969, while J. Boardman [Boa] and J. Mather [Ma3], [Ma4] were working out the details of the theory of smooth stability of smooth maps, Thom worked on realizing his program. In his landmark paper [T5] of 1969, he makes the daring steps of abandoning Whitney's ideas for semianalytic local triviality, and of considering instead discontinuous (but "controlled") vectorfields on a stratified set, showing that they have a continuous flow! (This idea is briefly mentioned in [T6] (1962).) He thus obtains topological local triviality of Whitney A and B stratifications. (This paper contains many more ideas: the modern definition of a Thom (A_f) mapping, proofs of the isotopy lemmas, definition of subanalytic sets, stratification of analytic maps, and another more complete outline of the proof that topologically stable maps are dense.)

Mather's 1970 Harvard notes [Ma1] are a detailed working out of some of these ideas, complete with multiple inductions and a number of valuable improvements and simplifications (for example, Thom's fonction tapis is replaced by a distance function). Although these notes are still the best reference on Whitney stratifications, they were not published because they were meant as the first chapter of a book on topological stability of smooth maps, which has not been completed.

O. Zariski (who was working since 1962 on algebraic notions of equisingularity) apparently followed these developments closely, as can be seen in his papers on equisingularity [Z2] (1965, 1966), [Z3] (1971).

Although we arbitrarily end this history at 1970, it should be noted that tremendous advances in both stratification theory and singularities of maps have been made since.

1.1. Decomposed Spaces and Maps

Each of the spaces in this paper (for example, the spaces A and B making up the local Morse data) have both a Whitney stratification and a coarser decomposition called an $\mathcal{S}$-decomposition. Here $\mathcal{S}$ is a partially ordered set which indexes the strata of Z, and the $\mathcal{S}$-decomposition of the local Morse data is given by its intersection with the strata of Z. The attaching maps and homeomorphisms of the theorems in Chapter 3 respect the $\mathcal{S}$-decompositions. Keeping track of this additional data is important when considering nonproper Morse functions on a space X, since our approach to this situation is to embed X as a union of strata in a space Z, i.e., as a union of pieces of the $\mathcal{S}$-decomposition.

The spaces which are derived from Z using the Morse function also have Whitney stratifications, which are finer than the stratification of Z. For example, the space $Z_{\leq c}$ has "boundary strata" where $f(z)=c$ (in other words, pieces of the $\mathcal{S}$-decomposition have codimension 1 strata). The local Morse data has boundary strata at the edge of the ball and where $f(z)=v\pm\varepsilon$ so it has "corner strata" (which are codimension two in the $\mathcal{S}$ pieces) where these intersect. The attaching maps and homeomorphisms of Theorems 3.5.4 and 3.7 do *not* respect these Whitney stratifications. This may already be seen in the nonsingular case: adding a handle to a manifold with boundary creates corners which are new strata. In Smale's theory of the diffeomorphism type of Morse data, these corners must be smoothed by "straightening the angle".

Definition. Let $\mathcal{S}$ denote a partially ordered set with order relation denoted by $<$. An *$\mathcal{S}$-decomposition* of a topological space Z is a locally finite collection of disjoint locally closed subsets called *pieces*, $S_i \subset Z$ (one for each $i \in \mathcal{S}$) such that

(1) $Z = \bigcup_{i \in \mathcal{S}} S_i$

(2) $S_i \cap \bar{S}_j \neq \phi \Leftrightarrow S_i \subset \bar{S}_j \Leftrightarrow i=j$ or $i<j$ (and we write $S_i < S_j$)

A decomposed map $f\colon A \to Z$ between two $\mathcal{S}$-decomposed spaces

$$A = \bigcup_{i \in \mathcal{S}} R_i \quad \text{and} \quad Z = \bigcup_{i \in \mathcal{S}} S_i$$

is a continuous map such that $f(R_i) \subset S_i$ for each $i \in \mathcal{S}$. Any subspace B of an $\mathcal{S}$-decomposed space A inherits an $\mathcal{S}$-decomposition and the inclusion $B \subset A$ becomes an $\mathcal{S}$-decomposed map. In this case we say that the pair of spaces (A, B) is *$\mathcal{S}$-decomposed.*

If (A, B) and Z are $\mathcal{S}$-decomposed spaces and if $f\colon B \to Z$ is an $\mathcal{S}$-decomposed map, then the adjunction space $Y = Z \cup_B A$ (which is obtained from the disjoint union $Z \cup A$ by identifying each $a \in B$ with $f(a) \in Z$) is canonically $\mathcal{S}$-

decomposed by subsets $Q_i = S_i \cup_{T_i} R_i$ for each $i \in \mathscr{S}$ (where S_i, T_i, and R_i are the pieces of the decomposition of Z, B, and A respectively).

If $\mathscr{T} \subset \mathscr{S}$ is a partially ordered subset and if Z is $\mathscr{S}$ decomposed, we define $Z \cap |\mathscr{T}|$ to be the corresponding $\mathscr{T}$-decomposed subset of Z, i.e.,

$$Z \cap |\mathscr{T}| = \bigcup_{i \in \mathscr{T}} S_i.$$

An $\mathscr{S}$-decomposed homeomorphism $f: Z_1 \to Z_2$ induces homeomorphisms

$$Z_1 \cap |\mathscr{T}| \to Z_2 \cap |\mathscr{T}|$$

for each subset $\mathscr{T} \subset \mathscr{S}$.

Suppose $F: \mathscr{S}' \to \mathscr{S}$ is a morphism of partially ordered sets (i.e., for each $i<j$ we have $F(i)<F(j)$). Let $\{S_i\}$ be an $\mathscr{S}$-decomposition of a topological space Z, and let $\{S'_i\}$ be an $\mathscr{S}'$ decomposition of Z. Then the $\{S'_i\}$ is called a *refinement* of the $\{S_i\}$ if, for each $i \in \mathscr{S}$ we have $S'_i \subset S_{F(i)}$.

1.2. Stratifications

Let Z be a closed subset of a smooth manifold M, and suppose that

$$Z = \bigcup_{i \in \mathscr{S}} S_i$$

is an $\mathscr{S}$-decomposition of Z, where $\mathscr{S}$ is some partially ordered set. This decomposition is a *Whitney stratification* ([T5]) of Z provided:

(1) Each piece S_i is a locally closed smooth submanifold (which may or may not be connected) of M.

(2) Whenever $S_\alpha < S_\beta$ then the pair satisfies Whitney's conditions A and B: suppose $x_i \in S_\beta$ is a sequence of points converging to some $y \in S_\alpha$. Suppose $y_i \in S_\alpha$ also converges to y, and suppose that (with respect to some local coordinate system on M) the secant lines $\ell_i = \overline{x_i y_i}$ converge to some limiting line ℓ, and the tangent planes $T_{x_i} S_\beta$ converge to some limiting plane τ. Then

(2a) $T_y S_\alpha \subset \tau$ and

(2b) $\ell \subset \tau$.

(It is easy to see [Ma1] that (2b)$\Rightarrow$(2a).) The Whitney conditions are important because:

(1) Any closed subanalytic subset of an analytic manifold admits a Whitney stratification ([Ha1], [Hi1]).

(2) Subanalytic maps admit "Whitney stratifications" (see Sects. 1.6, 1.7).

(3) Whitney stratifications are locally topologically trivial along the strata (see Sect. 1.4).

(4) Whitney stratified spaces can be triangulated ([G], [J], [Ve1]).

(5) The transversal intersection of two Whitney stratified spaces is again a Whitney stratified space, whose strata are the intersections of the strata of the two spaces [Ch].

1.3. Transversality

1.3.1. Definitions. If $Z_1 \subset M_1$ and $Z_2 \subset M_2$ are Whitney subsets of smooth manifolds and if $f\colon M_1 \to M_2$ is a smooth map, we shall say the restriction $f|Z_1\colon Z_1 \to M_2$ is *transverse* to Z_2 if, for each stratum A in Z_1 and for each stratum B in Z_2 the map $f|A\colon A \to M_2$ is transverse to B (i.e., $df(x)(T_x A) + T_{f(x)} B = T_{f(x)} M_2$). It follows ([Ma1], [Ma2]) that the map $f\colon M_1 \to M_2$ takes a neighborhood of Z_1 transversally to Z_2 and that

$$Z_1 \cap f^{-1}(Z_2) = (f|Z_1)^{-1}(Z_2)$$

is Whitney stratified by strata of the form $A \cap f^{-1}(B)$, where A is a stratum of Z_1 and B is a stratum of Z_2. (We will write $Z_1 \pitchfork Z_2$.)

1.3.2. Transversality is open and dense. Recall [Ma3] that if M_1 and M_2 are smooth manifolds, then the Whitney C^∞ topology on the space $C^\infty(M_1, M_2)$ of smooth maps, is the topology whose basis consists of the open sets

$$M(U) = \{f\colon M_1 \to M_2 \mid j^k f(M_1) \subset U\}$$

where $1 \le k < \infty$ and U is an open subset of the bundle $J^k(M_1, M_2) \to M_1 \times M_2$ of k-jets of maps.

Proposition ([G2], [Tr2]). *If $Z_1 \subset M_1$ and $Z_2 \subset M_2$ are closed subsets with Whitney stratifications, then*

$$T = \{f \in C^\infty(M_1, M_2) \mid f|Z_1 \text{ is transverse to } Z_2\}$$

is open and dense (in the Whitney C^∞ topology) in $C^\infty(M_1, M_2)$.

1.3.3. Remarks. We do not assume that Z_1 or Z_2 is compact. However, the stratifications must satisfy Whitney's condition A, otherwise the set T will fail to be open. This is in contradiction to [GG] Exercise 3, p. 59: see [Tr1] or [KT] for interesting counterexamples. In fact, it is shown in [Tr2] that the set T is open if and only if the stratifications of Z_1 and Z_2 satisfy Whitney's condition A.

1.3.4. Proof that T is open. Fix a point $(x_1, x_2, \xi) \in J^1(M_1, M_2)$, i.e., $\xi \in \mathrm{Hom}(T_{x_1} M_1, T_{x_2} M_2)$. Let us say that ξ is *not transverse* if $x_1 \in Z_1$, $x_2 \in Z_2$, and $\xi(T_{x_1} S_1) + T_{x_2} S_2 \neq T_{x_2} M_2$, where S_i is the stratum of Z_i which contains the point x_i. Then the set of nontransverse points is closed in $J^1(M_1, M_2)$. If $(x_1^{(i)}, x_2^{(i)}, \xi^{(i)})$ is a sequence of nontransverse points which converge to the point (x_1, x_2, ξ), then by taking subsequences if necessary, we can assume that (for $1 \le i < \infty$) the points $x_1^{(i)}$ all lie in the same stratum S_1 of Z_1, and the tangent spaces $T_{x_1} S_1$ converge to some limiting plane τ_1. We may also assume that $x_2^{(i)}$ all lie in the same stratum S_2 of Z_2, and the tangent spaces $T_{x_2} S_2$ converge to some limiting tangent space τ_2. It follows that $\xi(\tau_1) + \tau_2 \neq T_{x_2} M_2$. However, if S_1' and S_2' denote the strata of Z_1 and Z_2 which contain the limit points x_1 and x_2, then (by Whitney's condition A) we have $\tau_1 \supset T_{x_1} S_1'$ and $\tau_2 \supset T_{x_2} S_2'$. Thus, (x_1, x_2, ξ) is not transverse. It follows that the complement

U of the nontransverse jets is open in $J^1(M_1, M_2)$. Therefore the set

$$T = \{f \in C^\infty(M_1, M_2) \mid j^1 f(Z_1) \subset U\} = \{f \mid j^1 f(M_1) \subset U\}$$

is open.

Next we shall prove that the set T is dense, i.e. that any smooth function $f: M_1 \to M_2$ can be perturbed so as to take Z_1 transversally to Z_2. In fact, such a perturbation can be found within any "submersive family" of self maps of M_2:

1.3.5. Definition. A *family* (P, Φ) *of self maps* of a manifold M is a smooth manifold P together with a smooth mapping $\Phi: P \times M \to M$. For each $p \in P$ and $m \in M$ we associate the partial maps,

$$\Phi_p: M \to M \quad \text{by } \Phi_p(x) = \Phi(p, x)$$
$$\Phi^m: P \to M \quad \text{by } \Phi^m(y) = \Phi(y, m).$$

The family (P, Φ) of self maps will be called *submersive* if, for each $m \in M$ the differential

$$d\Phi^m(p): T_p P \to T_{\Phi(p, m)} M$$

is surjective.

1.3.6. Theorem. *Let (P, Φ) be a submersive family of self maps of M_2. Then the set*

$$U = \{p \in P \mid \Phi_p \circ f \mid Z_1 \text{ is transverse to } Z_2\}$$

is dense in P. If Z_1 is compact then U is also open.

1.3.7. Examples. Take $M_1 = M_2 = P = \mathbb{R}^n$ and f = identity, to see that any two Whitney stratified subsets of Euclidean space can be made transverse by an arbitrarily small translation. Take $M_1 = M_2 = \mathbb{CP}^n$, and P = the manifold of projective transformations, to see that any two projective algebraic varieties can be made transverse by a projective transformation. The same method works if we replace $\mathbb{CP}^n$ by any space which is homogeneous under the action of a Lie group G; such an action is automatically a submersive family of self maps. This gives the characteristic 0 part of the Kleiman transversality theorem ([K1], 1974). More generally, every smooth manifold M admits a submersive family of self maps: choose finitely many vectorfields $V_1, V_2, \ldots, V_r$ on M such that at each point $m \in M$ the vectors $V_1(m), V_2(m), \ldots, V_r(m)$ span the tangent space $T_m M$. Take P to be the r-dimensional vectorspace of formal linear combinations (with $\mathbb{R}$ coefficients) of the V_i and let Φ_p be the time = 1 flow of the corresponding vectorfield. Thus, Proposition 1.3.6 shows that the set T considered above is dense.

Proof. Our proof that U is dense follows the ingenious method of Abraham ([Ab]) and Morse ([Mo2]), as recalled in [AR] and [GG]. This is by now a standard technique. Consider the following diagram

$$\begin{array}{ccc} P \times M_1 & \xrightarrow{\Psi} & M_2 \\ \downarrow{\scriptstyle \pi} & & \uparrow \\ P & & Z_2 \end{array}$$

where $\pi(p, m)=p$ and $\Psi(p, m)=\Phi(p, f(m))=(\Phi_p \circ f)(m)$. Since Φ is a submersive family, each $\Psi_m: P \to M_2$ is transverse to Z_2. Thus $P \times Z_1$ is transverse to $\Psi^{-1}(Z_2)$ and the intersection $(P \times Z_1) \cap \Psi^{-1}(Z_2)$ can be Whitney stratified by strata of the form $(P \times A_1) \cap \Psi^{-1}(A_2)$, where A_1 is a stratum of Z_1 and A_2 is a stratum of Z_2.

For each such pair of strata we define

$$J(A_1, A_2)=\{p \in P \mid p \text{ is a critical value of } \pi \mid (P \times A_1) \cap \Psi^{-1}(A_2)\}$$
$$K(A_1, A_2)=\{(p, m) \in (P \times A_1) \cap \Psi^{-1}(A_2) \mid \{0\} \times T_m A_1 + T_{(p, m)} \Psi^{-1}(A_2) \neq T_p P \times T_m M_1\}.$$

It is easy to see that the following four statements are equivalent:

1. $p \in J(A_1, A_2)$ i.e. p is a critical value of $\pi \mid (P \times A_1) \cap \Psi^{-1}(A_2)$.
2. $p \in \pi(K(A_1, A_2))$ i.e. $\{p\} \times A_1$ is not transverse to $\Psi^{-1}(A_2)$.
3. The restriction $\Psi: \{p\} \times A_1 \to M_2$ is not transverse to A_2.
4. The partial map $\Psi_p: A_1 \to M_2$ is not transverse to A_2.

By Sard's theorem [Sa1], [Sa2] the set $J(A_1, A_2) \subset P$ has measure 0. Thus the set

$$J=\cup\{J(A_1, A_2) \mid A_k \text{ is a stratum of } Z_k\}$$

also has measure 0. By (4) above, $U=P-J$ is the set of parameter values p such that Ψ_p takes Z_1 transversally to Z_2. Thus, U is dense in P.

To see that U is open, it suffices to show that

$$K=\cup\{K(A_1, A_2) \mid A_k \text{ is a stratum of } Z_k\}$$

is closed in $P \times M_1$. But, this is the same argument (using Whitney's condition A) as 1.3.4. □

1.3.8. Remark. If the spaces P, Z_k, and M_k and the maps Φ and f are complex algebraic, and if Z_1 is compact, then the set K is Zariski closed so the set U is Zariski open in P.

1.4. Local Structure of Whitney Stratifications

Fix a point p in a Whitney stratified subset Z of some smooth manifold M. Let S denote the stratum of Z which contains p. Let N' be a smooth submanifold of M which is transverse to each stratum of Z, intersects the stratum S in the single point p, and satisfies $\dim(S)+\dim(N')=\dim(M)$. Choose a Riemannian metric on M and let $r(z)=\text{distance}(z, p)$ for each $z \in M$. Let $B_\delta(p)$ denote the closed ball

$$B_\delta(p)=\{z \in M \mid r(z) \leq \delta\}$$

with boundary

$$\partial B_\delta(p)=\{z \in M \mid r(z)=\delta\}.$$

By Whitney's condition B, if δ is sufficiently small then $\partial B_\delta(p)$ will be transverse to each stratum of Z, and it will also be transverse to each stratum in $Z \cap N'$. Fix such a $\delta>0$.

Definition. The *normal slice* $N(p)$ through the stratum S at the point p is the set
$$N(p) = N' \cap Z \cap B_\delta(p).$$
The *link* $L(p)$ of the stratum S at the point p is the set
$$L(p) = N' \cap Z \cap \partial B_\delta(p).$$

These spaces are canonically Whitney stratified since they are transverse intersections of Whitney stratified spaces. They are also canonically $\mathscr{S}$-decomposed by their intersection with the strata of X. The Whitney stratification is a refinement of the $\mathscr{S}$-decomposition. For δ sufficiently small, the topological type of the pair $(N(p), L(p))$ is independent of the choice of δ, the Riemannian metric, the choice of N', or the choice of the point p within a single connected component of the stratum S. (The proof follows from Thom's first isotopy lemma (Sect. 1.5) and the fact that any two choices of δ, metric, N', and p are connected by a one parameter family of such choices. See [Ma1]. In Chap. 7 we will prove a collection of similar independence results.)

There is a homeomorphism between the normal slice $N(p)$ and the cone
$$c(L(p)) = L(p) \times [0, 1]/(x, 0) \sim (y, 0) \quad \text{for all } x, y \in L(p)$$
which takes the point $p \in N(p) \cap S$ to the cone point. This homeomorphism is stratum preserving when $c(L(p))$ is stratified so that $L(p) \times \{1\}$ is a union of strata, the cone point is a stratum, and the remaining strata are of the form $A \times (0, 1)$ where A is a stratum of $L(p)$.

The point $p \in S$ has a (closed) neighborhood which is homeomorphic (by a stratum preserving homeomorphism) to $\mathbb{R}^s \times \operatorname{cone}(L_S)$ where $s = \dim(S)$. In fact, S has a (closed) neighborhood T_S in Z and a locally trivial projection $\pi: T_S \to S$ such that the fibre $\pi^{-1}(p)$ is homeomorphic to the cone over L_S ([T5], [Ma1]).

1.5. Stratified Submersions and Thom's First Isotopy Lemma

Suppose Z is a Whitney stratified subset of a smooth manifold M. Let $f: M \to N$ be a smooth map such that

(a) $f \mid Z$ is proper.

(b) for each stratum A of Z, the restriction $(f \mid A): A \to N$ is a submersion.

Such a map is called a (proper) *stratified submersion.* For each $t \in \mathbb{R}^n$, the set $Z \cap f^{-1}(t)$ is Whitney stratified by its intersection with the strata of Z.

Theorem (Thom's first isotopy lemma [T5], [Ma1], [Ma2]). *Let $f: Z \to \mathbb{R}^n$ be a proper stratified submersion. Then there is a stratum preserving homeomorphism,*
$$h: Z \to \mathbb{R}^n \times (f^{-1}(0) \cap Z)$$
which is smooth on each stratum and commutes with the projection to $\mathbb{R}^n$. In particular the fibres of $f \mid Z$ are homeomorphic by a stratum preserving homeomorphism.

Proof. The complete proof takes over fifty pages in Mather's 1970 notes [Ma1], but here is an outline: it is possible to find a system $\{T_A, \pi_A, \rho_A\}$ of *control data* on Z. This consists of, for each stratum A of Z, a tubular neighborhood T_A of A in M, and a tubular projection $\pi_A: T_A \to A$ and a tubular "distance function" $\rho_A: T_A \to [0, \varepsilon]$ such that $\rho_A^{-1}(0) = A$, and so that for any stratum $B > A$ we have:

(1) $(\pi_A, \rho_A) | B \cap T_A: B \cap T_A \to A \times (0, \varepsilon)$ is a smooth submersion.
(2) $\pi_A \circ \pi_B = \pi_B \circ \pi_A$ whenever both sides are defined.
(3) $\rho_A \circ \pi_B = \rho_A$ whenever both sides are defined.

Furthermore, such a system of control data can be found which is *compatible with f*, i.e., so that

(4) $f \circ \pi_A = f$ for each stratum A of Z.

Any vectorfield V on $\mathbb{R}^n$ has a *controlled lift* to a vectorfield W on Z. This means that W is tangent to the strata of Z, and whenever $A < B$ are strata we have

(1) $(\pi_A)_*(W | B \cap T_A) = W | A$.
(2) $d\rho_A(p)(W(p)) = 0$ for all $p \in B \cap T_A$ (i.e., W is tangent to surfaces of constant ρ_A).
(3) $f_*(W) = V$.

It turns out that the integral curves of such a controlled vectorfield W fit together (stratum by stratum) to give a continuous one parameter family of stratum preserving homeomorphisms of Z which commute with f. Furthermore, commuting vectorfields $V_1, V_2, \ldots, V_n$ on $\mathbb{R}^k$ can be lifted to commuting controlled vectorfields $W_1, W_2, \ldots, W_n$ on Z. This exhibits Z as a topological product $Z = f^{-1}(0) \cap Z \times \mathbb{R}^n$. □

1.6. Stratified Maps

Suppose $Y_1 \subset M_1$ and $Y_2 \subset M_2$ are Whitney stratified subsets of smooth manifolds M_1 and M_2. Let $f: M_1 \to M_2$ be a smooth map such that $f | Y_1$ is proper and $f(Y_1) \subset Y_2$. Then $f | Y_1$ is a *stratified map* provided that, for each stratum $A_2 \subset Y_2$, the preimage $f^{-1}(A_2)$ is a union of connected components of strata of Y_1, and f takes each of these components of strata submersively to A_2.

Such a stratified map has several local triviality properties along any connected component B of a stratum in Y_2. The restriction $f | f^{-1}(B)$ is a proper stratified submersion and hence (by Thom's first isotopy lemma) is topologically a locally trivial fibre bundle in a statum preserving way. Therefore, the fibres over any two points in B are homeomorphic. The homeomorphism type of the inverse image $f^{-1}(N(b))$ of a normal slice $N(b)$ through a point $b \in B$ is independent of the choices made in the definition of $N(b)$, and is also independent of b. However it is *not* true that the topological type of the map $f^{-1}(N(b)) \to N(b)$ is independent of the choices, nor is it independent of the point b. See [T6] and [Nak] for interesting counterexamples.

1.7. Stratification of Subanalytic Sets and Maps ([Ha1], [Ha2], [Hi1])

A *semianalytic subset* A of a real analytic manifold M is a subset which can be covered by open sets $U \subset M$ such that each $U \cap A$ is a union of connected components of sets of the form $g^{-1}(0) - h^{-1}(0)$, where g and h belong to some finite collection of real valued analytic functions in U. A *subanalytic* subset B of a real analytic manifold M is a subset which can be covered by open sets $V \subset M$ such that $V \cap B$ is a union of sets, each of which is a connected component of $f(G) - f(H)$, where G and H belong to some finite family $\mathcal{G}$ of semianalytic subsets of an analytic manifold M', and where $f\colon M' \to M$ is an analytic mapping such that the restriction $f \mid \text{closure}(\cup \mathcal{G})$ is proper. A *subanalytic map* between two subanalytic sets is one whose graph is subanalytic.

Theorem. *Suppose A and B are subanalytic (resp. real semialgebraic, resp. complex analytic, resp. complex algebraic) subsets of real analytic (resp. real algebraic, resp. complex analytic, resp. complex algebraic) smooth manifolds M and N. Suppose $f\colon A \to B$ is a proper subanalytic (resp. proper real algebraic, resp. proper complex analytic, resp. proper complex algebraic) map. Then there exist Whitney stratifications of A and B into subanalytic (resp. semialgebraic, resp. complex analytic, resp. complex algebraic) smooth manifolds, such that f becomes a stratified map. Furthermore, if $\mathcal{C}$ is a locally finite collection of subanalytic (resp. semialgebraic, resp. complex analytic, resp. complex algebraic) subsets of A, and if $\mathcal{D}$ is a locally finite collection of subanalytic (resp. semialgebraic, resp. complex analytic, resp. complex algebraic) subsets of B, then the stratification may be chosen so that each element of $\mathcal{C}$ and each element of $\mathcal{D}$ is a union of strata of the stratification.*

Proof. As mentioned in the introduction to this chapter, this theorem (as stated) has never appeared in print, although several proofs may be synthesized from the literature. The first outline of a proof is in [T5] Theorem 3.5.1, where he effectively defines subanalytic sets ("P.S.A" sets).

For $M = \mathbb{R}^m$ and $N = \mathbb{R}^n$ and $f\colon M \to N$ analytic, and A, B subanalytic subsets, see [Hi2] p. 215. His method easily globalizes to the situation where M and N are arbitrary analytic manifolds. (Alternatively one can choose embeddings of the analytic manifolds M and N as analytic submanifolds of Euclidean space and extend the map $f\colon M \to N$ to the ambient Euclidean spaces). The complex analytic case is covered by Theorem 1, Sect. 4 of [Hi2].

The second most complete reference is Theorem 2.1 and Applications 2.4, and Theorem 4.1 of [Ha2]. However, Hardt does not verify the Whitney conditions but mentions instead (in the introduction) that his stratifications can be refined so as to satisfy the Whitney conditions, using the method of [Lo]. This is true, but is not simple: one must use a double induction (on the dimension of the stratum and on the dimension of its projection), applying Hardt's Theorem 2.1 at each stage.

Another outline of proof appears in [V] Sect. 3.6, and similar results may be found in [Ma2], [Hi1], [Hi2], [Lo], [Ha2], [Gi], [DS], [DSW].

Caution. If the map f is algebraic and the sets A and B are real algebraic (resp. real analytic, resp. semianalytic) then we cannot conclude that the strata of the resulting stratifications are real algebraic (resp. real analytic, resp. semianalytic). A counterexample is given by $f: \mathbb{R} \to \mathbb{R}$, $f(x) = x^2$ in the first two cases, and by [Lo] p. 135 in the semianalytic case.

1.8. Tangents to a Subanalytic Set

Fix an analytic Whitney stratification of a subanalytic subset Z of some analytic manifold M. Suppose $p \in Z$. Let S be the stratum of Z which contains p.

Definition. A cotangent vector $\xi \in T_p^* M$ is *conormal* for Z at p if $\xi(T_p S) = 0$. The conormal bundle $T_S^* M$ of S in M is the subbundle of $(T^* M)|S$ consisting of all conormal covectors at points in S, i.e.,

$$T_S^* M = \bigcup_{q \in S} \{\xi \in T_q^* M \mid \ \xi | T_q S = 0\}.$$

By Whitney's condition A, the conormal covectors of Z (i.e., the union of the conormal bundles to all the strata of Z) form a closed subset of $T^* M$, which is also called the set of *characteristic covectors* of Z.

Remark. If M is a complex analytic manifold then there is a canonical isomorphism

$$T_p^* M = \mathrm{Hom}_{\mathbb{R}}(T_p M, \mathbb{R}) \cong \mathrm{Hom}_{\mathbb{C}}(T_p M, \mathbb{C})$$

and so, if $S \subset M$ is a complex analytic submanifold we may unambiguously refer to the conormal bundle $T_S^* M \to S$ as a complex vectorbundle.

Definition. A *generalized tangent space* Q at the point p is any plane of the form

$$Q = \lim_{p_i \to p} T_{p_i} R$$

where $R > S$ is a stratum of Z and $p_i \in R$ is a sequence converging to p. The cotangent vector ξ is *degenerate* if there exists a generalized tangent space Q $\neq T_p S$ such that $\xi(Q) = 0$.

Proposition (see [W1]). *The total space of the conormal bundle $T_S^* M$ is a manifold whose dimension is equal to $m = \dim(M)$. The degenerate covectors which are conormal to S form a conical subanalytic subvariety of codimension ≥ 1 in the conormal bundle $T_S^* M$. If Z is a complex analytic variety, then the set of degenerate covectors form a conical complex analytic subvariety of complex codimension ≥ 1 in the conormal bundle $T_S^* M$.*

Proof. Fix any stratum R which contains the stratum S in its closure. Let $r = \dim(R)$, and $m = \dim(M)$. By choosing local coordinates for M about the point p, we may replace M with R^m.

Consider the image of the Gauss mapping

$$g: R \to \mathbb{R}^m \times G_r(\mathbb{R}^m)$$

which assigns to each point $x \in R$ the pair $(x, T_x R)$. The image of g is a subanalytic variety of dimension $\leq r$, so its closure is obtained by adding some subanalytic variety of dimension $\leq r-1$. (This is the *Nash blowup* of R.) Thus, the set of generalized tangent spaces of dimension r at the points of S form a subanalytic variety V in $G_r(\mathbb{R}^m)$ of codimension $\geq r(m-r)-r+1$.

Let $\mathbb{P}(T_S^* M)$ denote the projective bundle of conormal directions. A direction $\xi \in \mathbb{P}(T_S^* M)$ is degenerate (for the stratum R) if $G_r(\ker \xi) \subset G_r \mathbb{R}^m$ has nonempty intersection with this variety V. Consider the Grassmann bundle of r-planes in the kernel bundle, KER over $\mathbb{P}(T_S^* M)$,

$$\begin{array}{ccc} G_r(\mathrm{KER}) & \xrightarrow{\phi} & G_r(\mathbb{R}^m) \\ \downarrow{\scriptstyle \pi} & & \uparrow \\ \mathbb{P}(T_S^* M) & & V \end{array}$$

The map ϕ is induced by the inclusion $\ker(\xi) \subset \mathbb{R}^m$ for each $\xi \in \mathbb{P}(T_S^* M)$, and it is a submersion. Thus, ϕ is transverse to (each stratum of a Whitney stratification of) V, so $\phi^{-1}(V)$ is a subanalytic subvariety in $G_r(\mathrm{KER})$, of dimension $\leq m-2$. Thus, the degenerate cotangent directions $\pi(\phi^{-1}(V))$ form a subanalytic subvariety of dimension $\leq m-2$ and so the degenerate covectors (for the stratum R) form a conical subvariety of dimension $\leq m-1$ in $T_S^* M$. The collection of all degenerate covectors is the union of the covectors which are degenerate for the various strata $R > S$, so this dimensional estimate holds for the set of all degenerate covectors. □

Discussion. One might expect the set of degenerate covectors in $T_S^* M$ to form a locally trivial fibre bundle (with conical fibre) over the stratum S. Unfortunately, this is not true. There may be certain "exceptional" points $p \in S$ where *every* conormal vector $\xi \in T_S^* M$ at p is degenerate! [OT1]. No Morse function is allowed to have such a point as a critical point, so from the point of view of Morse theory, they may be ignored.

Definition. Let Z be a Whitney stratified subanalytic set, and let S be a stratum of Z. A point $p \in S$ is *exceptional* if the degenerate conormal vectors at p form a codimension 0 subvariety of the conormal space at p.

Teissier [Te1] (Proposition 1.2.1, p. 461) has proven that a Whitney stratification of a complex analytic variety has no exceptional points, and every real analytic variety admits Whitney stratifications with no exceptional points (in fact, the strata need only be the real points of a Whitney stratification of the complexification of the variety). Theorem 4.1 of [O] states that the exceptional points are precisely the "$b_{\mathrm{cod}\,1}$ faults" in the sense of [OT1], [Tr3], [LT2]. If Z is subanalytic and either

(1) $\dim(S) = 1$
(2) $\dim(S) = \dim(Z) - 1$
(3) Kuo's ratio test (v) holds on S([Ku]), or
(4) Verdier's test (w) holds on S([V])

then S has no exceptional points. See [NT] in case (1), (3), or (4), and [Hi4] in case (2). For complex analytic varieties Z, Whitney's condition (b) implies Kuo's criterion (v) by [HM], which gives another proof of the nonexistence of exceptional points in this case. At present, there is no known example of exceptional points on a Whitney (b) stratification of a subanalytic set.

1.9. Characteristic Points and Characteristic Covectors of a Map

Let $f\colon M\to N$ be a smooth map between smooth manifolds and let $Z\subset M$ be a Whitney stratified subset. Suppose $f|Z$ is proper.

Definition. A cotangent vector $\xi\in T_q^*(N)$ is *characteristic* (for f and Z) if, for any $p\in Z\cap f^{-1}(q)$, the pullback $f^*(\xi)\in T_p^*M$ is characteristic for Z. A *point* $q\in N$ is characteristic for the map f, if there exists a nonzero characteristic covector $\xi\in T_q^*N$.

By Whitney's condition A, the characteristic covectors $\xi\in T^*N$ of the map f form a closed subset of T^*N, and the characteristic points $q\in N$ form a closed subset of N.

Remark. Suppose W is a submanifold of N and suppose that $q\in W$. Then, the map f takes Z transversally to W at the point q if and only if T_qW is *not* contained in the kernel of any characteristic covector ξ.

1.10. Characteristic Covectors of a Hypersurface

Let M be a smooth manifold and $p\in M$. Let $g\colon M\to\mathbb{R}$ be a smooth map with $dg(p)\neq 0$. Let $N=g^{-1}g(p)$ be the level hypersurface of g. Suppose $\pi\colon M\to\mathbb{R}^2$ is a smooth map such that $d\pi(p)\colon T_pM\to\mathbb{R}^2$ is surjective.

Proposition. *Suppose the restriction $\pi|N\colon N\to\mathbb{R}^2$ has a singularity at p, i.e., for some numbers $a, b\in\mathbb{R}$, the form $a\,du+b\,dv$ pulls back to 0 on T_p^*N. Then, $dg(p)$ is a multiple of the covector $\pi^*(a\,du+b\,dv)$.*

Proof. Let $\langle dg\rangle$ denote the subspace of T_p^*M which is spanned by $dg(p)$. Then we have an exact sequence

$$0\to\langle dg\rangle\to T_p^*M\to T_p^*N\to 0$$

(where the surjection is given by restriction, r), and an injection

$$\pi^*\colon T_{(u,v)}^*\mathbb{R}^2\to T_p^*M.$$

But, $r\pi^*(a\,du+b\,dv)=0$, so $\pi^*(a\,du+b\,dv)=k\,dg$ for some number k. □

1.11. Normally Nonsingular Maps

In this section we recall the definition and basic properties of normally nonsingular inclusions [FM1], [G2], [GM5]. The main theorem in this section has been stated without proof in the literature several times.

Definition. An inclusion of locally compact Hausdorff spaces, $i\colon X\to Y$ is *normally nonsingular* if there is a vectorbundle $E\to X$ and a neighborhood $U\subset E$ of the zero section (which we also denote by X), and a homeomorphism $j\colon U\to Y$

of U to an open subset $j(U)$ in Y, such that the composition $X \to U \to Y$ is equal to i.

Suppose M is a smooth manifold and $Y \subset M$ is a (closed) Whitney stratified subset of M. Let $N \subset M$ be a smooth submanifold and suppose that $N \pitchfork Y$, i.e., N is transverse to each stratum of Y. The space $X = N \cap Y$ is stratified by its intersection with the strata of Y.

Theorem. *The inclusion $X = N \cap Y \to Y$ is normally nonsingular. Moreover,*

(1) *The vectorbundle $E \to X$ in question is equivalent to the restriction to X, $(TM/TN)|X$ of the normal bundle of N in M.*

(2) *The neighborhood $U \subset E$ can be taken to be any sufficiently small ε-neighborhood E_ε of the zero section (with respect to any smoothly varying inner product on the fibres of TM/TN).*

(3) *The image neighborhood $j(U) \subset M$ can be taken to be the intersection $\phi(E_\varepsilon) \cap Y$, where $\phi: E_\varepsilon \to M$ is any smooth embedding of E_ε into a neighborhood of N in M, so that $\phi | N$ is the identity.*

(4) *These choices may be made so that the projection $j(U) \to E \to X$ is stratum preserving. Consequently, for any union Y' of strata in Y, the inclusion $X' = N \cap Y' \to Y'$ is also normally nonsingular.*

The following two diagrams illustrate this theorem. The Whitney object Y consists of three pages of a book which meet at the binding. It is embedded in the manifold $M = \mathbb{R}^3$. The manifold N is the plane which is illustrated in the diagram on the left, and the intersection X is shown in the figure on the right. The vectorbundle neighborhood U of X (in Y) is shaded, and the lines used in the shading represent the fibres of the vectorbundle.

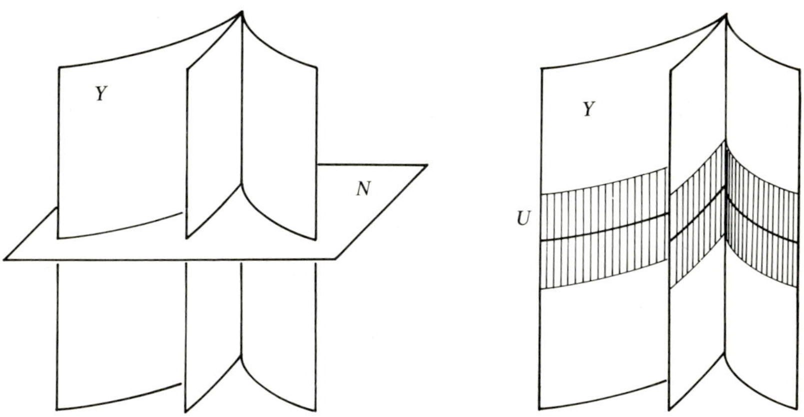

A normally nonsingular inclusion and its tubular neighborhood

Corollary. *Let A be any (not necessarily closed) algebraic subvariety of $\mathbb{CP}^n$. Let $B \subset \mathbb{CP}^n$ be a smooth manifold. Then there is a Zariski open subset V in the space T of projective transformations such that for any $f \in V$ the inclusion $f(B) \cap A \to A$ is normally nonsingular.*

Proof of Corollary. Choose a Whitney stratification of the closure $\bar{A}$ of A, so that A is a union of strata. If $f \in T$ is a projective transformation such that $f(B) \pitchfork \bar{A}$, then the inclusion $f(B) \cap A \to A$ is normally nonsingular by the preceding theorem. However, the set of such transformations f is Zariski open in the set of projective transformations, by the transversality theorem, Sect. 1.3. □

Discussion of Theorem 1.11. One might try to prove this theorem as follows: Define $g: M \to \mathbb{R}$ by $g(p) = \text{distance}(p, N)$. The restriction $g | Y: Y \to \mathbb{R}$ is a stratified submersion (in some neighborhood of $g^{-1}(0)$) and $g^{-1}(0) = X$. Therefore, there is a controlled lift V of the unit vectorfield $-d/dt$ on $\mathbb{R}$. The limits of the integral curves of V define a projection from some neighborhood of X to X. Unfortunately, these limits may not exist, and even if they do exist the resulting map to X may not be continuous.

If the normal bundle of N in M is trivial ($E \cong N \times \mathbb{R}^k$), then the projection $U \to \mathbb{R}^k$ of a tubular neighborhood U of N to the fibre, defines a stratified submersion $\pi: U \cap Y \to \mathbb{R}^k$, which is therefore topologically trivial. In fact, Thom's first isotopy lemma (Sect. 1.5) gives k commuting controlled vectorfields which are lifts of the unit vectorfields $\partial/\partial x_1, \partial/\partial x_2, \ldots, \partial/\partial x_k$ on $\mathbb{R}^k$. Using these vectorfields, it is easy to see that X has a product neighborhood $U \cong X \times \mathbb{R}^k$ in Y. Unfortunately, if the normal bundle of N in M is not trivial, then the controlled vectorfields which are constructed with respect to one local trivialization of E may not agree with the controlled vectorfields which are constructed with respect to a different local trivialization of E. Furthermore, it may not be possible to patch these vectorfields together using a partition of unity.

It is possible to give a proof of Theorem 1.11 by developing a theory of "controlled foliations" which is parallel to the theory of controlled vectorfields, but does not suffer from the patching difficulties described above. We will instead use a trick analogous to the "deformation to the normal bundle" [BFM].

Proof of Theorem 1.11. Choose a tubular neighborhood V of N in M. This consists of a smoothly varying inner product on the fibres of the normal bundle $\pi: E \to N$ of N in M, together with a smooth embedding $\phi: E_\varepsilon \to M$ of some neighborhood

$$E_\varepsilon = \{e \in E \mid \langle e, e \rangle < \varepsilon\}$$

of the zero section N, onto the neighborhood V; with the property that $\phi | N$ is the identity. Since N is transverse to Y we can (by shrinking ε if necessary) assume that ϕ is transverse to (each stratum of) Y. Now define the following one-parameter family of maps

$$\Psi: E_\varepsilon \times (-\delta, 1+\delta) \to M$$

by $\Psi(e, t) = \psi_t(e) = \phi(te)$. Clearly $\psi_1 = \phi$, and $\psi_0 = \phi \circ \pi$. If $\delta > 0$ is chosen sufficiently small, then each map ψ_t is also transverse to Y. This means that $\Psi^{-1}(Y) \subset (-\delta, 1+\delta)$ is a Whitney stratified set which projects to the interval $(-\delta, 1+\delta)$ as a stratified submersion (Sect. 1.5). By Thom's first isotopy lemma, there is a stratum preserving homeomorphism H between $\psi_0^{-1}(Y) = \pi^{-1}(\phi^{-1}(N \cap Y)) \cap E_\varepsilon$ and $\psi_1^{-1}(Y) = \phi^{-1}(Y \cap V)$. Furthermore, $\psi_0^{-1}(Y)$ is the disk bundle over $X = N \cap Y$ of the vectorbundle $E | X$. By defining $j: \psi_0^{-1}(Y)$

$\to Y \cap V$ to be the composition $j = \phi \circ H$, we have shown that the inclusion $X \to Y$ is normally nonsingular. Furthermore, the composite projection

$$Z \cap V \xrightarrow{\phi^{-1}} \psi_1^{-1}(Z) \xrightarrow{H} \psi_0^{-1}(Z) \xrightarrow{\pi} X \xrightarrow{\phi} M$$

is stratum preserving. □

Chapter 2. Morse Functions and Nondepraved Critical Points

2.0. Introduction and Historical Remarks

Classical Morse theory is concerned with the critical points of a class of smooth proper functions f from a manifold Z to the real numbers, called Morse functions. For our generalization, we will let Z be a closed Whitney stratified space in some ambient smooth manifold M. We will need analogues for the notions of smooth function, critical point, and Morse function for this setting. A smooth function on Z will be a function which extends to a smooth function on M. A critical point of a smooth function f will be a critical point of the restriction of f to any stratum S of Z. A proper function f is called Morse if (1) its restriction to each stratum has only nondegenerate critical points, (2) its critical values are distinct, and (3) the differential of f at a critical point p in S does not annihilate any limit of tangent spaces to a stratum other than S. This third condition is a sort of nondegeneracy condition normal to the stratum. If Z is subanalytic (which includes the real and complex analytic cases), then the set of Morse functions forms an open dense subset of the space of smooth functions, and Morse functions are structurally stable, just as in the classical case [P1].

Nondepraved critical points. The Morse functions described above are required to have nondegenerate critical points on each stratum. This condition is more restrictive than is required for the truth of our topological theorems. To provide the natural generality, we define a condition on a critical point called nondepraved (see Sect. 2.3). This is a kind of Whitney condition which is new even for functions on a smooth manifold. Having only nondegenerate critical points is a condition which holds off a set of codimension one in the space of smooth functions. Having only nondepraved critical points holds off a set of infinite codimension.

We do not need the generality of "nondepraved critical points" for any of the results or applications of stratified Morse theory. However, it is the natural generality for the techniques and theorems of stratified Morse theory, because the proofs are exactly the same whether we consider nondegenerate critical points or nondepraved critical points. The notion of a nondepraved critical point may also be of some independent interest to specialists in singularity theory since the nondepravity condition is like a "Whitney – B" condition on the values of the function.

The main results (Sect. 3.7) of this part apply to arbitrary functions with nondepraved critical points which are defined on arbitrary Whitney stratified spaces. However, in order to prove that such functions exist (Sect. 2.2.1) it is necessary to assume that the Whitney stratified space is subanalytic. This is the only place in Part I where the subanalyticity assumption is necessary.

Historical remarks on Morse functions and transversality. In [Mo4] (1925), M. Morse showed that any smooth function could be approximated by one which has only nondegenerate critical points (we will call this the Morse approximation lemma), and that such a function could be expressed (locally near a critical point) as a linear combination of squares of coordinates.

In 1936, H. Whitney [W7] proved that a smooth map $f: M \to N$ between two manifolds could be perturbed so as to miss a submanifold $W \subset N$, provided that $\dim(M) < \operatorname{codim}_N(W)$. This may be considered an early form of the transversality theorem. (The notion of general position had, of course, been considered much earlier by the Italian school of algebraic geometry (see the historical comments in [Fu]) and later by Lefschetz and other topologists studying intersections of chains.)

The program to study the singularities of a generic smooth map was pushed by Whitney, who gave a seminar in Strasbourg during 1950, in which he announced that folds and cusps were the generic singularities of a smooth map from the plane to the plane. These results were published in [W3] (1955), in which he used his 1936 version of transversality to give another proof of the Morse approximation lemma.

In 1953, R. Thom [T11] introduced the modern definition of transversality and used the theorem of Sard [Sa1] (1942) to show that transversality could be attained by arbitrarily small perturbations. (Sard's lemma was preceded by the result of A.P. Morse on the critical values of a real valued function.) Thom later extended his transversality lemma to include the case of transversality in the jet space [T7] (1956), [T1] (1955), in which he showed, using Sard's theorem, that locally any smooth function could be approximated by one which "encounters its own singularities transversally", i.e., which is transverse to naturally defined submanifolds of the space of all possible derivatives. He also mentions that the Morse approximation lemma is a special case of this transversality result. (See Haefliger's 1956 article [Hae1] for a beautiful exposition and globalization of these results, translated into the modern language of jet bundles.)

Until 1963, all versions of the transversality and Morse lemmas were proven by modifying the function locally by a polynomial, and then patching the modifications together using a partition of unity. (See, for example, Sternberg [Ste1] (1964) p. 65.) However, in 1963, R. Abraham [Ab], [AR] gave a simple, elegant, and powerful proof of the jet transversality lemma, which is still the model for modern extensions (eg., [GG] (1973), [Kl] (1974)), and is reproduced in this chapter. Abraham's paper was apparently unknown to Morse, who rediscovered the method in [Mo2] (1967).

Morse functions which are obtained as the distance from a generic point in Euclidean space were considered in [AF1], [Mo3], [Mi1].

Maps which are transverse to a "manifold collection" were also considered in a variety of papers from [T7] (1956) to [W5] (1968). Although Thom's transversality lemma shows that the smooth maps which are transverse to a manifold collection form a dense subset of the space of all smooth maps, it was apparently missed (see [Tr1], [Tr2]) that the stratification must satisfy Whitney's condition A in order that the transversal maps form an open set. In 1965, Whitney [W2] and Feldman [Fe1] showed that the smooth maps which are transverse to a Whitney (a)-regular stratification form an open and dense set.

Morse functions on a stratified space are defined (for the case of isolated singularities) in Lazzeri's important paper [La] (1973), where he indicates the essential reason why such Morse functions are dense in the space of all functions, provided the stratified set is analytic. In [Be] (1977), Benedetti established the density of Morse functions on a complex analytic space in order to complete the proofs of the finiteness theorems of [AG]. Pignoni [P1] (1979) generalized the density theorem to real analytic spaces and also showed that Morse functions on a stratified space are structurally stable. Recent work on Morse functions for Whitney stratified spaces includes [O], [Ot1], [Ot2], [Bru]. (Note Sect. 2.2.2 that the Whitney conditions are not sufficient to guarantee that the Morse functions on a stratified space are dense in the space of all smooth functions.)

2.1. Definitions

Throughout this chapter we fix a Whitney stratification of a subset Z of some smooth manifold M. We will consider a function $f\colon Z\to\mathbb{R}$ which is the restriction of a smooth function $\tilde{f}\colon M\to\mathbb{R}$. A *critical point* of such a function f is any point $z\in Z$ such that $d\tilde{f}(p)|T_pS=0$, where S is the stratum of Z which contains p. The corresponding critical value $v=f(p)$ will be said to be *isolated* provided there exists $\varepsilon>0$ such that $f^{-1}[v-\varepsilon, v+\varepsilon]$ contains no critical points other than p.

Definition. A *Morse function* ([Be], [La], [P1], [P2]) $f\colon Z\to\mathbb{R}$ is the restriction of a smooth function $\tilde{f}\colon M\to\mathbb{R}$ such that

(a) $f=\tilde{f}|Z$ is proper and the critical values of f are distinct.

(b) For each stratum S of Z, the critical points of $f|S$ are nondegenerate (i.e., if $\dim(S)\geq 1$, the Hessian matrix of $f|S$ is nonsingular).

(c) For every such critical point $p\in S$, and for each generalized tangent space (Sect. 1.8) Q at the point p, the following nondegeneracy condition holds: $d\tilde{f}(p)(Q)\neq 0$ except for the single case $Q=T_pS$ (i.e., the cotangent vector $d\tilde{f}(p)$ is characteristic but nondegenerate in the sense of Sect. 1.8).

Remarks. Whether or not a point $p\in Z$ is a critical point of f depends as much on the stratification of Z as on the function f. For example, every zero-dimensional stratum of Z is a critical point of f. If $p\in M$ is a critical point for the function $\tilde{f}\colon M\to\mathbb{R}$, and if p is a singular point of Z, then it is a degenerate critical point of f, so f is not a Morse function. The critical points of a Morse function are isolated. Morse functions which are defined on a real analytic set are topologically stable (see [P1], [Bru]).

For notational convenience we shall no longer distinguish between f and $\tilde{f}$.

2.2. Existence of Morse Functions

The results in Part I of this book apply to any Morse function which is defined on an arbitrary Whitney stratified set. However, we can only be guaranteed of the existence of Morse functions if the set is subanalytic.

2.2.1. Theorem ([Mo1], [Mo2], [T7], [W1], [Be], [P1], [P2], [O]). *Suppose Z is a closed Whitney stratified subanalytic subset of an analytic manifold M. Then the functions $\tilde{f}: M \to \mathbb{R}$ whose restrictions to Z are Morse, form an open and dense subset (in the Whitney C^∞ topology (Sect. 1.3.2)) of the space $C_p^\infty(M, \mathbb{R})$ of smooth proper maps.*

2.2.2. Example. The rapid spiral

$$Z = \{(x, y) \in \mathbb{R}^2 \mid r = \exp(-\theta^2),\ \theta \geq 0\}$$

is a Whitney stratified subset of $\mathbb{R}^2$ (see Thom [T5]) which has *no* Morse functions.

For many applications it is not enough to know that a given function can be approximated by a Morse function. One may have a certain collection of approximations available and it may be necessary to find a Morse function within this collection. For this reason, we give the following version of the density theorem for Morse functions:

Let Z be a closed subanalytic subset of an analytic manifold M. Let P be a finite-dimensional smooth manifold (the "parameter space") and let $F: P \times M \to \mathbb{R}$ be a smooth function. We think of F as a family of functions, $f_\alpha = F(\alpha, .): M \to \mathbb{R}$ which is parametrized by P. Define $\Phi: P \times M \to T^*M$ by $\Phi(\alpha, x) = d(f_\alpha)(x) \in T_x^*M$.

2.2.3. Theorem. *If the map Φ is a submersion, then for almost any $\alpha \in P$, the corresponding function $f_\alpha: M \to \mathbb{R}$ is a Morse function on Z. If Z is compact, then the parameter values α such that f_α is Morse form an open subset of P.*

Proof. We must find functions f_α which achieve two conditions, for each stratum S of Z:

(a) $f_\alpha \mid S$ has nondegenerate critical points.

(b) $df_\alpha(s)$ is a nondegenerate covector (Sect. 1.8) for each $s \in S$.

Let $D_S \subset T_S^*M \subset T^*M \mid S \subset T^*M$ denote the set of degenerate covectors at points $s \in S$, and define $\Phi_S: P \times S \to T^*S$ by $\Phi_S(\alpha, s) = d(f_\alpha \mid S)(s)$. Consider the following diagram:

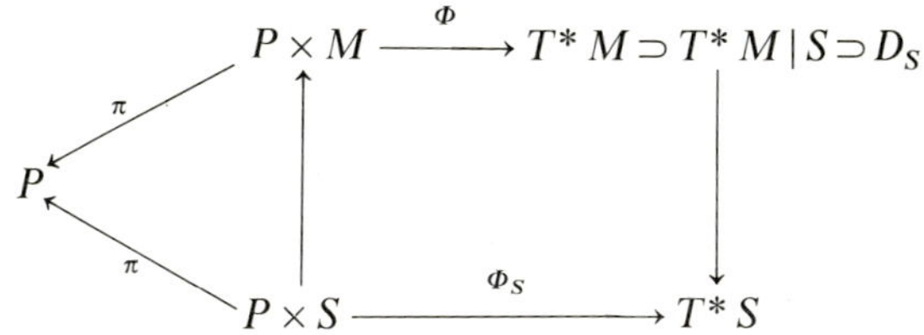

Since Φ is a submersion we have

(A) Φ_S is transverse to the zero section (which we also denote by S) in T^*S. (In fact, Φ_S is a submersion, as may be seen by expressing it in local coordinates, and differentiating.)

(B) Φ is transverse to each stratum of any Whitney stratification of D_S.

By (A), the set $V_1 = \Phi_S^{-1}(S)$ is a smooth submanifold of $P \times S$. Furthermore, $\alpha \in P$ is a critical value of $\pi | V_1 : V_1 \to P$ if and only if $f_\alpha | S$ has a degenerate critical point ($\Leftrightarrow d(f_\alpha | S): S \to T^*S$ is not transverse to S). By Sard's theorem, the set of such bad critical values α has measure 0 in P.

By (B) above, the set $V_2 = \Phi^{-1}(D_S)$ is a Whitney stratified subset of $P \times M$ whose dimension is less than or equal to $\dim(P) - 1$. (This is because $\dim(D_S) \leq \dim(M) - 1$ by Sect. 1.8.) Consequently, the image $\pi(V_2)$ has measure 0. But $\alpha \in \pi(V_2) \Leftrightarrow d(f_\alpha | S)$ kills some limiting tangent space.

We conclude that if $\alpha \notin \pi(V_1) \cup \pi(V_2)$, then f_α satisfies conditions (A) and (B) above. Since X has at most countably many strata, there is a dense subset of parameter values $\alpha \in P$ such that f_α is a Morse function.

Note that (B) implies that for any stratum $R > S$, the restriction $f_\alpha | R$ has *no* critical points in some neighborhood of S. It follows that f_α has finitely many critical points on any compact subset of Z. Thus, if Z is compact, the set of "Morse" parameter values $\alpha \in P$ is open. □

2.2.4. Examples. (1) Suppose $Z \subset \mathbb{R}^n$ and $F: \mathbb{R}^n \times \mathbb{R}^n \to \mathbb{R}$ is the function $F(p, q) = \text{distance}^2(p, q)$. It is easily seen that the corresponding map Φ is a submersion. We conclude ([P1], [Mo2], [Mi2]) that the distance from a generic point in Euclidean space is a Morse function on Z.

(2) Let $Z \subset \mathbb{R}^n$ and let $P = G_k(\mathbb{R}^n)$ denote the Grassmannian of all k-dimensional subspaces of $\mathbb{R}^n$. Take $F: P \times \mathbb{R}^n \to \mathbb{R}$ to be $F(Q, x) = \text{distance}^2(Q, x)$. The resulting map Φ is a submersion, so we conclude that the distance from a generic k-dimensional subspace of Euclidean space is a Morse function on Z.

2.3. Nondepraved Critical Points

The nondegenerate critical points of a Morse function are a particular kind of nondepraved critical point.

Definition. An isolated critical point $p \in \mathbb{R}^n$ of a smooth function $F: \mathbb{R}^n \to \mathbb{R}$ is *nondepraved* if it satisfies the following property: Suppose $p_i \in \mathbb{R}^n$ is a sequence of points converging to p. Suppose the vectors $v_i = (p_i - p)/|p_i - p|$ converge to some limiting vector v, and suppose the subspaces $\ker dF(p_i)$ converge to some limiting subspace τ. Suppose also that v is *not* an element of τ. Then for all i sufficiently large,

$$dF(p_i)(v_i) \cdot (F(p_i) - F(p)) > 0.$$

It follows that a critical point p is depraved if there exists a sequence $p_i \to p$ such that $v_i \to v$, $\ker dF(p_i) \to \tau$, $v \notin \tau$, and either (1) $dF(p_i)(v_i) < 0$ and $F(p_i) > F(p)$ for all i, or (2) $dF(p_i)(v_i) > 0$ and $F(p_i) < F(p)$ for all i.

The following diagram illustrates the second possibility. It consists of the level curves of an *isolated* depraved local maximum of a smooth function f: $\mathbb{R}^2 \to \mathbb{R}$. The points p_i are indicated by the sequence of dots in this figure.

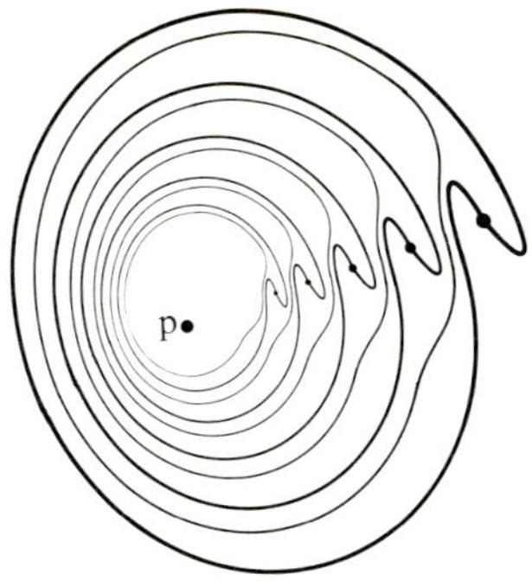

Level curves of a function with a depraved critical point

Remark. Although it is not needed for any of our applications, we will show in Sect. 2.6 that this condition is preserved under diffeomorphism. Thus, we may speak of nondepraved critical points of a smooth function $F: S \to \mathbb{R}$ which is defined on a smooth manifold S.

Now suppose (as above) that Z is a Whitney stratified subanalytic subset of an analytic manifold M, that $f: Z \to \mathbb{R}$ is the restriction of a smooth function defined on M, that $p \in Z$ is a critical point of f and that S is the stratum of Z which contains the point p.

Definition. A critical point $p \in Z$ of the function $f: Z \to R$ is *nondepraved* provided:

(a) the critical point p is isolated

(b) the restriction $f \mid S$ has a nondepraved critical point at p

(c) For each generalized tangent space Q at the point p, the following nondegeneracy condition holds: $d\tilde{f}(p)(Q) \neq 0$ except for the single case $Q = T_p S$ (i.e., the cotangent vector $df(p)$ is characteristic but nondegenerate for Z, in the sense of Sect. 1.8).

2.4. Isolated Critical Points of Analytic Functions

In this section we show that isolated critical points of an analytic function defined on Euclidean space are necessarily nondepraved. It follows that the critical points of Morse functions are nondepraved also.

Proposition. *If $F: \mathbb{R}^n \to \mathbb{R}$ is a (real) analytic function, and if $p \in \mathbb{R}^n$ is an isolated critical point of F, then it is a nondepraved critical point.*

Proof. Suppose the critical point p is depraved. Choose a sequence $p_i \to p$ which violates the nondepravity conditions, i.e., such that

(a) the vectors $v_i = (p_i - p)/|p_i - p|$ converge to some vector v

(b) $\ker dF(p_i)$ converge to some subspace τ

(c) $v \notin \tau$

(d) for all i, $dF(p_i)(v_i) \cdot (F(p_i) - F(p)) \leq 0$.

First let us assume that for all sufficiently large i, we have $F(p_i) \neq F(p)$. By taking a subsequence, if necessary, we may assume that for all i, $F(p_i) - F(p)$

will have a constant sign $s=\pm 1$, so $dF(p_i)(v_i)$ will have the opposite sign, $-s$. Let δ denote the angle between v and τ. Define Q to be the subset of $\mathbb{R}^n-\{p\}$ consisting of all points q which satisfy the following conditions:

(1) $dF(q)\left(\frac{q-p}{|q-p|}\right)\cdot(F(q)-F(p))\leq 0$

(2) $\operatorname{sign}\left(dF(q)\left(\frac{q-p}{|q-p|}\right)\right)=-s$

(3) $\operatorname{angle}\left(\left(\frac{q-p}{|q-p|}\right), \ker dF(q)\right)\geq\delta/2.$

Then Q is a nonempty semianalytic set which contains p in its closure. By the *curve selection lemma* ([Hi1] Lemma 1.3.3, [Mi2] Lemma 3.1, [BC], [Wa]), there is a (real) analytic curve $\alpha\colon[0,1]\to\bar{Q}$ such that $\alpha(0)=p$, $\alpha(t)\neq p$ unless $t=0$. The (germ near p of the) image of this curve satisfies Whitney's condition B. Choose $\varepsilon>0$ so small that

$$t\leq\varepsilon\Rightarrow\operatorname{angle}\left(\alpha'(t),\left(\frac{\alpha(t)-p}{|\alpha(t)-p|}\right)\right)<\delta/4.$$

Thus, angle $(\alpha'(t), \ker dF(\alpha(t)))>\delta/4$ for all $t<\varepsilon$. But this is a contradiction because

$$\begin{aligned}\operatorname{sign}(F(\alpha(\varepsilon))-F(\alpha(0)))&=\operatorname{sign}\int_{t=0}^{\varepsilon}F'(\alpha(t))\cdot\alpha'(t)\,dt\\&=\operatorname{sign}\int_{t=0}^{\varepsilon}F'(\alpha(t))\cdot\left(\frac{\alpha(t)-p}{|\alpha(t)-p|}\right)dt\\&=-s\end{aligned}$$

which is a contradiction.

The other possibility is that $F(p_i)=F(p)$ for all i sufficiently large. In this case the same argument gives a contradiction provided we replace the set Q by the set of points satisfying (1′) $F(q)=F(p)$ and also condition (3) above. □

Remark. The functions with degenerate critical points form a set of codimension 1 in the space of all smooth functions. The functions with depraved critical points form a set of infinite codimension in the space of all smooth functions. This follows from the fact that any finitely determined smooth function (i.e. a function with a k-sufficient jet, for some k) is locally C^∞-equivalent to an analytic function (its k-jet) so its critical points are nondepraved. But the set of functions whose jets are not k-sufficient for any k form a set of infinite codimension ([Ma4] Theorem 3.5, [T3], [DuP]). Nevertheless, there do exist smooth functions with depraved critical points.

2.5. Local Properties of Nondepraved Critical Points

In this section we prove several technical lemmas about nondepraved critical points which will be needed in Chapter 5.

2.5.1. Lemma. *If $F: M \to \mathbb{R}$ is a smooth function with a nondepraved critical point $p \in M$ and if no other critical point of F has the same critical value, then $F^{-1}F(p)$ is a smooth submanifold of M (except possibly at the point p), and the pair $(F^{-1}F(p)-\{p\}, \{p\})$ satisfies Whitney's condition B at the point p.*

Proof. Assume not. Then there is a sequence of points $p_i \in F^{-1}F(p)$ converging to p so that the secant lines $\overline{pp_i}$ converge to some limiting line ℓ, the tangent planes $T_{p_i}(F^{-1}F(p)) = \ker dF(p_i)$ converge to some limiting plane τ, and $\ell \not\subset \tau$. Let $v_i = (p_i - p)/|p_i - p|$ in some local coordinate system about p. Then, $dF(p_i)(v_i) \cdot (F(p_i) - F(p)) = 0$. Thus, the critical point is depraved. □

Suppose $p \in M$ is a nondepraved critical point of a smooth function $F: M \to \mathbb{R}$. Let $r: M \to \mathbb{R}$ denote the distance from the point p, with respect to some smooth Riemannian metric on M.

2.5.2. Lemma. *There is a neighborhood U of p in M with the following property: for each point $q \in U$ $(q \neq p)$, either*

(a) *$dF(q)$ and $dr(q)$ are linearly independent or*

(d) *$a\,dF(q) + b\,dr(q) = 0 \Rightarrow$ the sign of ab is opposite that of $F(q) - F(p)$.*

Proof. Choose a sequence of points $p_i \to p$ which fail both conditions. Choose a subsequence if necessary, so that

(1) the vectors $v_i = (p_i - p)/|p_i - p|$ converge to some limiting vector v

(2) the hyperplanes $\ker dF(p_i)$ converge to some plane τ.

Then, $v \notin \tau$ by condition (a) above, and there are numbers a_i and b_i so that $a_i\,dF(p_i) + b_i\,dr(p_i) = 0$. The nondepravity condition quarantees that $F(p_i) - F(p) \neq 0$ for sufficiently large i, so we may assume that $F(p_i) - F(p)$ has a constant sign $s = \pm 1$. For sufficiently large i, we have

(c) $dr(p_i)(v_i) > 0$ (since v_i points away from p)

(d) $dF(p_i)(v_i)$ has the same sign as $F(p_i) - F(p)$

by the nondepravity condition. However,

$$a_i\,dF(p_i)(v_i) = -b_i\,dr(p_i)(v_i) \quad \text{so} \quad b_i a_i = -a_i^2\,dF(p_i)(v_i)/dr(p_i)(v_i)$$

which has the opposite sign to $F(p_i) - F(p)$. This is a contradiction. □

2.5.3. Corollary. *Suppose local coordinates $x_1, x_2, \ldots, x_n$ on M have been chosen so that*

$$r^2 = x_1^2 + x_2^2 + \ldots + x_n^2$$

is the distance from a nondepraved critical point $p \in M$ of a smooth function $F: M \to \mathbb{R}$. Then at each point q in the neighborhood U described above, either

(a) *$dF(q)$ and $dr(q)$ are linearly independent, or*

(b) *$\dfrac{1}{x_i}\left(\dfrac{\partial F}{\partial x_i}\right)(q)$ has the same sign as $F(q) - F(p)$ (or is undefined).*

Proof. The equation $a\,dF(q) + b\,dr(q) = 0$ becomes in local coordinates,

$$\sum_{i=1}^{n} \left(a\,\frac{\partial F}{\partial x_i}(q) + b\,\frac{x_i}{r} \right) dx_i = 0$$

so for each i we have

$$\frac{1}{x_i}\left(\frac{\partial F}{\partial x_i}\right)(q) = -b/ar$$

which is negative when a and b have the same sign, and is positive when a and b have opposite signs $\square$.

2.6. Nondepraved is Independent of the Coordinate System

It is easy to check that the following is an equivalent statement of the nondepravity condition. (It uses the inner product in $\mathbb{R}^n$):

Suppose $F: (\mathbb{R}^n, 0) \to (\mathbb{R}, 0)$ is a smooth function with an isolated critical point at 0. This critical point is nondepraved provided the following occurs: suppose $p_i \to 0$ is a sequence of points and suppose that $F(p_i)$ has the same sign $s = \pm$ for all i. Suppose the unit vectors $p_i/|p_i|$ converge to a unit vector $\hat{\ell} \in S^{n-1}$, and suppose that the unit covectors $dF(p_i)/|dF(p_i)|$ converge to a unit covector λ. Suppose $\lambda(\hat{\ell}) \neq 0$. Then the sign of $\lambda(\hat{\ell})$ is equal to the sign s.

Proposition. *Suppose that $\phi: (\mathbb{R}^n, 0) \to (\mathbb{R}^n, 0)$ is a diffeomorphism. Then the point 0 is a nondepraved critical point of a smooth function $F: (\mathbb{R}^n, 0) \to (\mathbb{R}, 0)$ if and only if the point $\phi^{-1}(0)$ is a nondepraved critical point of the function $F \circ \phi$.*

Proof. We shall show, for example, that F nondepraved $\Rightarrow F \circ \phi$ is nondepraved. Suppose $q_i \in \mathbb{R}^n$ is a sequence of points converging to 0, and that

(a) the vectors $m_i = q_i/|q_i|$, converge to a limiting vector $\hat{m}$,

(b) the planes $\tau_i = \ker d(F \circ \phi)(q_i)$ converge to a limit τ,

(c) $\hat{m}$ is not contained in τ.

We will assume for simplicity (the other cases being entirely similar) that $F \circ \phi(0) = 0$, and that for all i, $F \circ \phi(q_i) > 0$. We must show that, for sufficiently large i,

$$d(F \circ \phi)(q_i)(\hat{m}) > 0.$$

Let $p_i = \phi(q_i)$. We may suppose that

(a) the unit vectors $\ell_i = p_i/|p_i|$ converge to some limiting unit vector $\hat{\ell}$,

(b) the unit covectors $\lambda_i = dF(p_i)/|dF(p_i)|$ converge to some limiting covector λ.

Since F is nondepraved, we have $\lambda(\hat{\ell}) > 0$. Thus, there is a neighborhood V of λ in the $n-1$ sphere of unit covectors, and a neighborhood U of $\hat{\ell}$ in the $n-1$ sphere of unit vectors so that

$$v \in V \quad \text{and} \quad u \in U \Rightarrow v(u) > 0.$$

We can find an integer I so large that

$$i \geq I \Rightarrow dF(p_i)/|dF(p_i)| \in V.$$

Since ϕ is smooth, the map $\psi: \mathbb{R}^n \to S^{n-1}$ given by

$$\psi(q)=d\phi(q)(\hat{m})/|d\phi(q)(\hat{m})|$$

is continuous and takes the origin to $\hat{\ell}$. Find a neighborhood W of the origin so that $\psi(W)\subset U$. Choose J so large that

$$i\geq J\Rightarrow q_i\in W.$$

Thus, if $i\geq \max(I, J)$, we have

$$\begin{aligned}\psi(q_i)&=\frac{d\phi(q_i)(\hat{m})}{|d\phi(q_i)(\hat{m})|}\in U\\&\Rightarrow\frac{dF(\phi(q_i))}{|dF(\phi(q_i))|}\,\frac{d\phi(q_i)(\hat{m})}{|d\phi(q_i)(\hat{m})|}>0\\&\Rightarrow dF(\phi(q_i))(d\phi(q_i))(\hat{m})>0\end{aligned}$$

as desired. □

Chapter 3. Dramatis Personae and the Main Theorem

3.0. Introduction

In this chapter the main objects of interest: local Morse data, normal Morse data, and tangential Morse data will be defined. The main theorems are stated in Sects. 3.7, 3.8, and 3.9.

Classical Morse theory studies the relation between the critical points of a Morse function and the topology of Z. Specifically, if $Z_{\leq c}$ denotes the set of all points in x where the value of f is $\leq c$, then as c varies, the topology of $Z_{\leq c}$ remains constant unless c passes a critical value, in which case the change in the topology is determined by the local behavior of f near the corresponding critical point. Exactly the same statement is true in our context. We want to study this change in the topology of $Z_{\leq c}$ as c passes the critical value v of a critical point p. We measure it by a pair of spaces (A, B) which we call local Morse data for f at p. The definition of local Morse data is as follows: Let B_δ be the intersection of Z with a small round ball in M centered at p. Then for small enough $\varepsilon > 0$,

$$A \text{ is } Z \cap B_\delta \cap f^{-1}[v-\varepsilon, v+\varepsilon] \quad \text{and} \quad B \text{ is } Z \cap B_\delta \cap f^{-1}(v-\varepsilon).$$

(The topological type of the pair (A, B) is independent of the choices of the ball and of ε.)

Theorem 3.5.4. *If v is the only critical value between $v-\varepsilon$ and $v+\varepsilon$, then $Z_{\leq v+\varepsilon}$ is obtained as a topological space from $Z_{\leq v-\varepsilon}$ by attaching the space A along the space B.*

For example, if Z has only one stratum (the classical case) then local Morse data is just

$$(D^\lambda \times D^{n-\lambda}, \partial D^\lambda \times D^{n-\lambda})$$

where λ denotes the Morse index and D^k is the k-dimensional disk.

The goal then becomes to study the local Morse data (A, B) at p. In order to do this, we define two auxiliary concepts pertaining to the critical point at p. The first, tangential Morse data, is just local Morse data for the restriction of f to the stratum S containing p. The second, normal Morse data, is local Morse data for the restriction of f to the intersection of Z with a smooth transversal slice to S at p. Tangential Morse data and normal Morse data are pairs of topological spaces that are independent of the choices involved in their definition. Together, they determine the local Morse data, and hence the change in the topology of $Z_{\leq c}$ as c passes the critical value v, by the following result whose proof occupies the bulk of the first part of this book:

The Main Theorem (see Sect. 3.7). *As pairs of topological spaces, the local Morse data is the product of the tangential Morse data with the normal Morse data.*

The relevance of theory just sketched to singular analytic varieties is clear, since they can be Whitney stratified. The relevance to noncompact varieties X is the following: Even if X is nonsingular, if the Morse function is not proper, then the topology of $X_{\leq c}$ can have anomalous changes as c varies, even though c does not pass a critical value. To remedy this, we suppose that X embeds as a union of strata in a larger analytic variety Z on which f is proper. (This can often be done. For example, if X is algebraic, it embeds as a union of strata in a compact variety.) Now we will have new critical values, arising from critical points on strata "off the edge" of X (in $Z-X$) which account for the anomalous changes in the topology of $X_{\leq c}$. We will be able to analyze the effect of these new critical points on the topology of X by observing that the two theorems stated above respect the decomposition of Z by strata (see Chap. 10).

Another generalization of classical Morse theory that interests us is the relative case: we suppose that we have a map π of X to Z; we wish to study the topology of X by proper Morse functions f defined on Z. It turnes out that this can be done if π is a stratified map. If we define $X_{\leq c}$ to be the set of points in X where the composition $f\pi$ is $\leq c$, then it is still true that the topological type of $X_{\leq c}$ changes only when c passes a critical value for f. The change in the topology occuring at the critical values can be analyzed through results analogous to the two theorems stated above (see Chaps. 9 and 11).

Notational Definition. Suppose that $B \subset A$ and $W \subset W'$ are topological spaces and that there exists an attaching map $h: B \to W$ such that the identity map $W \to W$ extends to a homeomorphism

$$W' \to W \cup_h A.$$

Then we shall say that W' was obtained from W by *attaching the pair* (A, B) and we will write

$$W' = W \cup (A, B).$$

We will supress the particular map h, although in all our applications h will be an embedding and B will be a closed subset.

3.1. The Setup

Throughout this chapter, we will assume that Z is a Whitney stratified subset of a smooth manifold M, and that the strata of Z are indexed by some partially ordered set $\mathcal{S}$. Let $f: M \to \mathbb{R}$ be a proper smooth function and suppose that $p \in Z$ is a nondepraved (Sect. 2.3) critical point of the restriction $f|Z: Z \to \mathbb{R}$. (For example, f might be a Morse function.) We shall use X to denote the stratum of Z which contains the critical point p, $v=f(p)$ to denote the critical value, and $Z_{\leq a}$ to denote the space $Z\cap f^{-1}(-\infty, a]$. Similarly, we shall denote the set $Z\cap f^{-1}[a,b]$ by $Z_{[a,b]}$. These spaces are canonically $\mathcal{S}$-decomposed (Sect. 1.1) by their intersection with the strata of Z.

3.2. Regular Values

In this section we will assume $f: Z\to \mathbb{R}$ is a proper function.

Proposition. *Suppose the interval* $[a,b]$ *contains no critical values of* $f|Z$. *Then* $Z_{\leq a}$ *is homeomorphic to* $Z_{\leq b}$ *by a homeomorphism which preserves the* $\mathcal{S}$*-decompositions.*

Proof. The proof will appear in Chap. 7.

3.3. Morse Data

Fix $\varepsilon>0$ so that the interval $[v-\varepsilon, v+\varepsilon]$ contains no critical values of $f|Z$ other than $v=f(p)$.

Definition. A pair (A, B) of $\mathcal{S}$-decomposed spaces is *Morse data* for f at the point p if there is an embedding $h: B\to Z_{\leq v-\varepsilon}$ such that $Z_{\leq v+\varepsilon}$ is homeomorphic to the space $Z_{\leq v-\varepsilon}\cup_B A$ (which is obtained from $Z_{\leq v-\varepsilon}$ by attaching A along B using the attaching map h) and the homeomorphism preserves the $\mathcal{S}$-decompositions (see Sect. 1.1). A pair (A', B') is *homotopy Morse data* if it is homotopy equivalent to some choice (A, B) of Morse data.

Remarks. If (A', B') is homotopy Morse data for f at p, then the space $Z_{\leq v+\varepsilon}$ is homotopy equivalent to the space $Z_{\leq v-\varepsilon}\cup(A', B')$ for some choice of attaching map.

Morse data is not a well-defined topological type. However, the homotopy type of the quotient A/B is well-defined and is the *Conley Morse index* (see [Co]). If (A, B) is Morse data for f at p for some choice of $\varepsilon>0$, then it is Morse data for f at p, for any other $\varepsilon'<\varepsilon$ (see Proposition 3.2).

Morse data satisfies the following *excision property:*

Lemma. *If* (A, B) *is Morse data for* f *at* p *and if* C *is a subset of* B *such that* $\bar{C}\subset$ *interior* (B), *then* $(A-C, B-C)$ *is Morse data for* f *at* p.

Proof. The space $Z\cup_B A$ has a basis for the topology, which consists of open sets U such that either $U\cap C=\phi$ or $U\subset B$. □

3.4. Coarse Morse Data

Suppose $f: Z\to\mathbb{R}$ is proper and the critical value $v=f(p)$ is isolated.

Definition. The *coarse* Morse data for f at p is the pair of $\mathcal{S}$-decomposed spaces

$$(A, B)=(Z\cap f^{-1}[v-\varepsilon, v+\varepsilon], Z\cap f^{-1}(v-\varepsilon))$$

where $\varepsilon>0$ is any number such that the interval $[v-\varepsilon, v+\varepsilon]$ contains no critical values other than $v=f(p)$. The coarse Morse data has a canonical $\mathscr{S}$-decomposition which is given by its intersection with the strata of Z.

Remark. The homeomorphism type of the coarse Morse data is independent of the choice of ε, (see the proof in Sect. 7 of Proposition 3.2) and the homeomorphisms may be chosen to preserve the $\mathscr{S}$-decomposition. The coarse Morse data is Morse data.

The following pair of spaces illustrates the coarse Morse data for the critical point p'_3 of the example in Sect. 1.2 of the introduction:

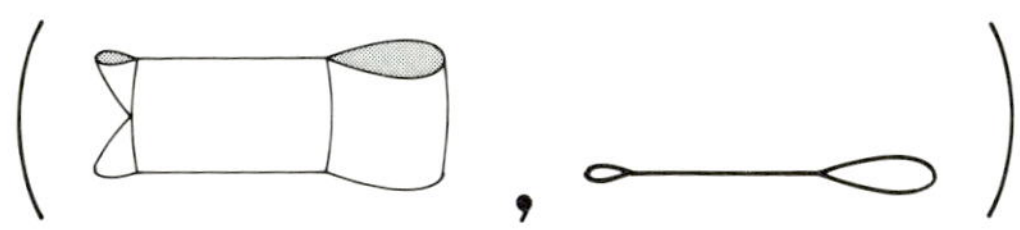

Coarse Morse data for p'_3

3.5. Local Morse Data

Suppose the point $p\in Z$ is a nondepraved critical point of the proper function $f: Z\to \mathbb{R}$. Choose a (smooth) Riemannian metric on M and let $B_\delta^M(p)$ denote the closed disk of radius δ in M, which is centered at p. For δ sufficiently small, the Whitney conditions imply that the sphere $\partial B_\delta^M(p)$ is transverse to each stratum of Z. Thus the transverse intersection

$$B_\delta=B_\delta^M(p)\cap Z$$

is canonically Whitney stratified, and its "boundary"

$$\partial B_\delta=\partial B_\delta^M(p)\cap Z$$

is a closed union of strata in B_δ.

3.5.1. Lemma. *There exists $\delta_0>0$ such that for any $0<\delta\leq\delta_0$ we have:*

(a) *$\partial B_\delta^M(p)$ is transverse to all the strata in Z and*

(b) *none of the critical points of $f\,|\,B_\delta$ have critical value v, except for the critical point p, i.e., for any stratum $S\subset B_\delta$ and for any critical point q of $f\,|\,S$, we have $f(q)\neq v$ unless $q=p$.*

Remark. Part (b) of this lemma is not completely obvious because B_δ contains (new) strata of the form $S=X\cap\partial B_\delta^M(p)$, where X is a stratum of Z. Even if $f\,|\,X$ has no critical points, it is possible that $f\,|\,S$ has critical points.

Proof. The proof will appear in Sect. 6.2.

3.5.2. Definition. Choose $\delta>0$ as in the above lemma. The *local Morse data* for f at p is the coarse Morse data for $f\,|\,B_\delta$ at p, i.e., it is the pair

$$(B_\delta\cap f^{-1}[v-\varepsilon, v+\varepsilon],\ B_\delta\cap f^{-1}(v-\varepsilon))$$

where $\varepsilon>0$ is so small that the interval $[v-\varepsilon, v+\varepsilon]$ contains no critical points of $f \mid B_\delta$ other than v. The local Morse data is $\mathscr{S}$-decomposed by its intersection with the strata of Z.

Example. Consider the critical point p_3' of the space $\mathscr{R}$ described in the introduction (Sect. 1.2). The local Morse data is the pair of spaces which is illustrated in the following diagram:

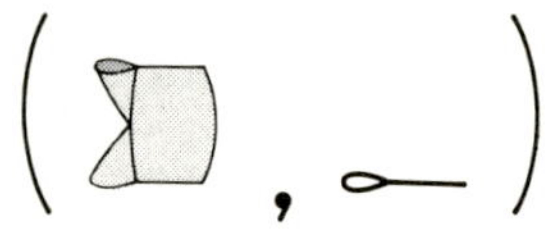

Local Morse data for p_3'

3.5.3. Proposition. *Suppose the function $f\colon Z\to\mathbb{R}$ has a nondepraved critical point at $p\in Z$. Then the local Morse data for f at p is well defined, i.e., independent (up to $\mathscr{S}$-decomposed homeomorphism of pairs) of the choice of the Riemannian metric and the choice of δ and ε provided these are chosen in accordance with the procedure of Sects. 3.4 and 3.5.1.*

3.5.4. Theorem. *If $v=f(p)$ is an isolated critical value, then the local Morse data for f at p is Morse data.*

Proof. The proofs will be delayed until Sects. 7.4 and 7.6.

Remark. It follows from the Morse-Bott-Thom-Smale theory of nondegenerate critical point analysis, that local Morse data at a critical point $p\in M$ of a smooth *Morse* function $f\colon M\to\mathbb{R}$ is diffeomorphic to the pair

$$(D^{s-\lambda}\times D_\lambda,\ D^{s-\lambda}\times\partial D^\lambda)$$

where D^k denotes the closed k-dimensional disk, ∂D^k is its boundary, λ is the Morse index of $f\mid M$ at the point p, and s is the dimension of M. It is shown [Mi1] that this pair is homotopy equivalent to the local Morse data for M at p (considering M to be stratified with a single stratum). In Sect. 4.5 we will show that local Morse data is homeomorphic to the above pair, using the powerful technique of moving the wall.

3.6. Tangential and Normal Morse Data

Let Z be a Whitney stratified subset of some smooth manifold M and let $f\colon Z\to\mathbb{R}$ be a proper function with a nondepraved critical point p. Let X denote the stratum of Z which contains the critical point p. Let N be a normal slice at p (see Sect. 1.4), i.e., $N=Z\cap N'\cap B_\delta^M(p)$, where $\delta>0$ is so small that $\partial B_\delta^M(p)\pitchfork(Z\cap N')$ and where N' is a submanifold of M which is transverse to Z and intersects the stratum X in the single point p, and satisfies $\dim(X)+\dim(N')=\dim(M)$.

3.6.1. Definition. The *tangential Morse data* for f at p is the local Morse data for $f|X$ at p. The *normal Morse data* for f at p is the local Morse data for $f|N$ at p.

The tangential Morse data is trivially $\mathscr{S}$-decomposed. The normal Morse data is $\mathscr{S}$-decomposed by its intersection with the strata of Z.

3.6.2. Proposition. *The $\mathscr{S}$-decomposed homeomorphism type of the tangential and normal Morse data are well defined (i.e., independent of the choices which were made in their construction).*

Proof. The proof will appear in Sects. 7.4 and 7.5.

Remarks. By [Mi1] the tangential Morse data is homotopy equivalent to the pair

$$(D^{s-\lambda}\times D^{\lambda},\ D^{s-\lambda}\times \partial D^{\lambda})$$

where λ is the Morse index of the restriction of f to the stratum which contains the critical point, and where s is the dimension of that stratum. We will show in Sect. 4.5 that the tangential Morse data is homeomorphic to this pair.

The normal Morse data is thus constructed as follows: choose $\delta_0>0$ so that for all $\delta\le\delta_0$, $\partial B_\delta^M(p)$ is transverse to each stratum of N, and none of the critical points of $f|N\cap B_\delta^M(p)$ have critical value v (except for the critical point p). Then choose $\varepsilon_0>0$ so that $f|N\cap B_\delta(p)$ has no critical values (other than v) in the interval $[v-\varepsilon_0,\ v+\varepsilon_0]$. Fix $0<\varepsilon\le\varepsilon_0$. The normal Morse data is the pair

$$(J_p,\ K_p)=(N\cap B_\delta^M(p)\cap f^{-1}[v-\varepsilon,\ v+\varepsilon],\ N\cap B_\delta^M(p)\cap f^{-1}(v-\varepsilon)).$$

The $\mathscr{S}$-decompositions of the tangential (resp. normal) Morse data are refined by canonical Whitney stratifications of the tangential (resp. normal) Morse data, which are given by strata of the form

$$\begin{array}{ll}
X\cap B_\delta^0(p)\cap f^{-1}(v-\varepsilon,\ v+\varepsilon) & (\text{resp. } A\cap N'\cap B_\delta^0(p)\cap f^{-1}(v-\varepsilon,\ v+\varepsilon))\\
X\cap B_\delta^0(p)\cap f^{-1}(v\pm\varepsilon) & (\text{resp. } A\cap N'\cap B_\delta^0(p)\cap f^{-1}(v\pm\varepsilon))\\
X\cap \partial B_\delta(p)\cap f^{-1}(v-\varepsilon,\ v+\varepsilon) & (\text{resp. } A\cap N'\cap \partial B_\delta(p)\cap f^{-1}(v-\varepsilon,\ v+\varepsilon))\\
X\cap \partial B_\delta(p)\cap f^{-1}(v\pm\varepsilon) & (\text{resp. } A\cap N'\cap \partial B_\delta(p)\cap f^{-1}(v\pm\varepsilon))
\end{array}$$

where $B_\delta^0(p)$ denotes the interior of the ball $B_\delta^M(p)$, and A denotes a stratum of Z. The tangential, local, and coarse Morse data are Whitney stratified in a similar manner.

3.7. The Main Theorem

For a fixed stratification of Z and a fixed function f with a nondepraved critical point $p\in Z$, there is a $\mathscr{S}$-decomposition preserving homeomorphism of pairs:

$$\text{Local Morse data}\cong(\text{Tangential Morse data})\times(\text{Normal Morse data}).$$

In other words, if $(P,\ Q)$ denotes the tangential Morse data and if $(J,\ K)$ denotes the normal Morse data, then the local Morse data is the pair $(P\times J,\ P\times K\cup Q\times J)$.

Proof. The proof will appear in Chapter 8.

Remark. Although both sides of this homeomorphism have canonical Whitney stratifications, the homeomorphism cannot be chosen so as to be stratum preserving, due to the extra strata in the tangential and normal Morse data which appear as "corners".

Remark. Similar results in the nonsingular case were obtained in [Ki1], [Ki2].

3.8. Normal Morse Data and the Normal Slice

Theorem. *The total space of the normal Morse data is homeomorphic (by an $\mathscr{S}$-decomposition preserving homeomorphism) to the normal slice.*

Proof. The proof will appear in Sect. 7.8.

3.9. Halflinks

The halflink is a real version of the "complex link" (which is a pair of spaces defined for any complex analytic variety [GM1], [GM3]). Like the complex link, it is a stratified space which is naturally associated to any Morse function, and it almost determines the link of the stratum X: In the complex case it is also necessary to know the monodromy; in the real case it is also necessary to know the lower halflink.

As above, let Z be a Whitney stratified subset of a manifold M, let f be a function with a nondepraved critical point $p \in Z$ and let X be the stratum of Z which contains p. Choose $\varepsilon \ll \delta \ll 1$ according to the procedure outlined in Sect. 3.5. Let $N = N' \cap Z \cap B_\delta^M(p)$ be a normal slice through the stratum X at the point p (see Sect. 1.4).

3.9.1. Definition. The upper *halflink* of Z at the point p (with respect to the function f) is the pair of spaces

$$(\ell^+, \partial \ell^+) = (N \cap B_\delta^M(p) \cap f^{-1}(v+\varepsilon),\ N \cap \partial B_\delta^M(p) \cap f^{-1}(v+\varepsilon))$$

and the lower halflink of Z is the pair of spaces

$$(\ell^-, \partial \ell^-) = (N \cap B_\delta^M(p) \cap f^{-1}(v-\varepsilon),\ N \cap \partial B_\delta^M(p) \cap f^{-1}(v-\varepsilon)).$$

These spaces are canonically $\mathscr{S}$-decomposed by their intersection with the strata of Z, and they are canonically Whitney stratified (by the same procedure as was used to stratify the normal Morse data in Sect. 3.6).

3.9.2. Proposition. *The upper and lower halflink are well defined, i.e., independent (up to $\mathscr{S}$-decomposition preserving homeomorphism of pairs) of the choices of normal slice, Riemannian metric and choice of ε and δ which were made in their definition.*

Proof. The proof will appear in Sect. 7.5.

Remark. The lower halflink for Z at p corresponding to a function f is precisely the upper halflink of Z at p which corresponds to the function $-f$.

3.9.3. Theorem. *If Z is a complex analytic variety, then the halflink is independent of the Morse function. If Z is a real analytic variety, then there are finitely many possibilities (up to stratum preserving homeomorphism) for the halflink at the point p.*

Proof. The proof will appear in Corollary 7.5.3 and 7.5.4.

3.10. The Link and the Halflink

Theorem. There is a $\mathscr{S}$-decomposition preserving homeomorphism between $\partial\ell^+$ and $\partial\ell^-$. The union $\ell^+ \cup_{\partial\ell} \ell^-$ is homeomorphic (by an $\mathscr{S}$-decomposition preserving homeomorphism) to the link of the stratum X which contains the critical point p.

Proof. The proof will appear in Sect. 7.7.

Remark. These homeomorphisms cannot be taken to be taken to be stratum preserving due to the "corners" in the stratification of the halflink.

3.11. Normal Morse Data and the Halflink

3.11.1. Theorem. *The normal Morse data for f at p is homeomorphic (by an $\mathscr{S}$-decomposition preserving homeomorphism) to the pair of spaces (cone $(\ell^+ \cup_{\partial\ell} \ell^-)$, ℓ^-), where the $\mathscr{S}$-decomposition of the cone is given as follows: the cone point is one piece (corresponding to the stratum X of Z) and the other pieces are of the form $A \times (0, 1]$ where A is a piece of the $\mathscr{S}$-decomposition of $\ell^+ \cup_{\partial\ell} \ell^-$.*

Proof. The proof will appear in Sect. 7.9.

3.11.2. Corollary. *The normal Morse data for f at p has the homotopy type of the pair*

$$(cone(\ell^-), \ell^-).$$

Proof. The proof follows directly from this result in homotopy theory (with $A = \ell^+$ and $B = \ell^-$):

3.11.3. Sublemma. *Suppose $(A, \partial A)$ and $(B, \partial B)$ are compact pairs of topological spaces. Suppose ∂A has a collared neighborhood $U \cong \partial A \times [0, 1]$ in A, and suppose there is a homeomorphism $h: \partial A \to \partial B$. Then there is a deformation retraction from the pair $(cone(A \cup_{\partial A} B), B)$ to the pair $(cone(B), B)$. Furthermore, if $\mathscr{S}$ is a partially ordered set and if A and B are $\mathscr{S}$-decomposed spaces such that the collaring $U \cong \partial A \times [0, 1]$ is an $\mathscr{S}$-decomposition preserving homeomorphism, then we obtain natural $\mathscr{S}$-decompositions of $cone(A \cup_{\partial A} B)$ and $cone(B)$, and the above deformation retractions may be chosen so as to preserve the $\mathscr{S}$-decompositions.*

Notation. A point in $\operatorname{cone}(A \cup B)$ will be denoted as a pair (x, r), where $x \in A \cup B$, and $r \in [0, 1]$ denotes the "distance" from the cone point. Points x in the collared neighborhood U will be denoted as pairs (y, s), where $s \in [0, 1]$ denotes the "distance" from the boundary ∂A, i.e., the points $(y, 0)$ lie in ∂A. We will use the variable $t \in [0, 1]$ to denote the "time" parameter in the homotopy.

Proof of sublemma. In fact, we will find a deformation retraction

$$H: (\operatorname{cone}(A \cup_\partial B, B) \times [0, 1] \to (\operatorname{cone}(B), B)$$

which is the identity on $\operatorname{cone}(B) \times [0, 1]$, is the identity at $t = 1$, and collapses the first space to the second space when $t = 0$, and which preserves the $\mathcal{S}$-decompositions. If $x \in \operatorname{cone}(B)$, define $H(x, r, t) = (x, r)$. If $x \in \operatorname{cone}(A - U)$ define $H(x, r, t) = (x, rt)$. If $(x, r, t) = (y, s, r, t) \in \operatorname{cone}(U)$, define H as follows:

$s = 1$: $\quad H(y, s, r, t) = (y, s, tr)$

$\frac{r}{2} \leq s \leq 1$: $\quad H(y, s, r, t) = \left(y, s - \frac{(1-t)\, r(1-s)}{2-r}, tr\right)$

$0 \leq s \leq \frac{r}{2}$: $\quad H(y, s, r, t) = (y, ts, r - 2s(1-t))$

$s = 0$: $\quad H(y, s, r, t) = (y, s, r)$.

This deformation is illustrated in the following diagram:

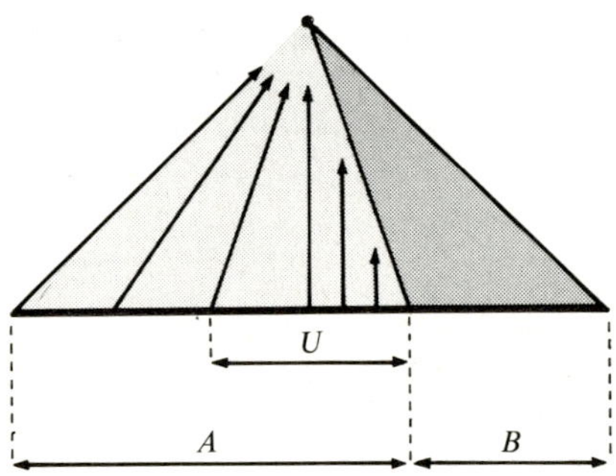

of the space $\operatorname{cone}(A \cup B)$. The cone on B is the dark shaded region, and the collared neighborhood U of ∂A is indicated. □

3.12. Summary of Homotopy Consequences

(See also Sects. 10.8 and 11.8 for generalizations to nonproper and relative Morse functions.) Suppose Z is a Whitney stratified space, $f: Z \to \mathbb{R}$ is a proper *Morse* function, and $[a, b] \subset \mathbb{R}$ is an interval which contains no critical values except for a single isolated critical value $v \in (a, b)$ which corresponds to a critical point p which lies in some stratum S of Z. Let λ be the Morse index of $f \mid S$ at the point p.

Theorem. *The space $Z_{\leq b}$ has the homotopy type of a space which is obtained from $Z_{\leq a}$ by attaching the pair* (*Sect.* 3.0)

$$(D^\lambda, \partial D^\lambda) \times (\text{cone}(\ell^-), \ell^-).$$

Proof. By 3.5.4, the local Morse data is Morse data. By Sect. 3.7, this is a product of normal and tangential Morse data. By remark 3.5.4, the tangential Morse data has the homotopy type of the pair $(D^\lambda, \partial D^\lambda)$. By 3.11.2, the normal Morse data has the homotopy type of the pair $(\text{cone}(\ell^-), \ell^-)$. □

3.13. Counterexample

The following example illustrates that the delicate estimates made in Chapter 6 have nontrivial content: local Morse data $\neq$ normal Morse data $\times$ tangential Morse data for arbitrary analytic functions.

Stratify $Z = \mathbb{R}^2$ with a singular stratum $S =$ the x-axis. Consider the function $f(x, y) = x^2 - y^2$. This has a nondegenerate (in the classical sense) critical point at the origin and the restriction $f|S$ has a nondegenerate critical point at the origin. However, the normal Morse data at (0, 0) is not well-defined, and this may be attributed to the fact that $df(0, 0)$ kills a limiting tangent plane from the large stratum.

Even if we make the assumption that the normal and tangential Morse data should be well-defined, it may still fail that the local Morse data is the product of the two: Let $Z = \mathbb{R}^2$, stratified with one singular one-dimensional stratum,

$$S = \{(x, y) \in \mathbb{R}^2 \mid y = x^2\}.$$

Let N be the normal slice through S at the origin,

$$N = \{(x, y) \in \mathbb{R}^2 \mid x = 0\}.$$

Define $f: Z \to \mathbb{R}$ by $f(x, y) = y^2 - x^6$. This has a singular point at the origin. The following facts are easy to verify:

(1) f is a real algebraic function.

(2) $f|S$ has a nondepraved critical point (a minimum) at the origin; the tangential Morse data is the pair (D^1, ϕ).

(3) $f|N$ has a minimum at the origin; the normal Morse data is (D^1, ϕ) and is independent of the choice of the normal slice N.

(4) The origin is a saddle point of f, i.e., the local Morse data is $(D^1 \times D^1, \partial D^1 \times D^1)$.

Thus, the local Morse data is not equal to the product of the normal and tangential Morse data. This failure can be attributed to the fact that $df(0)$ kills a limiting tangent plane from the large stratum.

Chapter 4. Moving the Wall

4.1. Introduction

This chapter and the next contain the main technical tools which will be used in Part I. Moving the wall is a rigorous but intuitive technique for verifying the hypotheses and expressing the conclusions of Thom's first isotopy lemma, which is particularly useful when the isotopy lemma is applied to a complicated geometric situation. The power of this method even in the nonsingular case is illustrated in Sect. 4.5, where we reprove the classical result in Morse theory: crossing a nondegenerate critical point corresponds to attaching a handle.

Many of the (pieces of) spaces which are considered in this book are constructed by projecting some Whitney stratified subset Z of some smooth manifold M to some auxiliary manifold N (by a proper smooth map $g: M \to N$) and then taking the counter-image $Z \cap g^{-1}(Y_0)$ in Z of some Whitney stratified region $Y_0 \subset N$.

$$\begin{array}{ccc} M & \xrightarrow{g} & N \\ \cup & & \cup \\ Z & & Y_0 \end{array}$$

(For example, local Morse data at a point $p \in Z$ is the preimage of a box

$$Y_0 = \{(r, f) \in \mathbb{R}^2 \mid 0 \leq r \leq \delta;\ v - \varepsilon \leq f \leq v + \varepsilon\}$$

in the two-dimensional space whose coordinates are f and the distance r from the point p.) It often happens that the restriction $g|Z: Z \to N$ is *not* a submersion on each stratum of Z, but that we nevertheless need a criterion which guarantees that $Z \cap g^{-1}(Y_0)$ is homeomorphic to $Z \cap g^{-1}(Y_1)$, where Y_0 and Y_1 are connected by some one-parameter family of regions Y_t. (By this, we mean that there is a Whitney stratified space $Y \subset \mathbb{R} \times N$ such that the projection to the first factor, $\pi: Y \to \mathbb{R}$ is a submersion on each stratum and such that $Y_0 = \pi^{-1}(0)$ and $Y_1 = \pi^{-1}(1)$).

Such a criterion is the following: for each $t \in [0, 1]$ the characteristic covectors of the map g must be nonzero on each stratum of Y_t. (Recall from Sect. 1.9 that a covector $\xi \in T_q^* N$ is characteristic for $g|Z$ if there is a point $z \in Z$ such that $g(z) = q$ and such that the preimage $g^*(\xi)$ vanishes on the tangent space $T_z S$ to the stratum S which contains the point z.)

Thus, the family Y_t is restricted not by the characteristic *points* of the map g, but by the characteristic covectors. This allows us to vary Y_0 in a family which may pass through singular values of the map g.

Although this criterion is simply a restatement of Thom's first isotopy lemma (the deformation Y_t gives rise to a controlled vectorfield on Z), the point of presenting our proofs in this language is that arguments involving complicated geometry in Z and difficult estimates on the components of controlled vectorfields are replaced by the simpler geometry of strata in the wall space and the calculation of the characteristic covectors of the map g.

Remarks. In most of our applications, N will be a Euclidean space $\mathbb{R}^n$ of low dimension ($n \leq 4$) and Y_t will be homeomorphic to a closed ball whose boundary ("the wall") is stratified in particular ways. For example the local Morse data

$$(B_\delta(p) \cap Z \cap f^{-1}[v-\varepsilon, v+\varepsilon],\ B_\delta(p) \cap Z \cap f^{-1}(v-\varepsilon))$$

has "boundary strata" at the edge $\partial B_\delta(p)$ of the ball and also where $f(z) = v \pm \varepsilon$, so it has a codimension two "corner stratum" where these intersect. These boundary and corner strata arise from boundaries and corners of a region in the wall space. The wall space defining tangential Morse data and the wall space defining normal Morse data are each two-dimensional, giving rise to codimension two corners in each. The product of tangential Morse data and normal Morse data, therefore, has codimension four corners and it is defined by a four-dimensional wall space. To prove the main theorem, we must move the wall in this four-dimensional wall space, avoiding characteristic covectors at odd angles. This accounts for the extraordinary complexity of the proof of the main theorem (see Chapters 6 and 8).

4.2. Example

Let P be the parabola in $\mathbb{R}^2$ given by $x = y^2$, and let $f: P \to \mathbb{R}$ denote the projection to the x axis. Let Y_t be the closed interval $[t, t+1]$. Then, the characteristic covectors of f are the elements of $T_0^* \mathbb{R}$. The topological type of $f^{-1}(Y_t)$ changes only at $t=1$ and $t=0$, i.e., when these covectors vanish on the strata at the endpoints of the interval $[t, t+1]$.

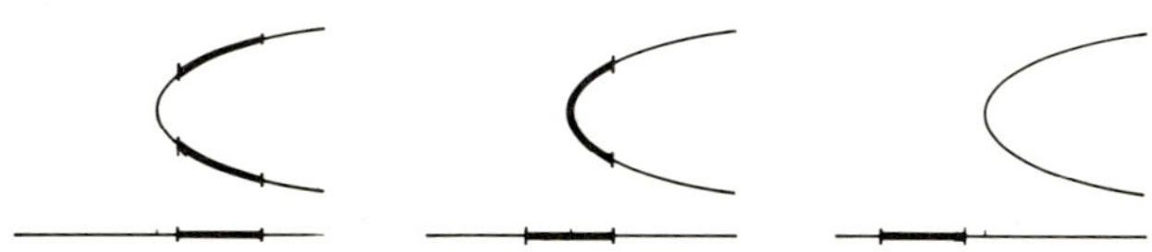

Characteristic covectors are an obstruction to moving the wall

4.3. Moving the Wall: Version 1

Let $f: M \to N$ be a smooth map between two manifolds. Let $Z \subset M$ be a Whitney stratified subset whose strata are indexed by some partially ordered set $\mathcal{S}$. Suppose $f|Z: Z \to N$ is proper. Let $Y \subset N \times \mathbb{R}$ be a (closed) Whitney stratified subset such that the projection to the second factor, $\pi: Y \to \mathbb{R}$ is a proper stratified submersion. Suppose that for each $t \in \mathbb{R}$ and for each $p \in f(Z) \cap Y_t$, and for each nonzero characteristic covector $\lambda \in T_p^* N$ of the map $f|Z: Z \to N$, the restriction of λ to the subspace

$$T_p S_t = (\ker(d\pi(p))) \cap T_p S$$

is nonzero, where S is the stratum of Y which contains the point (p, t) and $S_t = \pi^{-1}(t) \cap S$ is the stratum of $Y_t = \pi^{-1}(t) \cap Y$ which contains the point p. This is equivalent to the statement that, for each $t \in \mathbb{R}$, the restriction $f|Z$ is transverse to Y_t.

Thus, for each $t \in \mathbb{R}$, $Z \cap f^{-1}(Y_t)$ is $\mathcal{S}$-decomposed by its intersection with the strata of Z. This $\mathcal{S}$-decomposition is refined by the canonical Whitney stratification of $Z \cap f^{-1}(Y_t)$, which consists of strata of the form $A \cap f^{-1}(B)$, where A is a stratum of Z and B is a stratum of Y_t.

Theorem. *There is a homeomorphism*

$$Z \cap f^{-1}(Y_0) \cong Z \cap f^{-1}(Y_1)$$

which preserves the $\mathcal{S}$-decomposition of both sides, preserves the Whitney stratification of both sides, and is smooth on each stratum.

Proof. Consider the composition

$$\begin{array}{ccccc} M \times \mathbb{R} & \xrightarrow[f \times I]{} & N \times \mathbb{R} & \xrightarrow[\pi]{} & \mathbb{R} \\ \cup & & \cup & & \\ Z \times \mathbb{R} & & Y & & \end{array}$$

where I denotes the identity map. It is easy to see that $Z \times \mathbb{R}$ is transverse to $(f \times I)^{-1}(Y)$ and that the composition

$$\pi \circ (f \times I): Z \times \mathbb{R} \cap (f \times I)^{-1}(Y) \to \mathbb{R}$$

is a stratified submersion (unless $Y \cap f(Z) = \phi$, in which case the theorem is trivial). Thus, the first isotopy lemma (Sect. 1.5) can be applied to this projection. □

4.4. Moving the Wall: Version 2

Suppose Z is a Whitney stratified subset of $M \times \mathbb{R}$ such that the projection $p: Z \to \mathbb{R}$ to the second factor is a proper stratified submersion. Let $\mathcal{S}$ denote a partially ordered set which indexes the strata of Z. Let $F: M \times \mathbb{R} \to N \times \mathbb{R}$ be a one-parameter family of smooth functions, $F(x, t) = (f_t(x), t)$. Let $Y \subset N \times \mathbb{R}$ be a Whitney stratified subset and let $W \subset Y$ be a closed union of strata. Assume that the projections to the second factor $\pi: (Y, W) \to \mathbb{R}$ are proper stratified submersions, and let $(Y_t, W_t) = \pi^{-1}(t) \cap (Y, W)$. Suppose that for each $t \in \mathbb{R}$ and for each $p \in f_t(Z_t) \cap Y_t$ and for each nonzero characteristic covector $\lambda \in T_p^* N$ of the map $f_t: Z_t \to N$, the restriction of λ to the subspace

$$T_p S_t = (\ker d\pi(p)) \cap T_p S$$

is nonzero, where S is the stratum of Y which contains the point (p, t) and $S_t = \pi^{-1}(t) \cap S$ is the stratum of $Y_t = \pi^{-1}(t) \cap S$ which contains the point p. This is equivalent to the statement that for each $t \in \mathbb{R}$ the restriction $f_t | Z_t \colon Z_t \to N$ is transverse to each stratum of Y_t.

In this case, the intersections $Z_t \cap f_t^{-1}(Y_t)$ and $Z_t \cap f_t^{-1}(W_t)$ are canonically $\mathcal{S}$-decomposed by their inclusion in Z, and they are canonically Whitney stratified by a stratification which refines the $\mathcal{S}$-decomposition, whose strata are of the form $A_t \cap f_t^{-1}(B_t)$ where A_t is a stratum of Z_t and B_t is a stratum of Y_t.

Theorem. *There is a homeomorphism of pairs,*

$$Z_0 \cap f_0^{-1}(Y_0, W_0) \cong Z_1 \cap f_1^{-1}(Y_1, W_1)$$

which preserves the $\mathcal{S}$-decomposition of each side, and preserves the canonical Whitney stratifications of each side, and is smooth on each stratum.

Proof. The proof is exactly is exactly the same as above: $Z \cap F^{-1}(Y, W)$ is a Whitney stratified space such that the projection

$$p \colon Z \cap F^{-1}(Y, W) \to \mathbb{R}$$

is a proper stratified submersion. Thus, $p^{-1}(0)$ and $p^{-1}(1)$ are homeomorphic (by Thom's first isotopy lemma Sect. 1.5). □

4.5. Tangential Morse Data is a Product of Cells

To illustrate the power of the technique of moving the wall, we will now prove the (homeomorphism version of the) following classical result:

Proposition. *Suppose $f \colon M \to \mathbb{R}$ is a smooth function defined on an s-dimensional manifold M, and let $p \in M$ be a nondegenerate critical point of f with Morse index λ. Then, local Morse data (which equals the tangential Morse data) for f at p is homeomorphic to the pair*

$$(D^\lambda \times D^{s-\lambda}, \partial D^\lambda \times D^{s-\lambda})$$

where D^λ denotes the disk of dimension λ and ∂D^λ denotes its boundary sphere.

Remark. Our method gives the homeomorphism type of the local Morse data. It is a deeper result, due to Smale, that in fact the above pair is the diffeomorphism type of the local Morse data. See [Mi1] for a careful proof that the above pair has the homotopy type of the local Morse data.

Proof. By the Morse lemma, there exists a coordinate system $(x_1, \ldots, x_\lambda, y_1, \ldots, y_{s-\lambda})$ on M, centered at the point p, such that locally the Morse function is given by

$$f(x, y) = -\sum_{i=1}^{\lambda} x_i^2 + \sum_{i=s-\lambda}^{s} y_i^2.$$

The locally defined functions

$$F_1(x, y) = \sum_{i=1}^{\lambda} x_i^2 \quad \text{and} \quad F_2(x, y) = \sum_{i=s-\lambda}^{s} y_i^2$$

define a map $F = (F_1, F_2) \colon M \to \mathbb{R}^2$ which has *no* characteristic covectors (Sect. 1.8) except the origin, and every covector at the origin is characteristic. Since the local Morse data is independent of the Riemannian metric involved in its definition (Sect. 3.6.2), we may take the distance from the point p to be given by

$$r^2(x, y) = F_1(x, y) + F_2(x, y).$$

With these choices, the local Morse data is (by definition) the preimage under the map F of the pair (R, E) where R is the region

$$R = \{(F_1, F_2) \in \mathbb{R}^2 \mid F_1 + F_2 \leq \delta \text{ and } -\varepsilon \leq F_2 - F_1 \leq \varepsilon\}$$

(with $\varepsilon \ll \delta$) and E is the lower edge of the region R,

$$E = \{(F_1, F_2) \in \mathbb{R}^2 \mid F_1 + F_2 \leq \delta \text{ and } F_2 - F_1 = -\varepsilon\}.$$

These regions are illustrated in the following diagram of (F_1, F_2)-space:

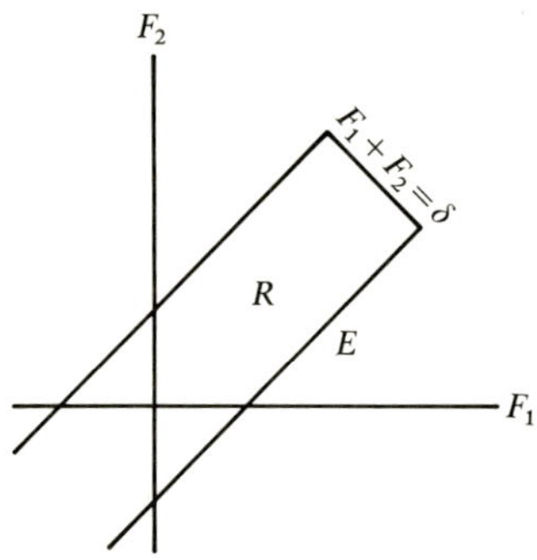

Local Morse data in the wall space

We will show by moving the wall that the pair $(F^{-1}(R), F^{-1}(E))$ is homeomorphic to the pair $(F^{-1}(R'), F^{-1}(E'))$ where

$$R' = \{(F_1, F_2) \mid F_1 \leq \eta \text{ and } F_2 \leq \nu\}$$
$$E' = \{(F_1, F_2) \mid F_1 = \eta \text{ and } F_2 \leq \nu\}$$

where $\eta = (\delta + \varepsilon)/2$ and $\nu = (\delta - \varepsilon)/2$. The regions are illustrated in the following diagram of (F_1, F_2)-space, with E' given by the right hand side of the box. For technical reasons (see below) we introduce an auxiliary stratum in R' consisting of the single point (ν, ν).

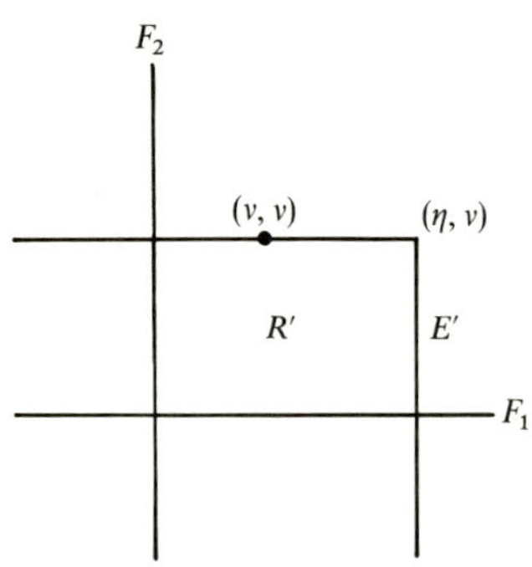

Diagram of (R', E')

Notice that the pair $(F^{-1}(R'), F^{-1}(E'))$ is precisely

$$([D_\eta^\lambda \times \mathbb{R}^{s-\lambda}] \cap [\mathbb{R}^\lambda \times D_v^{s-\lambda}],\ [\partial D_\eta^\lambda \times \mathbb{R}^{s-\lambda}] \cap [\mathbb{R}^\lambda \times D_v^{s-\lambda}])$$

which equals

$$(D_\eta^\lambda \times D_v^{s-\lambda},\ \partial D_\eta^\lambda \times D_v^{s-\lambda})$$

where D_η^λ denotes the closed ball of radius η in the space $\mathbb{R}^\lambda$.

The required deformation by moving the wall is a triviality to find because the function F has no characteristic covectors except at the origin (or when $s=1$, in which case the result is trivial). We will use Sect. 4.3 (Moving the wall, Version 1) with $f\colon M \to N$ of Sect. 4.3 replaced by $F=(F_1, F_2)\colon M \to \mathbb{R}^2$, and with $Z=M$. Since we are dealing with pairs of spaces (R, E) and (R', E') we will need a pair of one-parameter families of spaces (R_t, E_t) which interpolate between them. (These are called Y in the statement of Theorem 4.3). It is most convenient to construct this deformation in two steps:

Step 1: R_t varies between $R_0 = R$ and

$$R_1 = \{(F_1, F_2) \in \mathbb{R}^2 \mid F_1 + F_2 \le \delta,\ F_2 \le \eta,\ F_1 \le v\}.$$

The space E_t varies between $E_0 = E$ and

$$E_1 = E' = \{(F_1, F_2) \in \mathbb{R}^2 \mid F_1 = \eta,\ F_2 \le v\}.$$

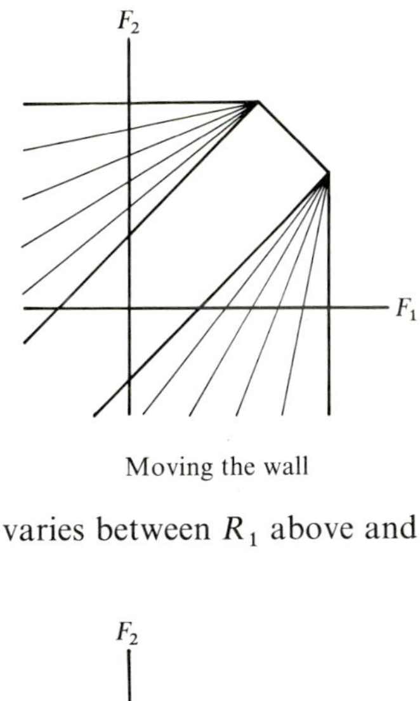

Moving the wall

Step 2. The space R_t varies between R_1 above and R'. The space $E_t = E_1 = E'$ does not vary.

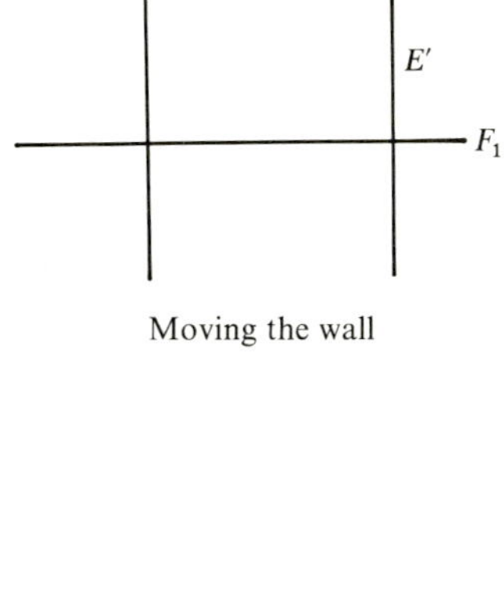

Moving the wall

Remark. During the deformation in Step 2, the boundary of the region R_t contains a zero-dimensional stratum at the "kink" in the top of the box. These form a one-dimensional stratum in the family

$$Y = \bigcup_t (R_t \times \{t\}) \subset \mathbb{R}^2 \times \mathbb{R}.$$

In order to satisfy the hypothesis (Sect. 4.3) that the projection $Y \to \mathbb{R}$ is a submersion on each stratum of Y, we must prolong this one-dimensional stratum throughout the deformation. This gives rise to the zero-dimensional stratum in R' which is indicated in the above figure of R'.

Chapter 5. Fringed Sets

The definition of local Morse data involves certain choices of allowable parameters ε and δ. The set of all such allowable ε and δ form a region in the (ε, δ) plane of a certain shape, which we call "fringed, of type $0<\varepsilon \ll \delta$". In this chapter we study fringed sets: these are open subsets of the first quadrant in $\mathbb{R}^2$ whose closure contains a segment of the x-axis ending at the origin, and which are unions of vertical segments. Fringed sets of this type will appear throughout the technical discussions in Part I.

We shall use the symbol $\mathbb{R}^+$ to denote the positive real numbers.

5.1. Definition

A set $A=\{(x, y)\} \subset \mathbb{R} \times \mathbb{R}^+$ is *fringed* over a subset $S \subset \mathbb{R}$ provided:

(1) The projection $\pi(A)$ of A to the first factor is equal to S

(2) A is open in $S \times \mathbb{R}^+$

(3) If $(s, y) \in A$ and if $0<y' \leq y$ then $(s, y') \in A$.

A set $A \subset \mathbb{R}^+ \times \mathbb{R}^+$ is of *type* $0<y \ll x$ if it is fringed over some interval $S=(0, x_0)$.

The intersection of finitely many sets of type $0<y \ll x$ is again a set of type $0<y \ll x$. An arbitrary union of such sets is again such a set.

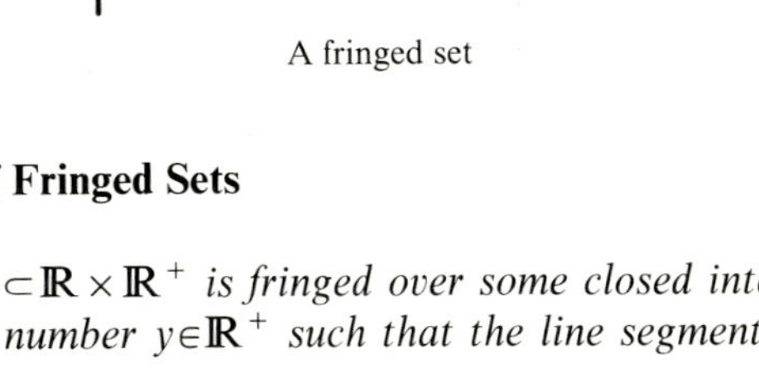

A fringed set

5.2. Connectivity of Fringed Sets

Lemma. *Suppose $A \subset \mathbb{R} \times \mathbb{R}^+$ is fringed over some closed interval $S=[\alpha, \beta] \subset \mathbb{R}$. Then there exists a number $y \in \mathbb{R}^+$ such that the line segment $[\alpha, \beta] \times \{y\}$ is contained in A.*

Proof. Suppose not. Then there is a sequence of values $y_i \to 0$ such that all line segments $[\alpha, \beta] \times \{y_i\}$ are not contained in A. Choose a sequence of points (α_i, y_i) which are not in A, such that $\alpha_i \in [\alpha, \beta]$ converge to some $\alpha_0 \in [\alpha, \beta]$ and so that $y_i \to 0$. By property (1) there is some point $(\alpha_0, y_0) \in A$, and since A is open there is a neighborhood U of α_0 such that $(u, y_0) \in A$ for all $u \in U$. It follows from property (3) that $U \times (0, y_0) \subset A$. This contradicts the statement that for all i, (α_i, y_i) is not contained in A. □

Proposition. *Suppose $A \subset \mathbb{R}^+ \times \mathbb{R}^+$ is a set of type $0 < y \ll x$. Then A is connected in the following strong sense: For any two points (x_1, y_1), (x_2, y_2) in A there exists a number $y' \leq \min(y_1, y_2)$ such that the following three straight line segments are each contained in A:*

(1) (x_1, y_1) *to* (x_1, y')
(2) (x_1, y') *to* (x_2, y')
(3) (x_2, y') *to* (x_2, y_2).

Proof. The set A is fringed over the interval $[x_1, x_2]$, so the preceding lemma gives a value $y' > 0$ such that the line segment $[x_1, x_2] \times \{y'\} \subset A$. □

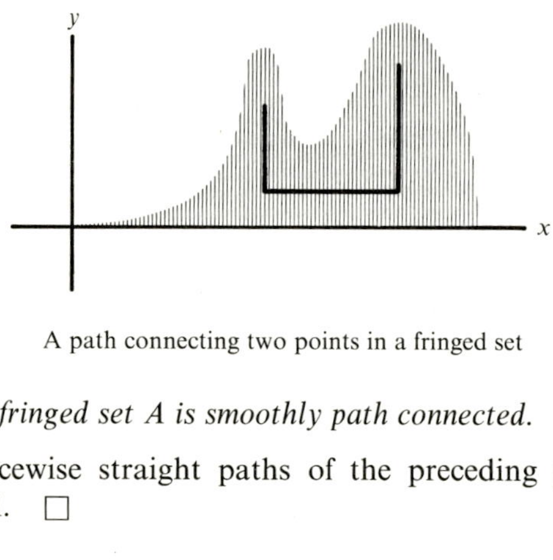

A path connecting two points in a fringed set

Corollary. *Any fringed set A is smoothly path connected.*

Proof. The piecewise straight paths of the preceding proposition can be smoothed within A. □

5.3. Characteristic Functions

To each fringed set A we can associate a characteristic function $f\colon S \to \mathbb{R}^+$ which is given by

$$f(s) = \sup\{y \mid (s, y) \in A\}.$$

This function is lower semicontinuous (since A is open in $S \times \mathbb{R}^+$).

Proposition. *Let A be a set of type $0 < y \ll x$ and suppose $(x_0, y_0) \in A$. Then there exists a positive monotone increasing k-times continuously differentiable function $f\colon \mathbb{R}^+ \to \mathbb{R}^+$ such that*

$$\{(x, y) \mid y < f(x) \text{ and } x \leq x_0\} \subset A$$

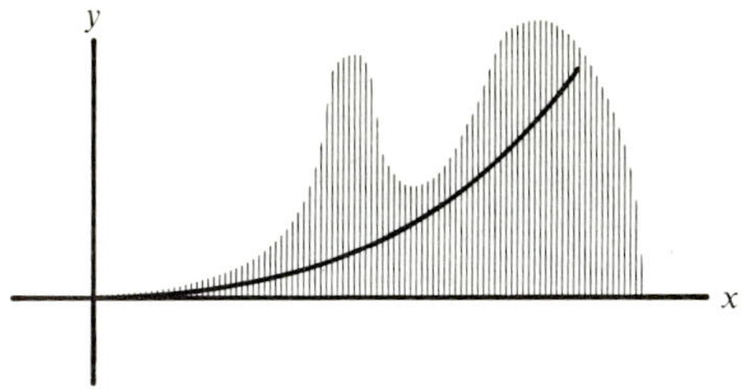

A fringed set contains the graph of a smooth positive increasing function

Proof. Define

$$f_0(x)=\sup\{y\in\mathbb{R}^+ \mid (x', y)\in A, \text{ for all } x'\in[x, x_0]\}.$$

The argument of Sect. 5.2 shows that $f_0(x)>0$ whenever $x>0$. Furthermore, f_0 is monotone increasing and satisfies

$$\{(x, y) \mid 0<x\leq x_0 \text{ and } y<f_0(x)\}\subset A.$$

Now define inductively

$$f_i(x)=\frac{1}{x_0}\int_0^x f_{i-1}(t)\,dt.$$

Each f_j is positive, monotone increasing, $j-1$ times continuously differentiable, and satisfies $f_j(x)\leq f_{j-1}(x)$ whenever $x\in(0, x_0]$. Thus, the function $f=f_{k+1}$ satisfies the requirements of the proposition. □

5.4. One Parameter Families of Fringed Sets

An *open* subset $B\subset\mathbb{R}^+\times\mathbb{R}^+\times[0, 1]$ is a one-parameter family of sets of type $0<y\ll x$ if, for each $z\in[0, 1]$ the set

$$B_z=\{(x, y)\in\mathbb{R}^+\times\mathbb{R}^+ \mid (x, y, z)\in B\}$$

is a set of type $0<y\ll x$.

Proposition. *Suppose B is a one-parameter family of sets of type* $0<y\ll x$. *Then B is path connected in the following strong sense: for any two points* (x_1, y_1, z_1), *and* (x_2, y_2, z_2) *in B, there are numbers* $x'\leq\min(x_1, x_2)$ *and* $y'\leq\min(y_1, y_2)$ *such that the following five segments are contained in B:*

(1) (x_1, y_1, z_1) *to* (x_1, y', z_1)
(2) (x_1, y', z_1) *to* (x', y', z_1)
(3) (x', y', z_1) *to* (x', y', z_2)
(4) (x', y', z_2) *to* (x_2, y', z_2)
(5) (x_2, y', z_2) *to* (x_2, y_2, z_2).

Proof. First choose x' as follows: Let $\pi_{xz}\colon B\to\mathbb{R}^+\times[0, 1]$ denote the projection to the xz coordinate plane and let $C=\pi_{xz}(B)$ denote the image of B. Thus, C is an open subset of $\mathbb{R}^+\times[0, 1]$ which contains the z-axis in its closure and which is a union of lines of the form (tx, z) where $0<t\leq 1$. Thus the set C is fringed over the interval $[0, 1]$. By Lemma 5.2 there exists $x'>0$ such that the line segment $\{x'\}\times[0, 1]$ is contained in C. Now consider the slice

$$D=B\cap\pi_{xz}^{-1}(\{x'\}\times[0, 1]).$$

This is an open subset of the flat

$$\{(x, y, z)\,|\,x=x', y>0, z\in[0, 1]\}$$

which contains the "z-axis", $\{(x', 0, z)\,|\,z\in[0, 1]\}$ in its closure, and which is fringed over $[0, 1]$. Thus, there exists a $y'>0$ such that the segment $(x', y')\times[0, 1]$ is contained in D (and hence also in B). Finally, diminish y' if necessary so as to guarantee that the segments from (x, y', z_1) to (x', y', z_1) and from (x', y', z_2) to (x_2, y', z_2) are also contained in D. □

5.5. Fringed Sets Parametrized by a Manifold

Let M be a smooth n-dimensional manifold. We shall say that a subset $B\subset M\times\mathbb{R}^+\times\mathbb{R}^+$ is a fringed set parametrized by M provided

(a) B is open

(b) for each point $p\in M$ the set

$$B_p=\{(x, y)\in\mathbb{R}^+\times\mathbb{R}^+\,|\,(p, x, y)\in B\}$$

is a nonempty fringed set of type $0<y\ll x$.

We shall use the symbol $\pi_1\colon B\to M$ to denote the projection to the first factor.

Proposition. *Suppose B is a fringed set parametrized by a manifold M. Then B has a smooth section, i.e., a smooth map $s\colon M\to B$ such that $\pi_1\circ s=\text{identity}$.*

Proof. Since B is open, for any compact set $K_1\subset M$ it is possible to find numbers $(x_1, y_1)\in\mathbb{R}^+\times\mathbb{R}^+$ which form a section over K_1, i.e., so that $(p, x_1, y_1)\in B$ for all $p\in K_1$. Now choose an exhaustive sequence of open sets $U_i\subset M$, with compact closures K_i, i.e.,

$$U_1\subset K_1\subset U_2\subset K_2\subset U_3\ldots$$

and find corresponding points $(x_i, y_i)\in\mathbb{R}^+\times\mathbb{R}^+$ which form a section over K_i. By the method of Sect. 5.2, there is for each i a smooth curve $(x_i(t), y_i(t))$ joining (x_i, y_i) (when $t=0$) to (x_{i+1}, y_{i+1}) (when $t=1$) such that, for each $t\in[0, 1]$ the point $(x_i(t), y_i(t))$ is a section of B over the set K_i. Use these curves to join the sections together as follows: Choose a smooth function $\phi_i\colon K_i\to[0, 1]$ such that

(1) $\phi_i(x)=0$ for all x in some neighborhood of K_{i-1}

(2) $\phi_i\,|\,K_i-U_i=1$

and define the section

$$s(p)=(x_i(\phi_i(p)), y_i(\phi_i(p)))\qquad\text{if } p\in K_i-K_{i-1}.$$

It is easy to check that this section is smooth and has the required properties. □

Chapter 6. Absence of Characteristic Covectors: Lemmas for Moving the Wall

This chapter contains the tools needed to prove the main theorems of Part I (Sects. 3.7, 3.10, 3.11). In order to carry out the "moving the wall" arguments which are needed in the proof of these theorems, it is necessary to know that there are no characteristic covectors which might impede the motion of the wall, i.e., that the wall is transverse to the strata of Z. In this chapter we prove that there are no such characteristic covectors.

6.1. The Setup

Throughout this chapter, Z will denote a Whitney stratified subset of some smooth manifold M, and $f\colon M \to \mathbb{R}$ will denote a smooth function which has an isolated nondegenerate (or nondepraved) critical point $p \in Z$ which lies in some stratum X of Z. Assume that $f(p)=0$. We choose a tubular neighborhood T_X of X in M, and a tubular projection $\pi\colon T_X \to X$ with the property that for any stratum Y of Z, either $Y \cap T_X = \phi$ or else the restriction

$$\pi \mid (Y \cap T_X)\colon Y \cap T_X \to X$$

is a submersion.

In this chapter we will fix a local coordinate system on M with coordinates $\{x_1, x_2, \ldots, x_m\}$ which are defined in some neighborhood $U \subset T_X$ of the critical point p, so that the point p becomes the origin, the stratum X is given by $x_{s+1}=0$, $x_{s+2}=0, \ldots, x_m=0$, and so that the tubular projection $\pi\colon U \to X$ is the linear projection

$$\pi(x_1, x_2, \ldots, x_m) = (x_1, x_2, \ldots, x_s, 0, \ldots, 0)$$

where $s=\dim(X)$. We will use the Euclidean metric in this coordinate system and denote by $r(z)$ the *square* of the distance between a point $z \in M$ and the critical point p. We will denote by $\rho(z)$ the square of the distance between z and $\pi(z)$. By Pythagorus,

$$r(z) = r\pi(z) + \rho(z).$$

We denote by $B_\delta(p)$ the ball

$$B_\delta(p) = \{z \in M \mid r(z) \leq \delta\}$$

and we denote its boundary by

$$S_\delta(p)=\partial B_\delta(p)=\{z\in M \mid r(z)=\delta\}.$$

We shall use (u, v) to denote coordinates in $\mathbb{R}^2$. If $A\subset\mathbb{R}^+\times\mathbb{R}^+$ is a set of type $0<v\ll u$, we denote by $A^\flat\subset\mathbb{R}\times\mathbb{R}$ the closure of the set obtained by adding to A its reflection about the u axis:

$$A^\flat=\{(u, v)\in\mathbb{R}^2 \mid \text{there exists a point } (u, \delta)\in A \text{ with } |v|\leq\delta\}.$$

Recall that a covector $\lambda=(u, v, a\,du+b\,dv)\in T^*\mathbb{R}^2$ is *characteristic* for the map $(r, f)\colon Z\to\mathbb{R}^2$ if there exists a point $z\in Z$ such that $(r(z), f(z))=(u, v)$ and $(r, f)^*(\lambda)=a\,dr(z)+b\,df(z)$ vanishes on $T_z Y$, the tangent space to the stratum which contains the point z (Sect. 1.8).

6.2. Lemma. *There exists a fringed set $A_1\subset\mathbb{R}^+\times\mathbb{R}^+$ of type $0<v\ll u$, such that the projection*

$$(r, f)\colon Z\to\mathbb{R}^2$$

has no characteristic covectors in the region $A_1^\flat$. Furthermore, there exists $r_0>0$ such that if $\lambda=(u, v, a\,du+b\,dv)\in T^*\mathbb{R}^2$ is a nonzero characteristic covector for the map $(r, f)\colon Z\to\mathbb{R}^2$ and if $u\leq r_0$ then $v\neq 0$, $a\neq 0$, $b\neq 0$, and the slope $-a/b$ of ker λ *has the same sign as v.*

Remarks. This result implies Lemma 3.5.1 and more: it says that the kernels of the characteristic covectors in $\mathbb{R}^2$ form a set of tangent lines outside a set $A^\flat$ which looks like this:

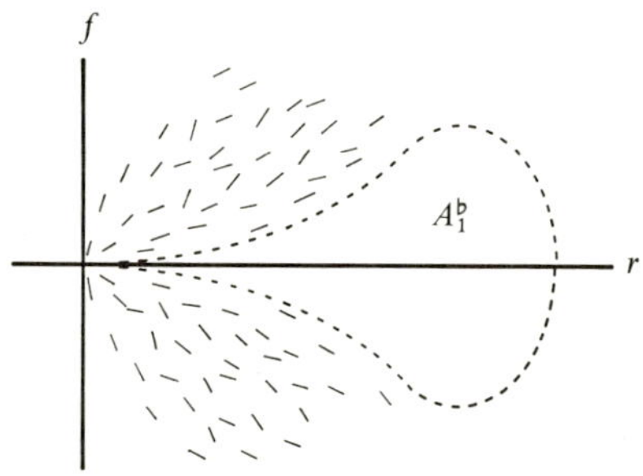

If f and r are real analytic functions, then this result is obvious because the map $(f, r)\colon Z\to\mathbb{R}^2$ may be stratified with analytic strata (i.e., with curves in $\mathbb{R}^2$) which contain the origin in their closures. The tangent vectors to these strata are the kernels of the characteristic covectors.

Proof. Step 1. First we show that the restriction $(r, f)|Z\colon Z\to\mathbb{R}^2$ has no characteristic covectors in some fringed set $A_1^\flat$. This will use the fact that f is nondepraved. The set of points $(u, v)\in\mathbb{R}^2$ which have nonzero characteristic covectors is closed. Thus, it suffices to show that there are no characteristic covectors over any point $(u, 0)\in\mathbb{R}^2$. Suppose this is false, i.e., that there is a sequence of points $p_i\in Z\cap f^{-1}(0)$ converging to p such that $df(p_i)$ and $dr(p_i)$ are linearly dependent when restricted to the stratum Y of Z which contains the point p_i. This means that f has a critical point on $Y \cap \partial B_{\delta_i}(p)$ where $\delta_i = r(p_i)$. By restricting to a subsequence if necessary, we may assume the points p_i are all contained in the same stratum Y of Z, that the tangent planes $T_{p_i} Y$ converge to some plane Q, and that the secant lines $\ell_i = \overline{p_i p}$ (in the fixed local coordinate system on M) converge to some limiting line ℓ. By Whitney's condition B,

$$Q = \lim_{i \to \infty} \ell_i \oplus T_{p_i}(Y \cap \partial B_\delta(p)).$$

This limit is a perpendicular direct sum, but $f(p) = f(p_i) = v$, so $df(p)(Q) = 0$. If $Y > X$, then this contradicts the assumption that $df(p)$ is a nondegenerate covector. We may therefore suppose that $Y = X$, so

$$\ker d(f \mid X)(p_i) = T_{p_i}(X \cap \partial B_\delta(p)) = T_{p_i}(X \cap f^{-1}(v)).$$

These kernels converge to a codimension one subspace τ of $Q = T_p X$, which is perpendicular to ℓ. However, $\ell \subset \tau$ since (by Sect. 2.5.1 and the fact that f is nondepraved), $f^{-1}(v) \cap X$ satisfies Whitney's condition B. This is a contradiction. □

Step 2. If $u > 0$ is sufficiently small, $v \neq 0$, and $\lambda = (u, v, a\,du + b\,dv) \in T^* \mathbb{R}^2$ is a characteristic covector for the map $(r, f) \mid X : X \to \mathbb{R}^2$, then the sign of the slope of $\ker(\lambda)$ is equal to the sign of v. This is simply a rewording of Lemma 2.5.2.

Step 3. We now study the slope of the nonzero characteristic covectors arising from a stratum $Y > X$ in Z. By our choice of local coordinates, we may assume the ambient space M is Euclidean space $\mathbb{R}^n$. Suppose there is a sequence of points $q_i \in Y$ which converge to $p \in X$ such that $d(f \mid Y)(q_i)$ and $d(r \mid Y)(q_i)$ are linearly dependent, i.e., $a_i d(f \mid Y)(q_i) + b_i d(r \mid Y)(q_i) = 0$. We must show that $a_i b_i$ has the opposite sign to $f(q_i) - f(p)$. By choosing a subsequence if necessary, we may assume that:

(a) $f(q_i) - f(p)$ has a constant sign $s = \pm 1$ (independent of i).

(b) The tangent planes $\tau_i = T_{q_i} Y$ converge to some limiting plane τ.

(c) The secant lines $\ell_i = \overline{q_i p}$ converge to some limiting line $\ell \subset \tau$, and the vectors

$$v_i = \text{projection to } \tau_i \text{ of } (q_i - p)/|q_i - p|$$

converge to some limiting vector $v \in \ell$.

(d) The subspaces $Q_i = \ker d(r \mid Y)(q_i) \subset \tau_i$ converge to some hyperplane $Q \subset \tau$.

Now consider the equation

$$a_i df(q_i)(v_i) + b_i dr(q_i)(v_i) = 0.$$

Clearly $dr(q_i)(v_i) > 0$, so it suffices to prove that, for sufficiently large i, $df(q_i)(v_i)$ has the same sign as $f(q_i) - f(p)$. Let us suppose, for example, that $f(q_i) - f(p) > 0$ for all i. Then

$$df(p)(v) = \lim (f(q_i) - f(p))/|q_i - p| \geq 0.$$

So it suffices to show that $df(p)(v) \neq 0$. However, we have a limiting direct sum, $\tau = Q \oplus \ell$. But $df(p)(Q) = 0$, while the assumption that $df(p)$ is nondegenerate implies that $df(p)(\tau) \neq 0$. Therefore, $df(p)(v) \neq 0$. □

6.3. Corollary. *The map* $(r\pi, f\pi)\colon T_X \cap Z \to \mathbb{R}^2$ *has no characteristic covectors in the region* $A_1^\flat$.

Proof. The map π is a submersion to the stratum X, and the restriction $(r, f)|X$ has no characteristic covectors in the region $A^\flat$. □

6.4. Lemma. *There exists* $r_1 > 0$ *such that if* $z \in Z$ *satisfies* $r(z) \le r_1$ *and* $z \notin X$, *then the differentials* $d\pi(z)$ *and* $df(z)$ *are independent, when restricted to the stratum* Y *of* Z *which contains the point* z *(i.e., their kernels are transversally intersecting subspaces of* $T_z Y$*).*

Proof. The point $p \in X$ is an isolated critical point of f, and $df(p)$ is a nondegenerate covector, i.e., it does not kill any limit of tangent spaces to strata $Y > X$. Thus, there is a neighborhood of p with the same property. □

Corollary. *If* z *satisfies the above conditions (i.e.,* $r(z) \le r_1$ *and* $z \notin X$*), and* $\pi(z) \ne p$, *the the differentials* $df(z)$ *and* $df\pi(z)$ *are linearly independent when restricted to the stratum* Y *which contains* z. *If* $r(z) \le r_1$, $z \notin X$, *but* $\pi(z) = p$, *then* $df(z) \ne 0$ *although* $df\pi(z) = 0$ *(when restricted to the stratum* Y*).*

6.5. Characteristic Covectors of Normal Slices

We now examine the restriction of the projection $(r, f)\colon Z \to \mathbb{R}^2$ to the normal slices through the stratum X (i.e., to the fibres $\pi^{-1}(x)$ of the tubular projection $\pi\colon T_X \to X$).

Lemma. *There exists a number* $r_2 > 0$ *and a region* $A_2 \subset \mathbb{R}^+ \times \mathbb{R}^+$ *of type* $0 < v \ll u$, *such that for any* $z \in Z$ *which satisfies* $r\pi(z) \le r_2$, *the map*

$$(\rho, f - f\pi)|\pi^{-1}(\pi(z))\colon \pi^{-1}(\pi(z)) \to \mathbb{R}^2$$

has no characteristic covectors in the region $A_2^\flat$. Furthermore, if $\lambda = (u, v, a\,du + b\,dv) \in T^*\mathbb{R}^2$ *is a nonzero characteristic covector of this map, then* $v \ne 0$, $a \ne 0$, $b \ne 0$, *and the slope* $-a/b$ *of* $\ker \lambda$ *has the same sign as* v.

Proof. The proof of this lemma is completely analogous to the proof of Lemma 6.2. It is only necessary to observe that if $df(p)$ is a nondegenerate characteristic covector for Z, then it is also a nondegenerate characteristic covector for $Z \cap \pi^{-1}(p)$ (with its stratification induced by transversal intersection). □

Remark. Lemma 6.5 remains valid for any smooth function $f\colon M \to \mathbb{R}$ such that $df(p)$ is a nondegenerate covector – it is not necessary to assume that the restriction $f|X$ has a nondepraved critical point at p.

Corollary. *If* $r\pi(z) \le r_2$, $z \notin X$, *and* $(\rho(z), f(z) - f\pi(z)) \in A_2^\flat$, *then the differentials* $d\rho(z)$, $df(z)$, *and* $d\pi(z)$ *are independent (i.e., their kernels are transversally intersecting subspaces of* $T_z Y$*). The same holds for the differentials* $dr(z)$, $df(z)$, *and* $d\pi(z)$. *Furthermore, if* $\pi(z) \ne p$, *then each of the following triples of covectors are linearly independent:*

(a) $d\rho(z)$, $df(z)$, *and* $df\pi(z)$
(b) $dr(z)$, $df(z)$, *and* $df\pi(z)$

(c) $d\rho(z)$, $df(z)$, *and* $dr\pi(z)$
(d) $dr(z)$, $df(z)$, *and* $dr\pi(z)$.

Proof. The lemma gives independence of $d\rho(z)$, $d(f-f\pi)(z)$, and $d\pi(z)$. But

$$\ker(d\rho(z))\cap\ker(d\pi(z))=\ker(dr(z))\cap\ker(d\pi(z))$$

since $r=\rho+r\pi$. Thus, $dr(z)$, $d(f-f\pi)(z)$, and $d\pi(z)$ are independent. Similarly, $dr(z)$, $df(z)$, and $d\pi(z)$ are independent, i.e.,

$$\ker(dr(z))+(\ker(df(z))\cap\ker(d\pi(z)))=T_z Y.$$

If $\pi(z)\neq p$, then $d(f\pi)(z)$ is nonzero and $\ker(d(f\pi)(z))\supset\ker(d\pi(z))$. Thus, $dr(z)$, $df(z)$, and $df\pi(z)$ are independent. A similar argument holds with $f\pi$ replaced by $r\pi$. □

6.6. Lemma. *For any $a>0$ and $b>0$ there exists a fringed set $A_3\subset\mathbb{R}^+\times\mathbb{R}^+$ of type $0<v\ll u$ such that whenever $z\in Z$ satisfies*

6.6.1. $(r(z), f(z))\in A_3^\flat$
6.6.2. $(r\pi(z), f\pi(z))\in A_3^\flat$
6.6.3. $z\notin X$
6.6.4. $\pi(z)\neq p$

then the covectors $a\,d\rho(z)+b\,dr\pi(z)$, $df(z)$, and $d(f\pi)(z)$ are linearly independent when restricted to the stratum Y which contains the point z.

Proof. Assume not. By Whitney's condition A, the points $z\in Z$ such that $a\,d\rho(z)+b\,dr\pi(z)$. $df(z)$, and $d(f\pi)(z)$ are linearly dependent when restricted to the stratum containing z, form a closed subset of Z. Thus, we may assume there is a sequence $z_i\in Z$ which converge to the critical point p, such that $f(z_i)=0$ and $f\pi(z_i)=0$, and such that the covectors $a\,d\rho(z_i)+b\,dr\pi(z_i)$, $df(z_i)$, $d(f\pi)(z_i)$ are linearly dependent, for all i. By choosing a subsequence if necessary, we may also assume that the points z_i all lie in the same stratum Y of Z, that the secant lines $\overline{pz_i}$ converge to some limiting line ℓ, that the tangent planes $T_{z_i}Y$ converge to some limiting plane τ, that the subspaces

$$\ker(a\,d\rho(z_i)+b\,dr\pi(z_i)),\qquad \ker(df(z_i)),\qquad \ker(d(f\pi)(z_i))$$

converge to subspaces A, B, and C of τ respectively.

We will now make use of the fact that f and π have smooth extensions to some neighborhood T_X of the stratum X in the manifold M. There exists a smooth Riemannian metric on M such that the square of the distance from the point p is given by

$$R(z)=a\rho(z)+br\pi(z).$$

Since ℓ is a limit of radial lines, we have ℓ is perpendicular to A (in this Riemannian metric), and so $\ell\not\subset A$. But the relation of linear dependence between $a\,d\rho(z)+b\,dr\pi(z)$, df, and $df\pi$ implies that $A\supset B+C$. We now claim that $\ell\subset B$ and $\ell\subset C$, which will be a contradiction.

Let $w=\lim(z_i-p)/|z_i-p|$ (computed in some local coordinate system on M) denote a unit vector in the limiting line ℓ. Since $df(p)\neq 0$ we have $B=\lim\ker(df(z_i))=\ker(df(p))$. But, $df(p)(w)=\lim(f(z_i)-p)/|z_i-p|=0$. Thus, $\ell\subset B$.

Similarly

$$\begin{aligned} C &= \lim \ker(df\pi(z_i)) \\ &= \lim d\pi(z_i)^{-1}(\ker(df(\pi(z_i))) \cap T_{\pi(z_i)} X) \\ &= d\pi(p)^{-1}(\lim \ker(df(\pi z_i))) \cap T_{\pi(z_i)} X) \\ &= d\pi(p)^{-1}(B \cap T_p X) \supset \ker(d\pi(p)) + (B \cap T_p X). \end{aligned}$$

Let ℓ_A be the limit of secant lines $\overline{z_i \pi(z_i)}$ and let ℓ_B be the limit of the secant lines $\overline{\pi(z_i) p}$. These are nontrivial lines by Conditions 6.6.3 and 6.6.4, and $\ell \subset \ell_A + \ell_B$. However, $\ell_A \subset A$ and $\ell_B \subset B \cap T_p X$. Thus, $\ell \subset (A + B \cap T_p X) \subset C$. □

6.7. Lemma. *There exists a fringed set $A_4 \subset \mathbb{R}^+ \times \mathbb{R}^+$ of type $0 < v \ll u$ such that for any $z \in Z$ which satisfies*

6.7.1. $(r(z), f(z)) \in A_4^\flat$
6.7.2. $(r(z), f\pi(z)) \in A_4^\flat$
6.7.3. $z \notin X$
6.7.4. $\pi(z) \neq p$

then the covectors

$$df(z), \quad df\pi(z), \quad dr(z),$$

are linearly independent, when restricted to the stratum Y of Z which contains the point z.

Proof. Let A_2 denote the fringed set of Sect. 6.5 such that $(\rho(z), f(z) - f\pi(z)) \in A_2^\flat$ implies $d\pi(z)$, $d\rho(z)$ and $d(f - f\pi)(z)$ are linearly independent. Let A_3 denote the fringed set of Sect. 6.6 (corresponding to parameters $a = 1$ and $b = 1$) such that $(r(z), f(z)) \in A_3$ and $(r\pi(z), f\pi(z)) \in A_3$ implies $dr\pi(z) + d\rho(z)$, $df(z)$ and $df\pi(z)$ are linearly independent. Fix $v_0 > 0$ so that the point $(0, v_0) \in A_2 \cap A_3$. For each $\delta > 0$ (but $\delta \leq v_0$) choose $M(\delta) > 0$ so that the rectangular box of base $\delta/2$ and height $2M(\delta)$ is contained in both of these sets, i.e., so that

$$B(\delta) = \{(u, v) \in \mathbb{R}^2 \mid \delta/2 \leq u \leq \delta \text{ and } |v| \leq 2M(\delta)\} \subset A_2^\flat \cap A_3^\flat.$$

Now define

$$A_4 = \{(\delta, \varepsilon) \in \mathbb{R}^+ \times \mathbb{R}^+ \mid \delta \leq v_0 \text{ and } \varepsilon \leq M(\delta)\}.$$

We claim this fringed set has the desired properties. Note that if $z \in Z$ satisfies $z \notin X$, $\pi(z) \neq p$, and either

(a) $(\rho(z), f(z) - f\pi(z)) \in A_2^\flat$ or
(b) $(r(z), f(z)) \in A_3^\flat$ and $(r\pi(z), f\pi(z)) \in A_3^\flat$

then (by Sects. 6.5 and 6.6) the covectors $dr(z)$, $df(z)$ and $df\pi(z)$ are linearly independent. So we need only show that one or other of the conditions (a) or (b) is implied by the relations (6.7.1)…(6.7.4).

Suppose $\rho(z) \leq r(z)/2$. Then $r\pi(z) = r(z) - \rho(z) \geq r(z)/2$. By 6.7.2 we have $(r\pi(z), f\pi(z)) \in B(r(z)) \subset A_3^\flat$ so condition (b) is satisfied. Similarly if $\rho(z) \geq r(z)/2$, then $(\rho(z), f - f\pi(z)) \in A_2^\flat$ so consition (a) is satisfied. □

6.8. Lemma. *Let A_4 be the set of type $0 < \varepsilon \ll \delta$ from Sect. 6.7 and let $r_1 > 0$ denote the number found in Lemma 6.4. Then for any $(\delta, \varepsilon) \in A_4$ such that $\delta \leq r_1$,* *the projection*

$$(f\circ\pi, f)\colon Z\cap B_\delta(p)\to\mathbb{R}^2$$

has no characteristic covectors in the square

$$|f|\leq\varepsilon, \qquad |f\pi|\leq\varepsilon$$

except for the following three types of covectors $\lambda=(u, v, a\,du+b\,dv)$:

1. *Those* λ *such that* $u=0$ *and* $v=0$.
2. *Those* λ *such that* $u=v$ *and* $a=b$. *These occur when* $z\in X$.
3. *Those* λ *such that* $u=0$ *and* $du=0$. *These characteristic covectors only vanish on the tangent spaces* $T_z Y$ *of points for which* $\pi(z)=p$. *(In other words,* $d(f\pi)(z)\neq 0$ *unless* z *lies in the* π*-fibre over the critical point* p.*)*

Proof. Let Y denote the stratum of $Y\cap B_\delta(p)$ which contains the point z. First suppose that $z\in Z\cap B_\delta^0(p)$, i.e., z is not a point on the boundary of the ball. By Lemma 6.4, if $z\notin X$ then $d\pi(z)$ and $df(z)$ are independent when restricted to $T_z Y$. Thus, $d(f\pi)(z)$ and $df(z)$ are also independent, provided $\pi(z)\neq p$.

We now examine points $z\in Z\cap\partial B_\delta(p)$. It is possible that there are no points $z\in Z\cap\partial B_\delta(p)$ such that $|f(z)|\leq\varepsilon$ and $|f\pi(z)|\leq\varepsilon$, in which case the lemma is true for trivial reasons. However, assuming such a point $z\in Z$ exists, it then satisfies the conditions (6.7.1 and 6.7.2) of Sect. 6.7. Thus, (assuming $\pi(z)\neq p$ and $z\notin X$) the differential forms $df(z)$, $dr(z)$, and $d(f\pi)(z)\in T_z^* Y$ are linearly independent. This implies that $df(z)$ and $d(f\pi)(z)$ are linearly independent when restricted to the tangent space $T_z^*(Y\cap\partial B_\delta(p))$ of the surface $r(z)=\delta$. □

6.9. Lemma. *Let* A_4 *denote the fringed set of Lemma 6.7. Fix* $(\delta, \varepsilon)\in A_4$ *and suppose* $z\in Z$ *satisfies* $\pi(z)\neq p$, $z\notin X$ *and*

6.9.1. $\delta/2\leq r(z)\leq\delta$
6.9.2. $\delta/2\leq\rho(z)\leq\delta$
6.9.3. $|f(z)|\leq\varepsilon$
6.9.4. $|f\pi(z)|\leq\varepsilon$.

Fix $a, b\in\mathbb{R}$ *with* $a\neq 0$. *Then the three covectors*

$$df(z), \qquad df\pi(z), \qquad a\,d\rho(z)+b\,dr\pi(z)$$

are linearly independent in $T_z Y$, *where* Y *is the stratum of* Z *which contains the point* z.

Proof. Since $|f-f\pi|\leq 2\varepsilon$, we have (see Sect. 6.7) $(\rho(z), f(z)-f\pi(z))\in B(\delta)\subset A_2^\flat$ (where A_2 is the set defined in Sect. 6.5), so Lemma 6.5 implies that $d\rho(z)$, $d(f-f\pi)(z)$, and $d\pi(z)$ are independent (i.e., their kernels are transverse subspaces of $T_z Y$, where Y is the stratum of Z which contains the point z), so $df(z)$ and $d\rho(z)$ are independent when restricted to $T_z Y\cap\ker d\pi(z)$. But $r=\rho+r\pi$, and $d(r\pi)(z)$ vanishes on $\ker d\pi(z)$. Thus, $d(f-f\pi)(z)$, $d\pi(z)$, and $a\,d\rho(z)+b\,dr\pi(z)$ are independent, provided $a\neq 0$. It follows that $d(f-f\pi)(z)$, $df\pi(z)$, and $a\,d\rho(z)+b\,dr\pi(z)$ are independent, since the only critical points of $df\pi(z)$ occur when $\pi(z)=p$. □

6.10. Lemma. *There exists a fringed set* $A_5 \subset \mathbb{R}^+ \times \mathbb{R}^+$ *with the following property: For any* $(\delta, \varepsilon) \in A_5$ *and for any* $a, b \in \mathbb{R}$ *with* $a \geq 0$ *and* $b > 0$, *and for any* $z \in Z$ *which satisfies* $z \notin X$, $\pi(z) \neq p$, *and*

6.10.1. $\delta/2 \leq r(z) \leq \delta$
6.10.2. $\delta/2 \leq r\pi(z) \leq \delta$
6.10.3. $|f(z)| \leq \varepsilon$
6.10.4. $|f\pi(z)| \leq \varepsilon$

then the three covectors

$$df(z), \quad df\pi(z), \quad a\,d\rho(z) + b\,dr\pi(z)$$

are linearly independent in $T_z^* Y$, *where* Y *is the stratum of* Z *which contains the point* z.

Proof. First we shall find a fringed set A_0 such that if $z \in Z$ satisfies

$$z \notin X, \quad \pi(z) \neq p, \quad (r(z), f(z)) \in A_0^\flat, \quad \text{and} \quad (r\pi(z), f\pi(z)) \in A_0^\flat$$

then $df(z)$, $df\pi(z)$, and $a\,d\rho(z) + b\,dr\pi(z)$ are linearly independent. By Lemma 6.4 there is a number $r_1 > 0$ such that if $\pi(z) \neq p$, and $z \notin X$ and $r(z) \leq r_1$ then $df(z)$ is nonzero on $\ker(d\pi(z)) | T_z Y$. By Corollary 6.3 there is a fringed set A_1 such that if $\pi(z) \neq p$ and $(r\pi(z), f\pi(z)) \in A_1^\flat$, then $dr\pi(z)$ and $df\pi(z)$ are linearly independent in $T_z Y$. Thus, if $r(z) \leq r_1$, $\pi(z) \neq p$, $z \notin X$, and $(r\pi(z), f\pi(z)) \in A_1^\flat$, then $df(z)$, $dr\pi(z)$ and $df\pi(z)$ are linearly independent. Since linear independence is an open condition, there exists a number $\lambda_0 > 0$ such that for any λ with $|\lambda| \leq \lambda_0$, the covectors

$$df(z), \quad dr\pi(z) + \lambda\,d\rho(z), \quad \text{and} \quad df\pi(z)$$

are linearly independent for the same set of choices of z. If $a/b \leq \lambda_0$, then we are done. Otherwise, by Lemma 6.6, for each value of $\lambda \in [\lambda_0, a/b]$, there is a fringed set $A_\lambda \subset \mathbb{R}^+ \times \mathbb{R}^+$ such that if $z \in Z$ satisfies $z \notin X$, $\pi(z) \neq p$, $(r(z), f(z)) \in A_\lambda^\flat$, and $(r\pi(z), f\pi(z)) \in A_\lambda^\flat$ then the covectors

$$dr\pi(z) + \lambda\,d\rho(z), \quad df(z), \quad \text{and} \quad df\pi(z)$$

are linearly independent. By Sect. 4.5 there is a uniform choice A_∞ of a fringed set such that $A_\infty \subset A_\lambda$ for all $\lambda \in [\lambda_0, a/b]$. Now choose a fringed set $A_0 \subset A_\infty \cap A_1$, and shrink A_0 if necessary to ensure that every point $(\delta, \varepsilon) \in A_0$ has $\delta \leq r_1$. This set A_0 has the desired properties: If $a/b \leq \lambda_0$, then the first argument applies while if $a/b \geq \lambda_0$, then the second argument applies.

We now repeat the method of Sect. 6.7: For each $\delta > 0$ choose a number $M_2(\delta) > 0$ so that the rectangular box of base $\delta/2$ and height $M_2(\delta)$ is contained in $A_0^\flat$, i.e., so that

$$B_2(\delta) = \{(u, v) \in \mathbb{R}^2 \mid \delta/2 \leq u \leq \delta \text{ and } |v| \leq M_2(\delta)\} \subset A_0^\flat.$$

Define

$$A_5 = \{(\delta, \varepsilon) \in \mathbb{R}^+ \times \mathbb{R}^+ \mid \delta \leq r_1 \text{ and } \varepsilon \leq M_2(\delta)\}.$$

Fix $(\delta, \varepsilon) \in A_5^\flat$. If $z \in Z$ satisfies (6.10.1) to (6.10.4) then

$$(r(z), f(z)) \in B_2(\delta) \subset A_0^\flat$$

and

$$(r\pi(z), f\pi(z)) \in B_2(\delta) \subset A_0^\flat$$

which implies that $df(z)$, $df\pi(z)$, and $a\,d\rho(z)+b\,dr\pi(z)$ are linearly independent. □

6.11. Lemma. *There exists a fringed set* $A_6 \subset \mathbb{R}^+ \times \mathbb{R}^+$ *with the following property: Fix* $(\delta, \varepsilon) \in A_6$. *Suppose a point* $z \in Z$ *satisfies* $z \notin X$, $\pi(z) \neq p$, *and*

6.11.1. $\rho(z)=\delta$
6.11.2. $r\pi(z)=\delta$
6.11.3. $|f(z)| \leq \varepsilon$
6.11.4. $|f\pi(z)| \leq \varepsilon$.

Then the covectors

$$d\rho(z), \quad dr\pi(z), \quad df(z), \quad df\pi(z)$$

are linearly independent when restricted to $T_z Y$, *where* Y *is the stratum of* Z *which contains the point* z.

Proof. We claim that $A_6 = A_1 \cap A_2$ has the required properties (where A_1 and A_2 are the fringed sets defined in Sects. 6.2 and 6.5). Fix $(\delta, \varepsilon) \in A_6$ and suppose $z \in Z$ satisfies 6.11.1 through 6.11.4. Then

$$(\rho(z), f(z) - f\pi(z)) \in A_2^\flat$$

so Corollary 6.5 implies that $d\rho(z)$, $df(z)$, and $d\pi(z)$ are independent. Similarly

$$(r\pi(z), f\pi(z)) \in A_1^\flat$$

so Corollary 6.3 implies that $dr\pi(z)$ and $df\pi(z)$ are independent. Together, these imply that $d\rho(z)$, $df(z)$, $dr\pi(z)$ and $df\pi(z)$ are independent. □

Chapter 7. Local, Normal, and Tangential Morse Data are Well Defined

This chapter contains the proofs of Theorems 3.5.3, 3.6.2, 3.9.2, 3.9.3, 3.10, and 3.11.

7.1. Definitions

Throughout this chapter we will assume that Z is a Whitney stratified subset of some smooth manifold M, that $f: M \to \mathbb{R}$ is a smooth function which has a nondepraved (or nondegenerate) critical point $p \in Z$, which lies in some stratum X of Z. For simplicity we shall assume the critical value $f(p)=0$. As in Chapter 6, we will assume a tubular neighborhood T_X of X in M has been chosen, with tubular projection $\pi: T_X \to X$, and normal distance function $\rho: T_X \to \mathbb{R}$. We also assume a Riemannian metric has been chosen on the ambient manifold M, and we denote the square of the distance from the critical point p by r. As in Sect. 6.1, for any set $A \subset \mathbb{R}^+ \times \mathbb{R}^+$ of type $0 < v \ll u$, we denote by $A^\flat$ the closure of the set obtained by adding to A its reflection about the u axis.

7.2. Regular Values

This section contains the proof of Proposition 3.2: if $[a, b]$ contains no critical values of $f: Z \to \mathbb{R}$ then there is an $\mathcal{S}$-decomposition preserving homeomorphism between $Z_{\leq a}$ and $Z_{\leq b}$.

Proof. We wish to pass from the set $(-\infty, a]$ to $(-\infty, b]$ by moving the wall. Let Y denote the following subset of $\mathbb{R}^2$:

$$Y = \{(t, f) \in \mathbb{R} \times \mathbb{R} \mid f \leq a + (b-a)t\}.$$

This is stratified as a manifold with boundary, and satisfies the hypotheses in Sect. 4.3 (moving the wall). Thus, we obtain a stratum preserving homeomorphism between $Z \cap Y_0 = Z_{\leq a}$ and $Z \cap Y_1 = Z_{\leq b}$. □

Remark. This proposition can also be proven using controlled vectorfields instead of the moving wall. A sketch of such a proof follows: Choose $\varepsilon > 0$ so that the interval $[a-2\varepsilon, b+2\varepsilon]$ contains no critical values of f. Choose a smooth vectorfield V on $[a-2\varepsilon, b+2\varepsilon]$ whose time 1 flow maps the interval $[a-\varepsilon, b]$ homeomorphically onto $[a-\varepsilon, a]$ and which vanishes on $[a-2\varepsilon, a-\varepsilon]$.

Thom's first isotopy lemma (Sect. 1.5) provides a lift of this vectorfield to a controlled vectorfield V' on $Z_{[a-\varepsilon, b+\varepsilon]}$ such that $f_* V' = V$. Thus, the time 1 flow of V' restricts to the desired homeomorphism on $Z_{[a-\varepsilon, b]}$ (and is the identity on $Z_{\leq a-\varepsilon}$).

7.3. Local Morse Data, Tangential Morse Data, and Fringed Sets

Let $A \subset \mathbb{R}^+ \times \mathbb{R}^+$ denote the fringed set from Proposition 6.2 of type $0 < v \ll u$ such that the projection $(r, f)\colon Z \to \mathbb{R}^2$ contains no characteristic covectors in the region $A^\flat$. Fix $(\delta, \varepsilon) \in A$ and let

$$\mathrm{Box}(\delta, \varepsilon) = \{(x, y) \in \mathbb{R}^2 \mid x \leq \delta \text{ and } |y| \leq \varepsilon\}.$$

This subset is stratified by its interior, the interiors of its three sides,

$$\mathrm{Top}(\delta, \varepsilon) = \{(x, \varepsilon) \mid x \leq \delta\}$$
$$\mathrm{Bottom}(\delta, \varepsilon) = \{(x, -\varepsilon) \mid x \leq \delta\}$$
$$\mathrm{RS}(\delta, \varepsilon) = \{(\delta, y) \mid |y| \leq \varepsilon\}$$

and its two corners (δ, ε) and $(\delta, -\varepsilon)$. Define $F\colon Z \to \mathbb{R}^2$ by $F(z) = (r(z), f(z))$. Observe that the local Morse data (Sect. 3.5) is the pair

$$(F^{-1}(\mathrm{Box}(\delta, \varepsilon)), F^{-1}(\mathrm{Bottom}(\delta, \varepsilon)))$$

and the tangential Morse data is the pair

$$(X \cap F^{-1}(\mathrm{Box}(\delta, \varepsilon)), X \cap F^{-1}(\mathrm{Bottom}(\delta, \varepsilon))).$$

Claim. *For any $(\delta, \varepsilon) \in A$, the map $F\colon Z \to \mathbb{R}^2$ is transverse to (each stratum of) $\mathrm{Box}(\delta, \varepsilon)$ in $\mathbb{R}^2$, i.e., no stratum of $\mathrm{Box}(\delta, \varepsilon)$ is tangent to the kernel of any characteristic covector of the map F.*

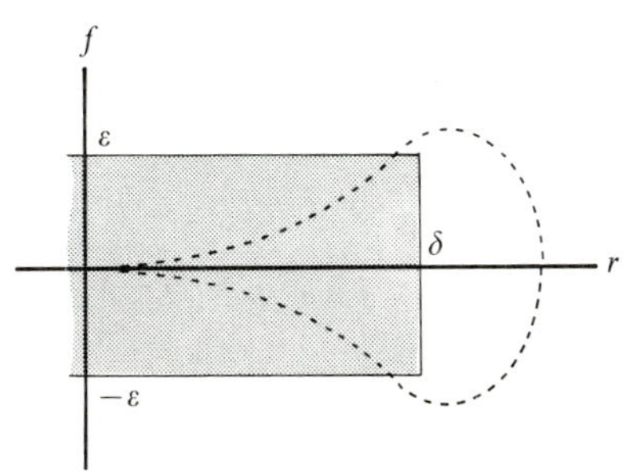

Diagram of Box (δ, ε) and fringed set

Proof of Claim. Any map is transverse to the interior of $\mathrm{Box}(\delta, \varepsilon)$. The map F is transverse to $\mathrm{Top}(\delta, \varepsilon)$, $\mathrm{Bottom}(\delta, \varepsilon)$ and $\mathrm{RS}(\delta, \varepsilon)$, because by Proposition 5.2 the kernels of the characteristic covectors have positive slope above the δ axis and negative slope below the δ axis. The map is transverse to the corners (δ, ε) and $(\delta, -\varepsilon)$ because these are points in the set $A^\flat$ which contains no characteristic covectors of the map F.

Corollary. *Local Morse data (resp. tangential Morse data) is Whitney stratified with strata of the form* $A \cap F^{-1}(B)$ *(resp.* $X \cap F^{-1}(B)$*), where* A *is a stratum of* Z *and* B *is a stratum of* Box(δ, ε).

7.4. Local and Tangential Morse Data are Independent of Choices

Suppose as above that Z is a Whitney stratified subset of a Riemannian manifold M (with Riemannian metric g_0), that $p \in Z$ is a nondepraved critical point of a function f_0, and that values of ε_0 and δ_0 have been chosen in accordance with the rules of Sect. 3.5. This gives rise to a particular construction of the local and tangential Morse data. Now suppose $f_1: M \to \mathbb{R}$ is another function which also has a nondepraved critical point at p. Choose another Riemannian metric g_1 on M, and values of ε_1 and δ_1 in accordance with the procedure in Sect. 3.5. This gives rise to another construction of local and tangential Morse data.

7.4.1. Theorem. *Suppose that the functions* f_0 *and* f_1 *are connected by a smooth one-parameter family of functions* $f: M \times \mathbb{R} \to \mathbb{R}$ *with a uniformly isolated nondepraved critical point* p *(i.e., there is a neighborhood* U *of* p*, such that for each* $t \in \mathbb{R}$ *the function* $f_t = f(*, t)$ *has no critical points in* U *except for the single critical point* p*, and this critical point is nondepraved). Then the local (resp. tangential) Morse data for* f_0 *at* p *(as constructed with respect to the first Riemannian metric and allowable choices of* ε_0 *and* δ_0*) is homeomorphic (by an* $\mathcal{S}$*-decomposition preserving and stratum preserving homeomorphism of pairs) to the local (resp. tangential) Morse data for* f_1 *at* p *(as constructed with respect to the new Riemannian metric and parameter values* ε_1 *and* δ_1*).*

Proof. We want to pass from Box$(\delta_0, \varepsilon_0)$ to Box$(\delta_1, \varepsilon_1)$ by moving the wall.

The local Riemannian metrics g_0 and g_1 are connected by a smooth one-parameter family of metrics, $g_t = t g_1 + (1-t) g_0$. These give rise to distance functions $r_t(z) = \text{distance}^2(p, z)$ (as measured by the metric g_t). We claim that there is a uniform chice of fringed set $A \subset \mathbb{R}^+ \times \mathbb{R}^+$ so that for each $t \in [0, 1]$ the map

$$(r_t, f_t): Z \to \mathbb{R}^2$$

has no characteristic covectors in the region $A^\flat$. This is easily verified using the same argument as in Sect. 6.2, plus the fact that p is a uniformly isolated critical point. Thus, it is possible to find a one parameter family of fringed sets $A_t \subset \mathbb{R}^+ \times \mathbb{R}^+$ such that

(a) $(\delta_0, \varepsilon_0) \in A_0$
(b) $(\delta_1, \varepsilon_1) \in A_1$
(c) the map

$$(r_t, f_t): Z \to \mathbb{R}^2$$

has no characteristic covectors in $A_t^\flat$.

(d) the total space of the family

$$\bigcup_{t\in[0,1]} \{t\}\times A_t \subset [0,1]\times\mathbb{R}^+\times\mathbb{R}^+$$

forms an open subset of $\mathbb{R}^+\times\mathbb{R}^+$.

By Proposition 5.4 there is a smooth one-parameter family of points $(\delta_t, \varepsilon_t)\in A_t$ which connect $(\delta_0, \varepsilon_0)$ to $(\delta_1, \varepsilon_1)$. Define $F: Z\times[0,1]\to\mathbb{R}^2\times[0,1]$ by

$$F(z,t)=(r_t(z), f_t(z)).$$

Let

$$Y_t=\operatorname{Box}(\delta_t, \varepsilon_t)\times\{t\}\subset\mathbb{R}^2\times[0,1]$$
$$W_t=\operatorname{Bottom}(\delta_t, \varepsilon_t)\times\{t\}\subset\mathbb{R}^2\times[0,1].$$

Then

$$(Y, W)=\bigcup_{t\in[0,1]}(Y_t, W_t)$$

forms a stratified fibre bundle over $[0,1]$ and each $F_t: Z\to\mathbb{R}^2$ is transverse to (Y_t, W_t). Thus, Lemma 4.4 (Moving the wall, Version 2) applies, giving us an $\mathscr{S}$-decomposition preserving homeomorphism between $F_0^{-1}(Y_0, W_0)$ and $F_1^{-1}(Y_1, W_1)$ (which also takes strata to strata and is smooth on each stratum). □

7.5. Normal Morse Data and Halflinks are Independent of Choices

Suppose as above that Z is a Whitney stratified subset of a Riemannian manifold M, that $p\in Z$ is a nondepraved critical point of a function f_0 and N_0' is a submanifold of M which meets the stratum X transversally at the point p. Assume that values of ε_0 and δ_0 have been chosen in accordance with the rules of Sect. 3.5. This gives rise to a particular construction of the normal Morse data, and the upper and lower halflinks. Now suppose $f_1: M\to\mathbb{R}$ is another function which also has a nondepraved critical point at p. Choose another Riemannian metric on M, another transversal N_1' through p, and values of ε_1 and δ_1 in accordance with the procedure in Sect. 3.5. This gives rise to another construction of normal Morse data, and the upper and lower halflinks.

7.5.1. Theorem. *Suppose the covectors $df_0(p)$ and $df_1(p)$ lie in the same connected component of the set of nondegenerate covectors (Sect. 1.8). Then the normal Morse data (resp. upper and lower halflinks) for f_0 at p (constructed with respect to the first Riemannian metric, normal slice N_0, and given choices of ε_0 and δ_0) is homeomorphic (by a homeomorphism H of pairs which preserves the $\mathscr{S}$-decompositions and preserves the stratifications) to the normal Morse data (resp. upper and lower halflinks) for f_1 at p (constructed with respect to the second Riemannian metric, normal slice N_1, and choices of parameter values, ε_1 and δ_1). Furthermore, if $df_0(p)=df_1(p)$ then there is a canonical choice up to isotopy for the homeomorphism H (cf.* [Bru]).

Proof. The proof is essentially identical to the proof of Theorem 7.4. It is possible to find a smooth one-parameter family $N'\subset M\times[0,1]$ of transversals

N_t' connecting N_0' and N_1', a smooth one-parameter family of Riemannian metrics connecting the two given metrics, and a smooth one-partameter family of functions f_t such that each $df_t(p)$ is a nondegenerate covector (we do not assume that each f_t has a nondepraved critical point at p). In the proof of Theorem 7.4, it is necessary to replace the function $F(z, t)=(r_t(z), f_t(z))$ with the function $F: (Z \times [0, 1]) \cap N' \to \mathbb{R}^2$ which is given by

$$F(z, t)=(r_t(z), f_t(z)-f_t(p)).$$

Lemma 6.5 guarantees the existence of a fringed set A_t and points $(\delta_t, \varepsilon_t) \in A_t$ such that $F_t: N_t' \cap Z \to \mathbb{R}^2$ has no characteristic covectors in the region $A_t^\flat$, and so that the normal Morse data is given by

$$(F_t^{-1}(\mathrm{Box}\,(\delta_t, \varepsilon_t), F_t^{-1}(\mathrm{Bottom}\,(\delta_t, \varepsilon_t)).$$

Since each $df_t(p)$ is nondegenerate, the singular point p is uniformly isolated in the normal slice. Thus, there is a uniform choice of fringed set A_∞ such that $A_\infty \subset A_t$ for all t. Thus, we can assume the points $(\delta_t, \varepsilon_t)$ form a smooth curve in $\mathbb{R}^2$ connecting $(\delta_0, \varepsilon_0)$ and $(\delta_1, \varepsilon_1)$. Defining (Y, W) as in the proof of Theorem 7.4.1, Lemma 4.4 (Moving the wall, Version 2) gives an $\mathscr{S}$-decomposition preserving homeomorphism between $F_0^{-1}(Y_0, W_0)$ and $F_1^{-1}(Y_1, W_1)$.

Proof of the "furthermore" part. Suppose we have two different one-parameter families of normal slices connecting N_0' and N_1', two different one-parameter families of Riemannian metrics connecting r_0 and r_1, and two different one-parameter families of ε and δ connecting $(\delta_0, \varepsilon_0)$ and $(\delta_1, \varepsilon_1)$. This gives rise to two different homeomorphisms G and H between the normal Morse data (resp. upper and lower halflinks). It is easy to find a two-parameter family of normal slices $N_{(s,t)}$ and distance functions $r_{(s,t)}$ which connect these two one-parameter families. Proposition 5.5 then provides the appropriate two-parameter family of choices $(\delta_{(s,t)}, \varepsilon_{(s,t)})$ (so that, for each value of s and t the corresponding map

$$(r_{(s,t)}, f_{(s,t)}): N_{(s,t)} \to \mathbb{R}^2$$

has no characteristic covectors in the region

$$\{(u, v) \mid u=\delta_{(s,t)}, |v| \leq \varepsilon_{(s,t)}\}).$$

This gives rise (by moving the wall as above) to a one-parameter family of homeomorphisms between G and H, i.e., G and H are isotopic. □

7.5.2. Corollary. *There is an $\mathscr{S}$-decomposition preserving and stratum preserving homeomorphism of pairs between the upper (resp. lower) halflink at p for the function f_0 with Riemannian metric g_0 and choice of parameters N_0', δ_0, ε_0, and the upper (resp. lower) halflink for the function f_1 constructed with respect to the Riemannian metric g_1 and parameters N_1', δ_1, and ε_1.*

Proof. The upper halflink is a union of strata in $F_t^{-1}(Y_t)$. □

7.5.3. Corollary. *If Z is a Whitney stratified subanalytic set, then there are finitely many possibilities (up to stratum preserving homeomorphism) for the normal Morse data of a function at the point p, and also for the upper and lower halflinks at p.*

Proof. By Proposition 1.8 the set of nondegenerate covectors has finitely many connected components. □

7.5.4. Corollary. *If Z is a complex analytically Whitney stratified complex analytic variety, then the normal Morse data at a point p is independent of the Morse function f. Furthermore, there is a stratum preserving homeomorphism between the upper and lower halflinks, and these spaces are independent of the function f.*

Proof. The set of nondegenerate covectors for Z at p is connected (see Sect. 1.8) since it is the complement of a subset of complex codimension one in the set of all characteristic covectors. □

7.5.5. Remarks. In Corollary 7.5.3 we are implicitly assuming that p is a nonexceptional point (see Sect. 1.8) by supposing that there exists a Morse function at the point p. See Sect. 13.2 for an example where the halflink is not well-defined in a family which passes through an exceptional point.

7.6. Local Morse Data is Morse Data

In this section we prove Proposition 3.5.4. We consider the situation outlined in Sect. 3.5, with a nondepraved critical point p of a function $f: Z \to \mathbb{R}$, with critical value $f(p)=0$, and a choice of parameter values δ and ε. Let $A^\flat \subset \mathbb{R}^2$ denote the fringed set of noncharacteristic points of the map $F(z)=(r(z), f(z))$ (as in Sect. 7.4) and choose $(\delta, \varepsilon) \in A$. Define the manifold with corners

$$\operatorname{Step}(\delta, \varepsilon)=\{(u, v) \in \mathbb{R}^2 \mid v \leq -\varepsilon\} \cup \{(u, v) \in \mathbb{R}^2 \mid u \leq \delta \text{ and } v \leq \varepsilon\}.$$

By moving the wall, we will find an $\mathscr{S}$-decomposition preserving homeomorphism between $Z_{\leq \varepsilon}$ and

$$F^{-1}(\operatorname{Step}(\delta, \varepsilon))=Z_{\leq -\varepsilon} \cup_Q P$$

where

$$(P, Q)=F^{-1}(\operatorname{Box}(\delta, \varepsilon), \operatorname{Bottom}(\delta, \varepsilon))$$

is the local Morse data. Note that the $\mathscr{S}$-decomposition of $Z_{\leq -\varepsilon} \cup_Q P$ defined as an attaching space (Sect. 1.1) coincides with the decomposition of $F^{-1}(\operatorname{Step}(\delta, \varepsilon))$ which is given by its intersection with the strata of X.

Choose a one-parameter family $Y_t \subset \mathbb{R}^2$ of manifolds with corners such that $Y_0=\operatorname{Step}(\delta, \varepsilon)$ and $Y_1=\{(u, v) \mid v \leq \varepsilon\}$ and so that for each t, the corners of Y_t are contained in the set $A^\flat$ and so that for large values of r the boundary of Y_t is horizontal. For example.

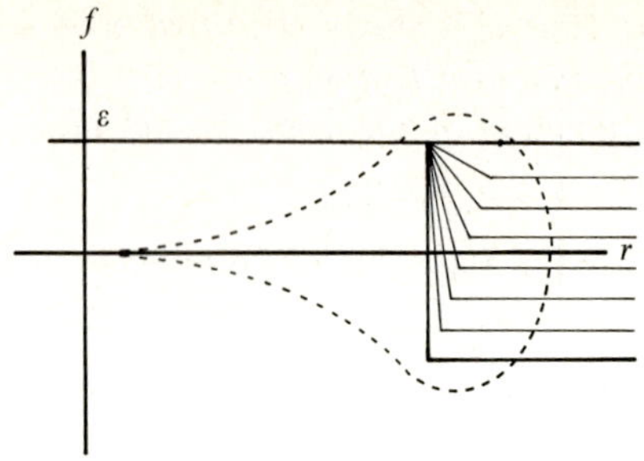

Moving the wall to show that local Morse data is Morse data

It follows that the tangent spaces to the boundary of Y_t are never contained in any characteristic covector of the map F, so moving the wall, version 1 gives an $\mathcal{S}$-decomposition preserving and stratum preserving homeomorphism between

$$F^{-1}(Y_0)=Z_{\leq -\varepsilon}\cup_Q P \quad \text{and} \quad F^{-1}(Y_1)=Z_{\leq \varepsilon}.$$

7.7. The Link and the Halflink

In this section we give the proof of Theorem 3.10. We shall consider the situation outlined in Sect. 3.9, with choices of a nondepraved critical point p, a normal slice N through the stratum X at the point p, a choice of Riemannian metric on M. We consider the map $F: Z\cap N \to \mathbb{R}^2$ which is given by $F(z)=(r(z), f(z)-f(p))$ and let A denote the set of type $0<v\ll u$ so that $F|N$ has no characteristic covectors in the region $A^\flat$. Choose a point $(\delta, \varepsilon)\in A$ and note that the upper and lower halflinks are given by

$$(F^{-1}(\mathrm{Top}(\delta, \varepsilon)), F^{-1}(\delta, \varepsilon))=(\ell^+, \partial\ell^-)$$
$$(F^{-1}(\mathrm{Bottom}(\delta, \varepsilon)), F^{-1}(\delta, -\varepsilon))=(\ell^-, \partial\ell^-).$$

Since F is a fibration over each point in the region $A^\flat$ we see immediately that any curve $(\delta_t, \varepsilon_t)\subset A^\flat$ determines (via the first isotopy lemma) a stratified homeomorphism $F^{-1}(\delta_0, \varepsilon_0)\cong F^{-1}(\delta_1, \varepsilon_1)$. In particular, it gives a homeomorphism between $\partial\ell^+$ and $\partial\ell^-$. In fact if we take the curve

$$\mathrm{RS}=\{(u, v)\in\mathbb{R}^2\,|\,u=\delta \text{ and } |v|\leq\varepsilon\}$$

then we obtain a homeomorphism

$$F^{-1}(\mathrm{RS})\cong\partial\ell^+\times[-\varepsilon, +\varepsilon]$$

which commutes with the projection to $[-\varepsilon, +\varepsilon]$.

We now find a homeomorphism between $\ell^+\cup_{\partial\ell}\ell^-$ and the link of the stratum X. First embed the space $\ell^+\cup_{\partial\ell}\ell^-$ into $Z\cap N$, by identifying it with $F^{-1}(Y_0)$, where Y_0 is the boundary of the triangle in the following diagram:

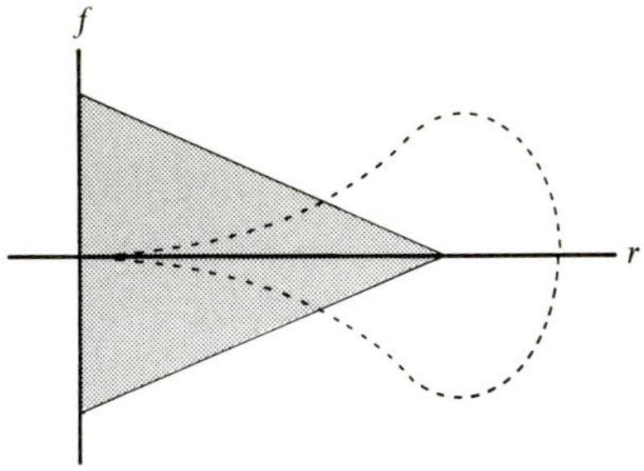

Upper and lower halflink joined along their boundary

Since $f(\{z \in Z \mid r(z) < 1\})$ is compact, we may assume it is contained in some interval $(v-a, v+a)$. Now consider the one-parameter family Y_t of stratified spaces

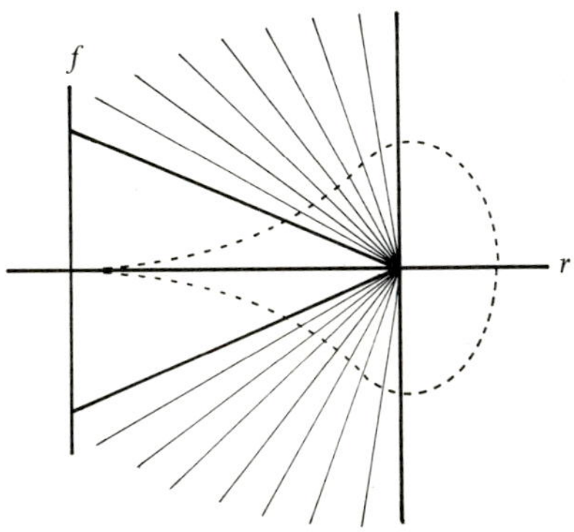

Moving the wall to show that the link is the union of the two halflinks

which interpolate between Y_0 and $Y_1 = \{(u, v) \in \mathbb{R}^2 \mid u = \delta\}$. By Lemma 6.5 the slope of the kernels of the characteristic covectors of F have the same sign as f, so no stratum of Y_t is tangent to the kernel of any characteristic covector. Thus, the map $F: Z \cap N \to \mathbb{R}^2$ is transverse to each Y_t. We obtain (by Moving the wall, Version 1) an $\mathcal{S}$-decomposition preserving homeomorphism between $F^{-1}(Y_0) = \ell^+ \cup_{\partial \ell} \ell^-$ and $F^{-1}(Y_1)$ which is the link of the stratum X, at the point p. $\square$

Remark. This homeomorphism identifies $\ell^+ \cup_{\partial \ell} \ell^-$ in a stratum preserving way with a refinement of the usual stratification of the link L of the point p: "fake" strata of the form $F^{-1}(\delta, 0) \cap L$ have been added.

7.8. Normal Morse Data is Homeomorphic to the Normal Slice

In this section we prove Proposition 3.8. Using the notation of the preceding section, the total space of the normal Morse data is given by $F^{-1}(\text{Box}(\delta, \varepsilon))$, while the normal slice (Sect. 1.4) $N = N' \cap Z \cap B_\delta(p)$ is given by

$F^{-1}(\text{Halfspace}(\delta))$, where

$$\text{Halfspace}(\delta)=\{(u,v)\in\mathbb{R}^2 \mid u\leq\delta\}.$$

We stratify Halfspace(δ) with its interior, Halfspace$^0(\delta)$, and the interiors of its three segments,

$$\begin{aligned}\text{RS}(\delta,\varepsilon)&=\{(u,v)\in\mathbb{R}^2 \mid u=\delta,\ |v|\leq\varepsilon\}\\ \text{RS}^+(\delta,\varepsilon)&=\{(u,v)\in\mathbb{R}^2 \mid u=\delta,\ v\geq\varepsilon\}\\ \text{RS}^-(\delta)&=\{(u,v)\in\mathbb{R}^2 \mid u=\delta,\ v\leq-\varepsilon\}\end{aligned}$$

and the two corners (δ,ε) and $(\delta,-\varepsilon)$.

Theorem. *There is an $\mathcal{S}$-decomposition preserving and stratum preserving homeomorphism between the normal Morse data,* $F^{-1}(\text{Box}(\delta,\varepsilon))$ *and the normal slice,* $F^{-1}(\text{Halfspace}(\delta))$, *which takes the interior*

$F^{-1}(\text{Box}^0)$ *to the interior* $F^{-1}(\text{Halfspace}^0)$, *and takes*
$F^{-1}(\text{Top}(\delta,\varepsilon))$ *to* $F^{-1}(\text{RS}^+(\delta,\varepsilon))$
$F^{-1}(\text{Bottom}(\delta,\varepsilon))$ *to* $F^{-1}(\text{RS}^-(\delta,\varepsilon))$
$F^{-1}(\text{RS}(\delta,\varepsilon))$ *to* $F^{-1}(\text{RS}(\delta,\varepsilon))$

and is the identity on the corners $F^{-1}(\delta,\pm\varepsilon)$.

Proof. Move the wall as follows:

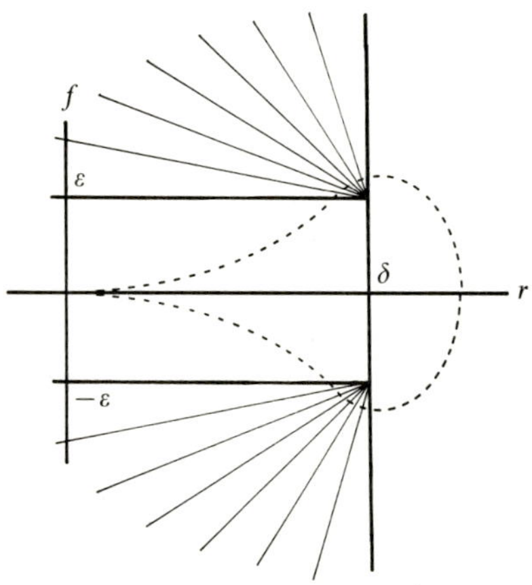

Moving the wall to show that Morse data is the normal slice

7.9. Normal Morse Data and the Halflink

In this section we prove Proposition 3.11: there is a homeomorphism of pairs between the normal Morse data and $(\text{cone}(\ell^+\cup_{\partial\ell}\ell^-),\ell^-)$.

Proof. The argument of Sect. 7.8 gives a homeomorphism between the normal Morse data and the pair

$$
\begin{aligned}
&F^{-1}(\mathrm{Box}(\delta, \varepsilon), \mathrm{Bottom}(\delta, \varepsilon)) \\
&\quad \cong (F^{-1}(\mathrm{Halfspace}(\delta)), F^{-1}(\mathrm{RS}^{-}(\delta, \varepsilon)) \\
&\quad \cong (\mathrm{cone}(F^{-1}(\mathrm{RS}^{+} \cup \mathrm{RS} \cup \mathrm{RS}^{-})), F^{-1}(\mathrm{RS}^{-})) \\
&\quad \cong (\mathrm{cone}(F^{-1}(\mathrm{Top} \cup \mathrm{RS} \cup \mathrm{Bottom})), F^{-1}(\mathrm{Bottom})) \\
&\quad \cong (\mathrm{cone}(\ell^{+} \cup (\partial\ell \times [0, 1]) \cup \ell^{-}), \ell^{-}) \\
&\quad \cong (\mathrm{cone}(\ell^{+} \cup_{\partial\ell} \ell^{-}), \ell^{-}). \quad \square
\end{aligned}
$$

Remark. The projection of a cone line under the map F will not necessarily be a straight line in $\mathbb{R}^2$. In other words, there is no particular relationship between this homeomorphism, the obvious conical structure of Box(δ, ε), and the axial lines in cone$(\ell^{+} \cup_{\partial\ell} \ell^{-})$.

Chapter 8. Proof of the Main Theorem

In this chapter we prove Theorem 3.7: the local Morse data is the Cartesian product of the tangential Morse data with the normal Morse data.

8.1. Definitions

Throughout this chapter, Z will denote a Whitney stratified subset of a smooth manifold M, and $f: Z \to \mathbb{R}$ will denote a function with an isolated nondepraved critical point $p \in Z$. We will use the symbol $\mathcal{S}$ to denote a partially ordered set which indexes the strata of Z. The stratum containing the point p will be denoted X, and $\pi: T_X \to X$ will denote the projection to X of a tubular neighborhood of X in M. We assume that T_X is sufficiently small that the restriction $\pi | Y$ to each stratum $Y > X$ is a submersion. As in Sect. 6.1 we choose a local Riemannian metric on M, with

$$r(z) = r\pi(z) + \rho(z)$$

where $r(z)$ is the square of the distance between p and z, and $\rho(z)$ is the square of the distance between z and $\pi(z)$. We shall consider a certain neighborhood $B_\delta = \{z \in Z \,|\, r(z) \leq \delta\}$ of p, where δ is chosen as follows:

(a) δ is so small that $f | X$ has no critical points in $B_{2\delta}$ or in $\pi(B_{2\delta})$, other than p.

(b) For all $z \in B_{2\delta}$, $df(z)(T_z Y) \neq 0$, where Y is the stratum of Z which contains z.

(c) $2\delta \leq r_1$ (of Sect. 6.4), and $2\delta \leq r_2$ (of Lemma 6.5).

(d) The point $(2\delta, 0)$ is an element of the set A_1 of Lemma 6.2, the set A_2 of Lemma 6.5, the set A_3 of Lemma 6.6, the set A_4 of Lemma 6.7, the set A_5 of Lemma 6.10, and the set A_6 of Lemma 6.11.

8.2. Embedding the Morse Data

In this section we show how to embed the product of pairs (normal Morse data) $\times$ (tangential Morse data) into the neighborhood B_δ defined above.

Choose $\varepsilon \ll \delta$ so that the point (δ, ε) is an element of the set $A_6^\flat$ of Lemma 6.11. Define

$$V_1=\{z\in Z \mid |f(z)-f\pi(z)|\leq \varepsilon/4,\ r\pi(z)\leq\delta, \text{ and } \rho(z)\leq\delta\}$$
$$V_2=\{z\in Z \mid f(z)-f\pi(z)=-\varepsilon/4,\ r\pi(z)\leq\delta, \text{ and } \rho(z)\leq\delta\}$$
$$W_1=\{z\in Z \mid |f\pi(z)|\leq 3\varepsilon/4 \text{ and } r\pi(z)\leq\delta\}$$
$$W_2=\{z\in Z \mid f\pi(z)=-3\varepsilon/4 \text{ and } r\pi(z)\leq\delta\}.$$

These spaces are $\mathscr{S}$-decomposed by their intersection with the strata of Z.

Proposition. *The pair of spaces* (V_1, V_2) *and* (W_1, W_2) *intersect transversally. Their intersection*

$$(V_1, V_2)\cap(W_1, W_2)=(V_1\cap W_1,\ V_1\cap W_2\cup V_2\cap W_1)$$

is homeomorphic (by a $\mathscr{S}$*-decomposition preserving homeomorphism of pairs) to the pair*

$$(\textit{normal Morse data at } p)\times(\textit{tangential Morse data at } p).$$

Proof. Let $D=\{x\in X \mid r(x)\leq\delta\}$ denote the disk of radius δ in the stratum X. We will show that the pair (V_1, V_2) fibres over D and is homeomorphic to $D\times$(Normal Morse data), and that the pair (W_1, W_2) fibres over the tangential Morse data (which is a subset of D) and is homeomorphic to the product (Tangential Morse data) $\times$ (normal slice).

Consider the open subset Z_1 of Z which is given by

$$Z_1=T_X\cap\{z\in Z \mid r\pi(z)\leq\delta\}.$$

By Proposition 6.5, the map $F\colon Z_1\to\mathbb{R}^2$ which is given by $F(z)=(\rho(z), f(z)-f\pi(z))$ is transverse to each stratum of $\mathrm{Box}(\delta, \varepsilon/4)$ (see Sect. 7.3). The pair (V_1, V_2) is the preimage under F of the pair $(\mathrm{Box}(\delta, \varepsilon/4), \mathrm{Bottom}(\delta, \varepsilon/4))$. By Lemma 6.5, for each $x\in D$, the restriction $F|\pi^{-1}(x)\colon \pi^{-1}(x)\to\mathbb{R}^2$ is transverse to the strata of $\mathrm{Box}(\delta, \varepsilon/4)$, which means that for each stratum S of $\mathrm{Box}(\delta, \varepsilon/4)$ we have a transversal intersection $F^{-1}(S)\pitchfork\pi^{-1}(x)$. Thus, $\pi|F^{-1}(S)$ has surjective differential. By Thom's first isotopy lemma (Sect. 1.5) this implies that the projection

$$\pi|F^{-1}(\mathrm{Box}(\delta, \varepsilon/4))\colon F^{-1}(\mathrm{Box}(\delta, \varepsilon/4))\to X$$

is a stratified fibre bundle over the (contractible) region $D\subset X$. Furthermore, the fibre over the critical point p is

$$\pi^{-1}(p)\cap(F^{-1}(\mathrm{Box}(\delta, \varepsilon/4)), F^{-1}(\mathrm{Bottom}(\delta, \varepsilon/4)))$$

which is precisely the normal Morse data as defined in Sects. 3.6 and 7.5.

Now consider the function $G\colon X\to\mathbb{R}^2$ given by $G(x)=(r(x), f(x))$. Note that δ and ε have been chosen so that the pair

$$(G^{-1}(\mathrm{Box}(\delta, 3\varepsilon/4)), G^{-1}(\mathrm{Bottom}(\delta, 3\varepsilon/4)))\subset X$$

is the tangential Morse data for f, and that this set is contained in the region D over which $\pi|(V_1, V_2)$ is a (trivial) stratified fibre bundle. Consequently the pair

$$(W_1, W_2)=\pi^{-1}G^{-1}(\mathrm{Box}(\delta, 3\varepsilon/4), \mathrm{Bottom}(\delta, 3\varepsilon/4))$$

is transverse to the pair (V_1, V_2). Therefore, the intersection $(V_1, V_2) \cap (W_1, W_2)$ is homeomorphic (by a stratum preserving homeomorphism) to

$$[\pi^{-1}(p) \cap (V_1, V_2)] \times \pi(W_1, W_2)$$

which is precisely the product (normal Morse data) × (tangential Morse data). □

8.3. Diagrams

From the five functions $f, f \circ \pi, r, \rho, r \circ \pi$, which are defined on Z, we obtain ten maps to $\mathbb{R}^2$ by projecting to any of the coordinate planes. A subset of Z can be designated by specifying a region in one of these coordinate planes.

Definition. A *picture* P is a Whitney stratified region in $\mathbb{R}^2$ together with a choice of two of the above functions (say, g and h). The picture is *allowable* if the map $(g, h): M \to \mathbb{R}^2$ takes each stratum of Z transversally (within $\mathbb{R}^2$) to each stratum of P. A *diagram* is a pair (P_a, P_b) of pictures (together with their two sets of functions, g_1, h_1, g_2, h_2). For any diagram D we define the realization $Z(D)$ to be the set

$$Z(D) = Z \cap (g_1, h_1)^{-1}(P_a) \cap (g_2, h_2)^{-1}(P_b)$$

which consists of all points $z \in Z$ such that $(g_i(z), h_i(z)) \in P_i$ (for $i = 1, 2$). The diagram is *allowable* if each of the pictures is allowable and if the map

$$(g_2, h_2) | (g_1, h_1)^{-1}(P_a) : (g_1, h_1)^{-1}(P_a) \to P_b$$

is transverse to each stratum of P_b (in $\mathbb{R}^2$). This definition is symmetric and is equivalent to saying that in the above formula for $Z(D)$, all the intersections of strata are transverse.

Remark. In Sect. 8.2 we found a homeomorphism between the product (normal Morse data) × (tangential Morse data) and the realization of the following diagram:

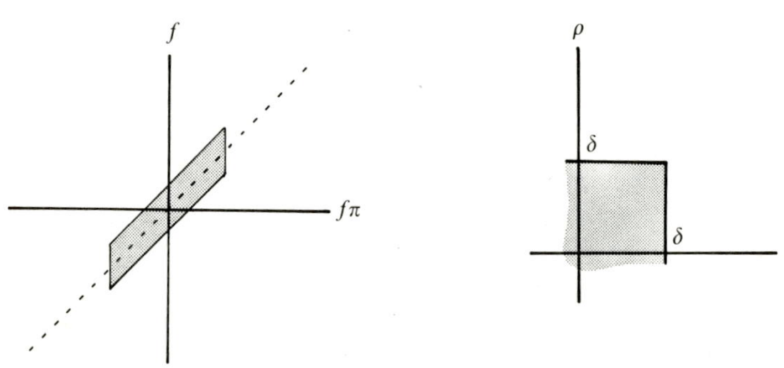

Diagram D_{11}: Normal Morse data × tangential Morse data

8.4. Outline of Proof

In this section we give a succession of diagrams which interpolate between the box diagram (Sect. 7.3) which defines local Morse data, and the above diagram D_{11}, which defines (normal Morse data) $\times$ (tangential Morse data). Each diagram will represent a pair of spaces, and the subspace will be explained in a caption below the diagram. Each diagram is obtained from the preceding by "moving the wall". The actual description of the motion of the wall, and the proof that there are no characteristic covectors to impede this motion will be given in Sect. 8.5.

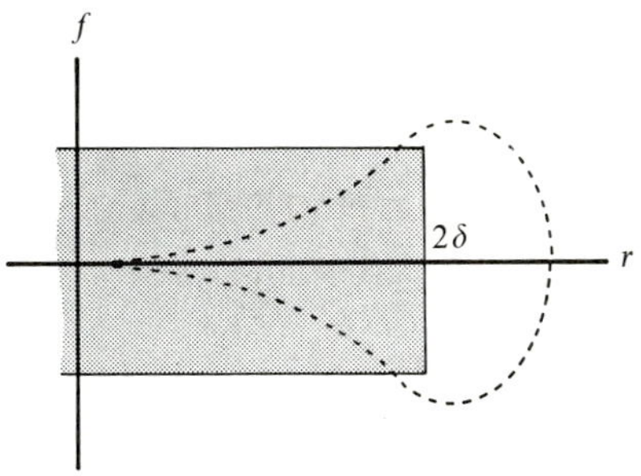

Diagram D_0: Local Morse data. The subspace is the bottom of the box

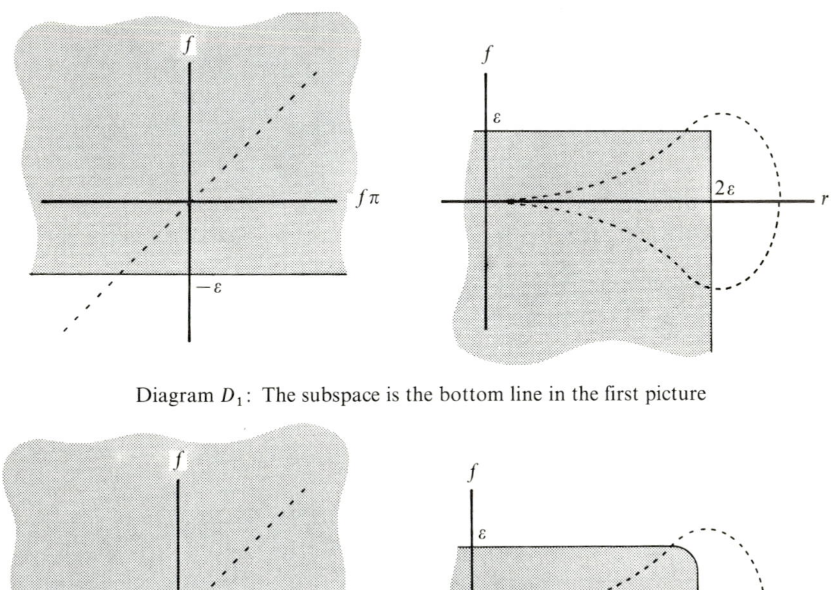

Diagram D_1: The subspace is the bottom line in the first picture

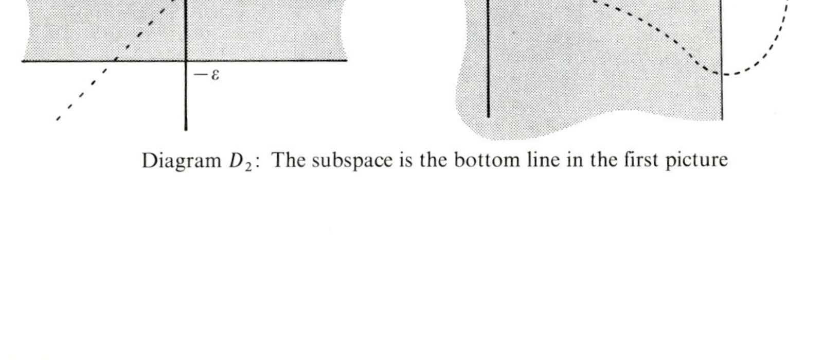

Diagram D_2: The subspace is the bottom line in the first picture

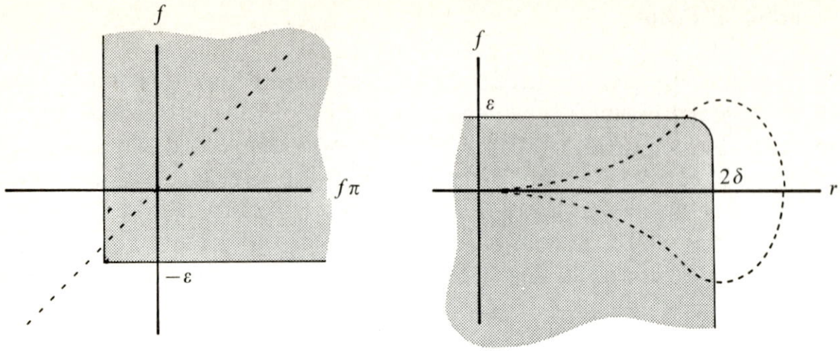

Diagram D_3: The subspace is the bottom and left hand sides in the first picture

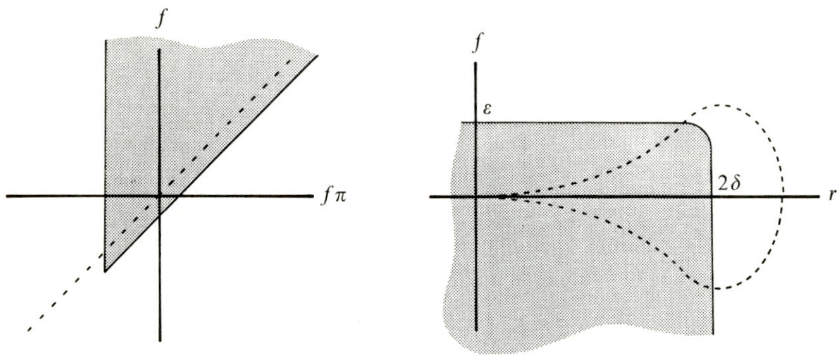

Diagram D_4: The subspace is the two edges shown in the first picture

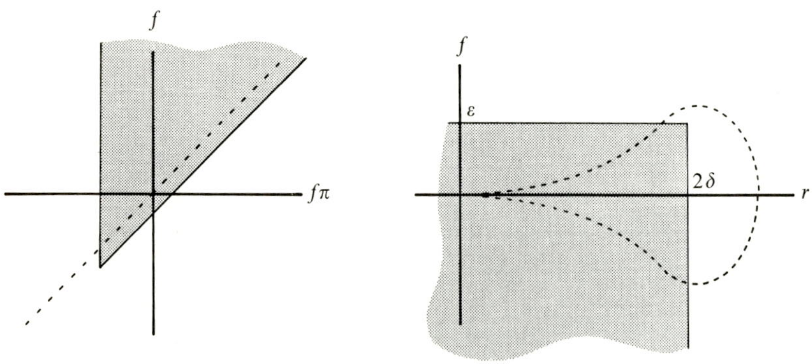

Diagram D_5: The subspace is the two edges shown in the first picture

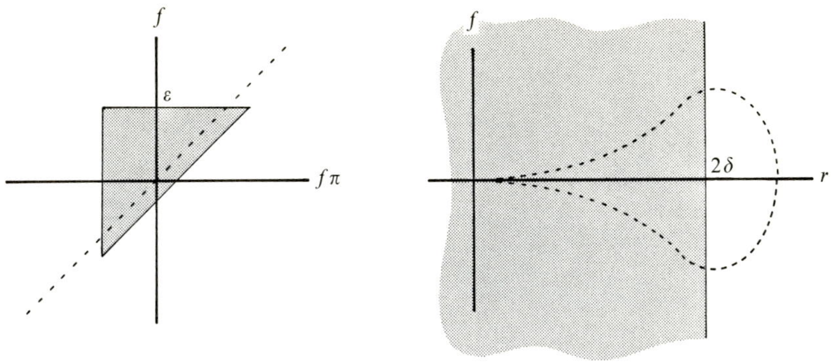

Diagram D_6: The subspace is the left edge and right diagonal in the first picture

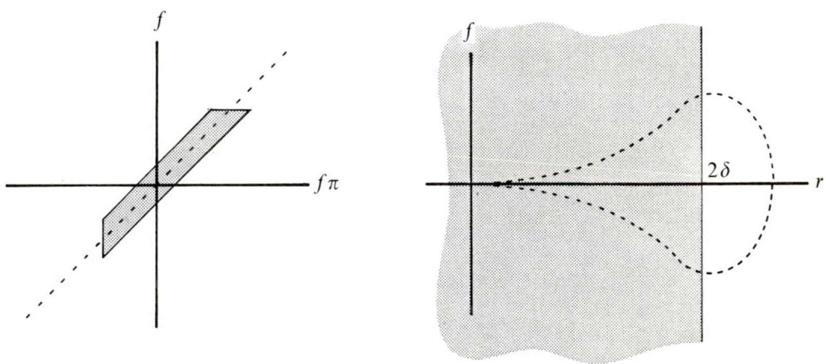

Diagram D_7: The subspace is the left vertical edge and bottom diagonal in the first picture

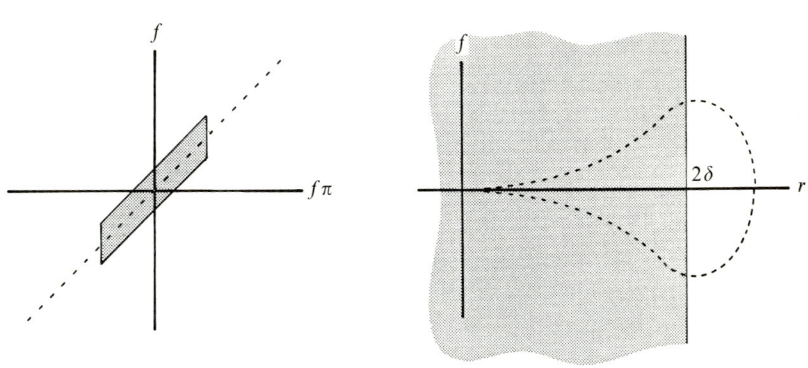

Diagram D_8: The subspace is the left vertical edge and bottom diagonal in the first picture

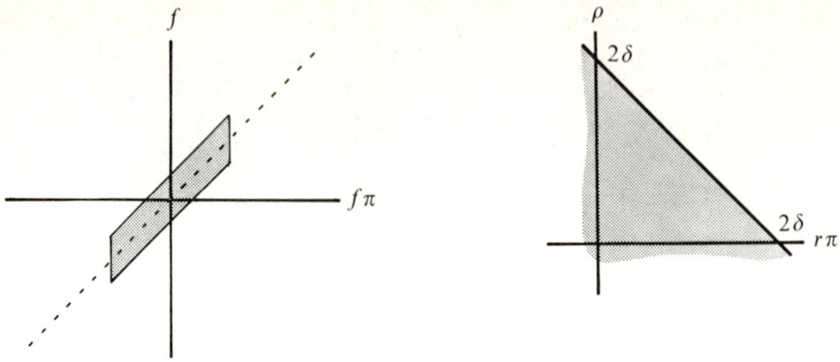

Diagram D_9: The subspace is the left vertical edge and bottom diagonal in the first picture

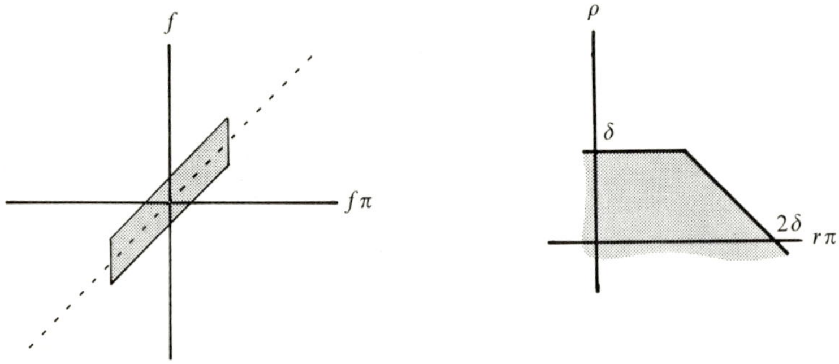

Diagram D_{10}: The subspace is the left vertical edge and bottom diagonal in the first picture

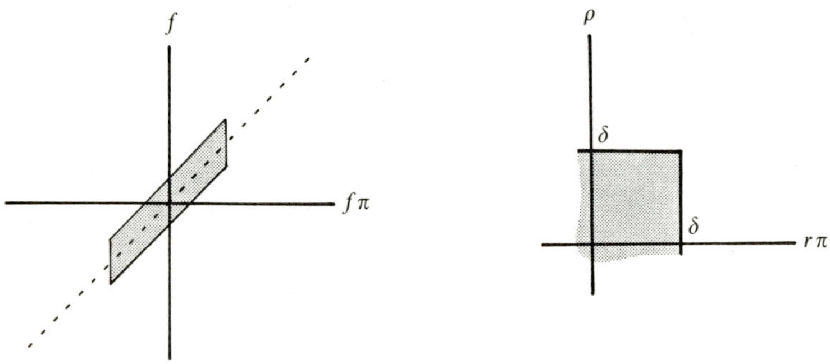

Diagram D_{11}: The subspace is the left vertical line and bottom diagonal in the first picture

8.5. Verifications

In this section we show how to move the wall and obtain homeomorphisms (of pairs) between the realizations of the preceding list of diagrams. The reader may check that in each case, the subspace is a union of strata of the realization, and that the subspace is taken to the subspace.

8.5.0. The realizations $Z(D_0)$ and $Z(D_1)$ are equal.

8.5.1. Since the map $(r, f)\colon Z \to \mathbb{R}^2$ is a submersion over the region $A_1^{\flat}$ of Sect. 6.2, the corner in the second picture of diagram D_2 can be rounded by moving the wall. Although this argument is straightforward, we will give it here in full because the remainder of this chapter consists of arguments along similar lines.

Consider the following motion of the wall: a one-parameter family of pictures $P_b(t)$ which interpolate between the second picture of diagram D_1 and the refinement which is obtained from the second picture of diagram D_2 by adding a fake zero-dimensional stratum to the rounded corner:

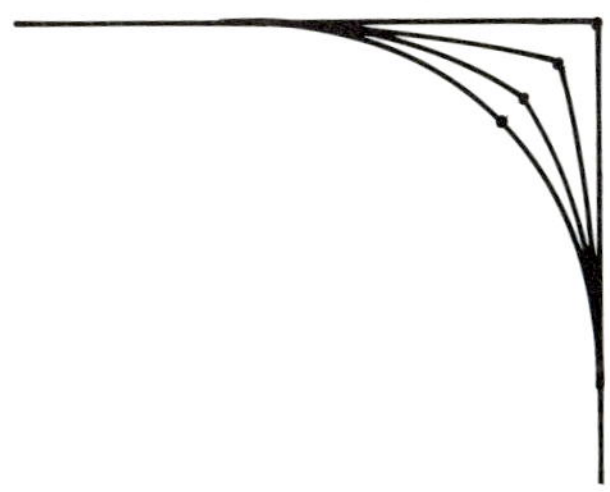

Rounding the corner

This one-parameter family $P_b(t)$ of pictures gives rise to a stratified space $Y \subset M \times \mathbb{R}$ as follows:

$$Y = \{(z, t) \in M \times \mathbb{R} \mid (r, f)(z) \in P_b(t)\}.$$

The allowability of the picture $P_b(t)$ implies that (each stratum of) each slice Y_t is transverse to (each stratum of) Z. Furthermore, $(f\pi, f)^{-1}(P_a) = f^{-1}[-\varepsilon, \infty)$ is transverse to each stratum in the intersection $Y_t \cap Z$ (where P_a is the first picture in the diagram D_1). Thus, Lemma 4.3 (Moving the wall) gives a stratum preserving homeomorphism between the set

$$Z(D_1) = Z \cap (f\pi, f)^{-1}(P_a) \cap (r, f)^{-1}(P_b(0))$$

and

$$Z(D_2) = Z \cap (f\pi, f)^{-1}(P_a) \cap (r, f)^{-1}(P_b(1)).$$

8.5.2. The wall motion from D_2 and D_3 and from D_3 and D_4 involve the most delicate arguments in this paper. We must first find all the characteristic covectors

$$\lambda = (u, v, a\,du + b\,dv) \in T^* \mathbb{R}^2$$

of the projection

$$(f\pi, f)\colon V \to \mathbb{R}^2$$

where $V = (r, f)^{-1}(P_b)$ and where the set P_b is stratified with two strata which we will denote by P^0 (for the interior) and ∂P (for the boundary).

Proposition. *Suppose the point $z \in V$ gives rise to the characteristic covector $\lambda = (u, v, a\,du + b\,dv)$. Then one of the following possibilities holds:*

(a) $z = p$ (*so* $u = v = 0$)

(b) $z \neq p$, *but* $z \in X$ (*so* $u = v$ *and* $a = b$)

(c) $z \notin X$ *but* $\pi(z)=p$ (*so* $u=0$ *and* $b=0$)
(d) $z \notin X$ *but* $(r, f)(z) \in \partial P$.

Proof. We will find the characteristic covectors arising from each of the strata of V. The strata of V are of 4 types:

(1) $(r, f)^{-1}(P^0) \cap X$
(2) $(r, f)^{-1}(P^0) \cap Y$
(3) $(r, f)^{-1}(\partial P) \cap X$
(4) $(r, f)^{-1}(\partial P) \cap Y$

where X is the stratum of Z which contains the critical point p, and Y is any other stratum of Z (with $Y > X$).

Type 1 and Type 3 strata. The image of the stratum X under $(f, f\pi)$ is precisely the diagonal $f = f\pi$. The Type 1 and Type 3 strata therefore give rise to characteristic covectors of type (a) and (b).

Type 2 strata. By Corollary 6.4 the differentials $df(z)$ and $d(f\pi)(z)$ are linearly independent unless $\pi(z)=p$, in which case $d(f\pi)(z)=0$ and $f\pi(z)=0$. These are characteristic covectors of type (c).

Type 4 strata. These give rise to characteristic covectors of type (d). □

We now analyze the covectors of type (d). Choose local coordinates about the point z in the stratum Y as follows: let $x_1, x_2, \ldots, x_s$ denote local coordinates in the stratum X, such that the critical point p corresponds to the *origin*. Extend these to coordinates on all of T_X (the tubular neighborhood of X in M) so that $x_i = x_i \circ \pi$, for each i. By the implicit function theorem, the functions $x_1, x_2, \ldots, x_s$ and f may be completed to a local coordinate system on Y about the point z by adding some functions $y_1, y_2, \ldots, y_{r-s-1}$ (where $r = \dim(Y)$). Let F denote the restriction of the function f to the stratum X. Thus,

$$\pi(x_1, x_2, \ldots, x_s, f, y_1, y_2, \ldots, y_{r-s-1}) = (x_1, x_2, \ldots, x_s)$$
$$f(x_1, x_2, \ldots, x_s, f, y_1, y_2, \ldots, y_{r-s-1}) = f$$
$$f\pi(x_1, x_2, \ldots, x_s, f, y_1, y_2, \ldots, y_{r-s-1}) = F(x_1, x_2, \ldots, x_s)$$

and we may choose the Riemannian metrics so that

$$r(z) = r\pi(z) + (f(z) - f\pi(z))^2 + g(y)$$

i.e.,

$$r(x_1, x_2, \ldots, x_s, f, y_1, y_2, \ldots, y_{r-s-1}) = \sum x_i^2 + (f - F(x))^2 + g(y)$$

where g is some function which depends only on $y_1, y_2, \ldots, y_{r-s-1}$.

At each point $(u, v) \in \partial P$ there are *nonnegative* numbers $\alpha(u, v)$ and $\beta(u, v)$ such that $\alpha\, df + \beta\, dr = 0$.

Claim. *There are no characteristic covectors of type (d) in the region* $|u| \leq \varepsilon$, $|v| < \varepsilon$. *If* λ *is a characteristic covector of type (d), arising from a point* $z \in (r, f)^{-1}(\partial P) \cap Y$, *then the slope of* $\ker(\lambda)$ *is:*

$$\frac{-a}{b} = \frac{f(z) - f\pi(z) - M}{\dfrac{\alpha}{\beta} + (f(z) - f\pi(z))}$$

where for some i,

$$\frac{1}{M} = \left(\frac{1}{x_i}\right)\left(\frac{\partial F(x)}{\partial x_i}\right).$$

Proof of claim. By Lemma 6.7, 6.8, and the assumption 8.1(d) (that $(2\delta, \varepsilon) \in A_4^\flat$, the covectors $df(z)$, $df\pi(z)$, and $dr(z)$ are linearly independent provided $|f(z)| \leq \varepsilon$ and $|f\pi(z)| \leq \varepsilon$. Thus there are no characteristic covectors in the shaded square. By Sect. 1.10, λ is a characteristic covector of the projection of a type (4) stratum if $\alpha df(z) + \beta dr(z)$ is a multiple (which we may take to be 1) of $adf\pi(z) + bdf(z)$. Thus,

$$\alpha df + \beta[\sum x_i dx_i + (f - f\pi)(df - df\pi) + dg] = adf\pi + bdf$$

where $df\pi = \sum \frac{\partial F}{\partial x_i} dx_i$. Equating coefficients of df, and coefficients of dx_i we obtain:

$$dg = 0$$

$$\alpha + \beta(f(z) - f\pi(z)) = b \qquad (\female 1)$$

$$\beta x_i - \beta(f - f\pi)\frac{\partial F}{\partial x_i} = a\frac{\partial F}{\partial x_i} \qquad (\female 2).$$

Dividing equations ($\female$2) by ($\female$1) gives the desired formula. □

The following diagram illustrates the possible values for the slope of $\ker(\lambda)$, where λ is a characteristic covector of the map $(f\pi, f)\colon V \to \mathbb{R}^2$ which arises from strata of V *other than* the stratum X which contains the critical point (i.e., covectors of types (c) and (d)). There are no characteristic covectors in the shaded regions and allowable slopes are designated by line segments in the circles.

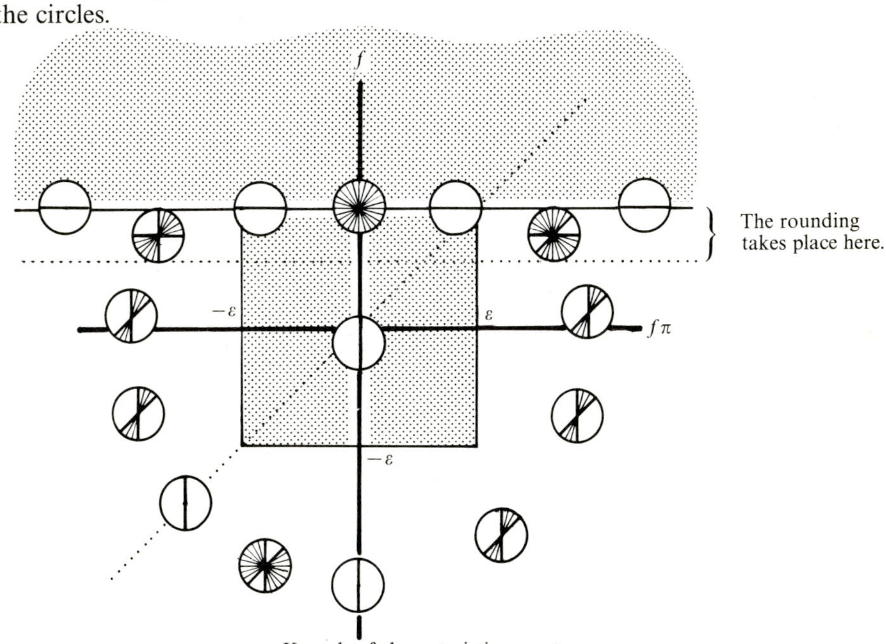

Kernels of characteristic covectors

Moving the wall. The homeomorphism between $Z(D_2)$ and $Z(D_3)$ is obtained by moving the wall in the first picture P_a. Let $P_a(t)$ denote the following set:

$$P_a(t)=\{(u, v)\in \mathbb{R}^2 \mid v \geq -\varepsilon \text{ and } (v+\varepsilon) \geq (u+\tfrac{3}{4}\varepsilon)\, t/(1-t)\}.$$

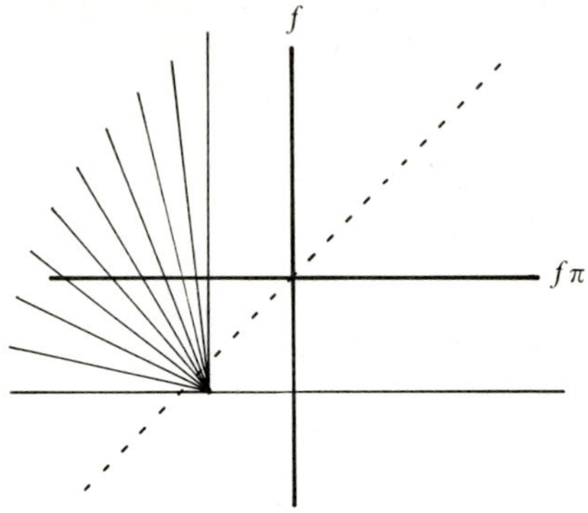

Moving the wall from diagram D_2 to diagram D_3

As t varies from 0 to 1 we obtain a one-parameter family of stratified spaces which interpolate between $P_a(0)$ and $P_a(1)$. We must check that the corresponding diagrams are allowable, i.e., that at no time t is any stratum of $P_a(t)$ contained in the kernel of any characteristic covector of the projection

$$(f, f\pi) \mid V\colon V \to \mathbb{R}^2.$$

The proposition and claim above guarantee that this wall motion is allowable because (a) the point $(-3\varepsilon/4, -\varepsilon)$ is not a characteristic point, and (b) in the region of the moving wall, the nondepravity condition guarantees (Sect. 2.5.3) that $M<0$, so the slope of $\ker(\lambda)$ is positive, for any characteristic covector λ.

8.5.3. We want to get from $Z(D_3)$ to $Z(D_4)$ by moving the wall. Let $P_a(t)$ denote the following one parameter family of pictures which interpolate between the first picture $P_a(0)$ in diagram D_4 and the first picture $P_a(1)$ in diagram D_5:

$$P_a(t)=\{(u, v)\in \mathbb{R}^2 \mid u \geq -3\varepsilon/4, \text{ and } v \geq t(u+3\varepsilon/4)-\varepsilon\}.$$

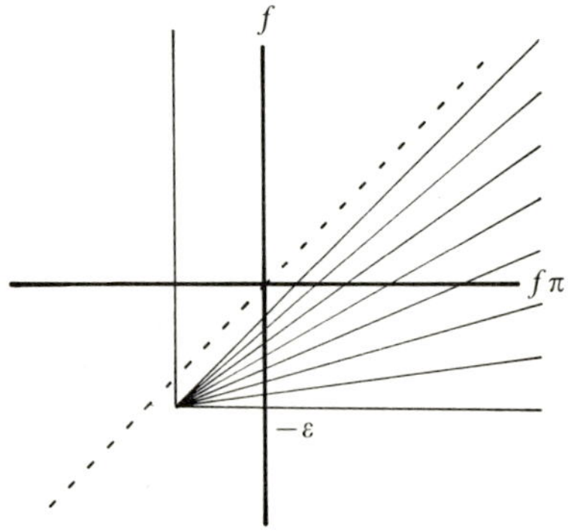

Moving the wall from diagram D_3 to diagram D_4

We now verify that this motion of the wall is allowable, i.e., that no stratum in $P_a(t)$ is contained in the kernel of any characteristic covector of the map $(f\pi, f)\colon V \to \mathbb{R}^2$ (where $V=(r,f)^{-1}(P_b)$). By lemma Sect. 6.8 (and assumption 8.1(d)) there are no characteristic covectors in the region $|f|<\varepsilon$ and $|f\pi|\le\varepsilon$. In the region of the moving wall, the claim and proposition of 8.5.2 above imply that the slope of the kernel of any characteristic covector λ is *not* in the interval $(0, 1]$ (since $\alpha\ge 0$, $\beta\ge 0$, $f-f\pi<0$, $f\pi>0$ and, by 2.5.3, $M>0$). □

8.5.4. Unrounding the corner. This is the same argument as 8.5.1.

8.5.5. The sets $Z(D_5)$ and $Z(D_6)$ are identical.

8.5.6. The homeomorphism between $Z(D_6)$ and $Z(D_7)$ is obtained by moving the wall through a one-parameter family of replacements $P_a(t)$ for the first picture as follows:

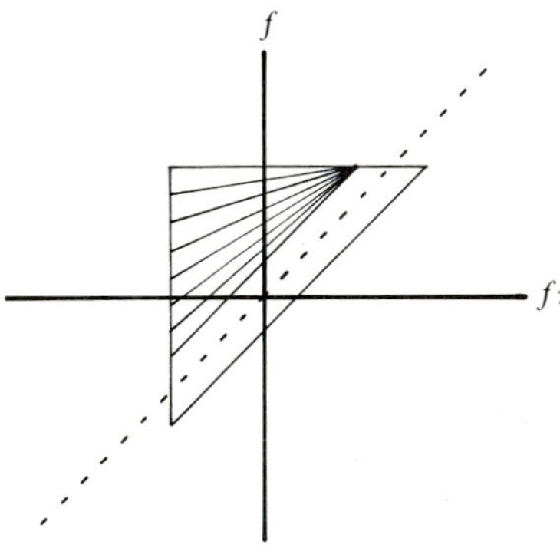

Moving the wall from diagram D_6 to diagram D_7

These diagrams are allowable because (by Lemma 6.7 and 6.8) there are no characteristic covectors of the map $(f\pi, f)\colon V\to\mathbb{R}^2$ in the region $|f|\le\varepsilon$, $|f\pi|\le\varepsilon$, where $V=(r,f)^{-1}(P_b)=\{z\in Z \mid r(z)\le 2\delta\}$.

8.5.7. Move the wall in the first picture P_a as follows:

$$P_a(t)=\left\{(u,v)\in\mathbb{R}^2 \mid u\ge -3\frac{\varepsilon}{4},\ u-\frac{\varepsilon}{4}\le v\le u+\frac{\varepsilon}{4},\ u\le\frac{(v-\varepsilon)(t-1)}{t}+3\frac{\varepsilon}{4}\right\}.$$

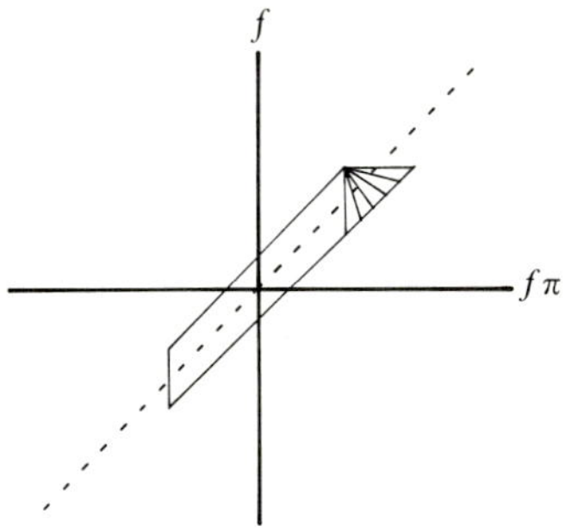

Moving the wall from diagram D_7 to diagram D_8

By Corollary 6.4, the covectors $df(z)$ and $df\pi(z)$ are linearly independent for all $z \in Z(D_7)$ (assuming $\pi(z) \neq p$). If $r(z) = 2\delta$, then (by Lemma 6.7 and Assumption 8.1(d)) the covectors $dr(z)$, $df(z)$ and $df\pi(z)$ are linearly independent. Thus, each diagram in this one-parameter family is allowable, and Lemma 4.3 (moving the wall) gives the desired homeomorphism between $Z(D_7)$ and $Z(D_8)$.

8.5.8. The sets $Z(D_8)$ and $Z(D_9)$ are identical since $r = \rho + r\pi$.

8.5.9. A homeomorphism between $Z(D_9)$ and $Z(D_{10})$ is given as follows: first refine the stratification of $Z(D_9)$ by adding a "fake" stratum of the form

$$Y \cap \{z \in Z \mid \rho(z) = \delta,\ r\pi(z) = \delta\}$$

whenever Y is a stratum of Z, i.e., by adding a zero-dimensional stratum $(\delta, \delta) \in \mathbb{R}^2$ to the second picture P_b in diagram D_9. The resulting diagram D_9' is allowable, because by Lemma 6.11 (and Assumption 8.1(d)) for any $z \in Z(D_9)$, if $\pi(z) \neq p$, then the covector $df(z)$, $df\pi(z)$, $d\rho(z)$ and $dr\pi(z)$ are linearly independent when restricted to $T_z Y$ (where Y is the stratum of Z which contains the point z). Now move the wall as follows: consider the one-parameter family of diagrams

$$P_b(t) = \{(u, v) \in \mathbb{R}^2 \mid v \leq 2\delta - u \text{ and } v \leq (t-1)(u-\delta) + \delta\}.$$

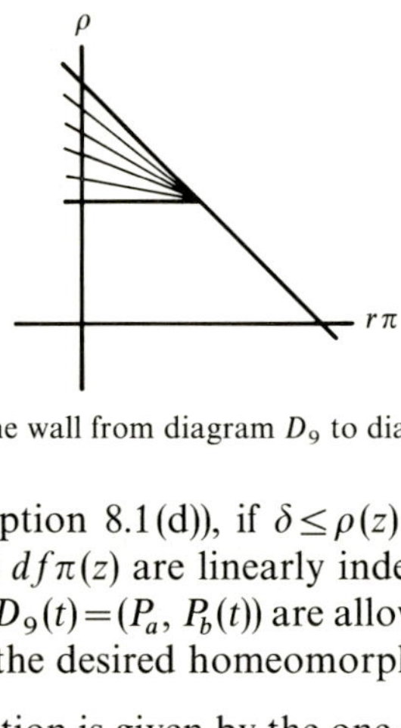

Moving the wall from diagram D_9 to diagram D_{10}

By Lemma 6.9 (and Assumption 8.1(d)), if $\delta \leq \rho(z) \leq 2\delta$ and $\delta \leq r(z) \leq 2\delta$, then $a\,d\rho(z) + b\,dr\pi(z)$, $df(z)$, and $df\pi(z)$ are linearly independent provided $a \neq 0$ and $\pi(z) \neq p$. Thus, the diagrams $D_9(t) = (P_a, P_b(t))$ are allowable and Lemma 4.3 (Moving the wall) applies, giving the desired homeomorphism.

8.5.10. The final wall motion is given by the one-parameter family of pictures

$$P_b(t) = \{(u, v) \in \mathbb{R}^2 \mid v \leq \delta/2 \text{ and } u \leq (t-1)(v-\delta/2) + \delta/2\}$$

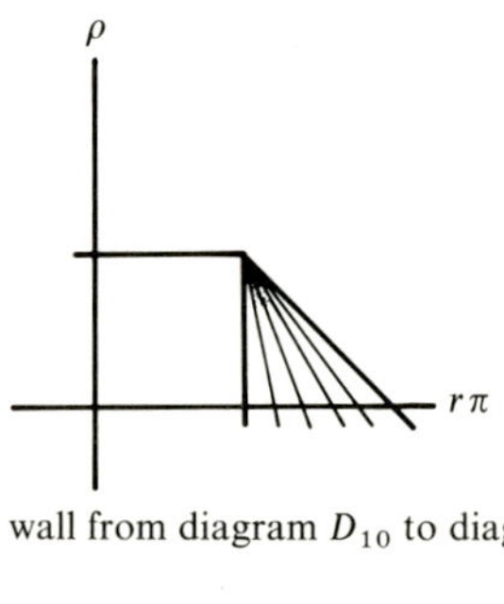

Moving the wall from diagram D_{10} to diagram D_{11}

which interpolate between D_{10} and D_{11}. According to Lemma 6.10 (and Assumption 8.1(d)), if $\delta \leq r\pi(z) \leq 2\delta$ and if $\delta \leq r(z) \leq 2\delta$ and $|f(z)| \leq \varepsilon$, $|f\pi(z)| \leq \varepsilon$, then the covectors $a\,d\rho(z) + b\,dr(z)$, $df(z)$ and $df\pi(z)$ are linearly independent (provided $a \geq 0$ and $b > 0$). Thus, the diagrams $D_{10}(t) = (P_a, P_b(t))$ are allowable so we can move the wall as illustrated.

Chapter 9. Relative Morse Theory

9.0. Introduction

The reader who is interested only in Morse theory for singular spaces or for nonproper Morse function may skip this chapter. We will consider the Morse theory of a composition

$$X \xrightarrow{\pi} Z \xrightarrow{f} \mathbb{R}$$

which will eventually be used (in Part II, Sects. 5.1 and 5.1*, with $Z = \mathbb{CP}^n$) to prove a conjecture of Deligne [D1] concerning Lefschetz hyperplane theorems for a variety X and an algebraic map $\pi: X \to \mathbb{CP}^n$. We will approximate the function f by a Morse function, although the composition $f\pi: X \to \mathbb{R}$ is not Morse (or even Morse-Bott) in any reasonable sense. All attempts to prove Deligne's conjecture by approximating $f\pi$ by a Morse function seem to end in failure because one loses curvature estimates on the Morse index of $f\pi$. Instead, we are forced to "relativize" the Morse theory of f. Our main result is stated in Sect. 9.8.

9.1. Definitions

Suppose $X \subset M'$ and $Z \subset M$ are Whitney stratified (closed) subsets of smooth manifolds M' and M. Let $\pi: X \to Z$ be a proper surjective stratified map (Sect. 1.6), i.e., π is the restriction of a smooth map $\tilde{\pi}: M' \to M$ and π takes each stratum of X submersively to a stratum of Z. We will assume that the strata of X are indexed by a partially ordered set $\mathcal{S}$. Let $f: Z \to \mathbb{R}$ be a Morse function (Sect. 2.1) with a nondegenerate critical point (or a smooth function with a nondepraved critical point) $p \in Z$. The preimage $\pi^{-1}(p)$ is called a *critical fibre* and $v = f(p)$ is called its critical value. As in Sect. 3.1 we consider the following spaces:

$$X_{\leq a} = \{x \in X \mid f\pi(x) \leq a\}$$
$$X_{<a} = \{x \in X \mid f\pi(x) < a\}$$
$$X_{[a,b]} = \{x \in X \mid a \leq f\pi(x) \leq b\}.$$

These spaces are $\mathcal{S}$-decomposed by their intersection with the strata of X. If a and b are regular values, then the above spaces are canonically Whitney stratified so that the a- and b-level sets of $f\pi$ are unions of strata.

9.2. Regular Values

Suppose $X \xrightarrow{\pi} Z \xrightarrow{f} \mathbb{R}$ is a relative Morse function on X. Let $[a, b] \subset \mathbb{R}$ be a closed interval which contains no critical values of $f\pi$. Then, there is an $\mathscr{S}$-decomposition preserving (in fact, a stratum preserving) homeomorphism

$$h^\pi : X_{\leq a} \cong X_{\leq b}.$$

Proof. The same proof as Sects. 3.2 and 7.2 works in the relative case because if f is a submersion over the interval $[a, b]$ then so is $f\pi$. □

9.3. Relative Morse Data

For a relative Morse function $X \xrightarrow{\pi} Z \xrightarrow{f} \mathbb{R}$ we define the relative local Morse data (A^π, B^π) over a critical point $p \in Z$ to be the preimage (under π) of the local Morse data (A, B) of Sect. 3.3. Similarly we define the relative normal Morse data, relative link L^π, relative (upper and lower) halflinks $(\ell^{\pi+}, \ell^{\pi-})$ at p to be the preimage (under π) of the normal Morse data, the link of p in Z, and the (upper and lower) halflinks of p in Z. Each of these spaces is $\mathscr{S}$-decomposed by its intersection with the strata of X. They are also canonically Whitney stratified by a stratification which refines the $\mathscr{S}$-decomposition.

Remark. In each case these definitions involve a choice of Riemannian metric on M (the ambient analytic manifold which contains Z), a choice of an $\varepsilon > 0$ and a $\delta > 0$ and a choice of a normal slice $N \subset M$ through the stratum of Z which contains the critical point p. However, these choices do not involve data or geometry on X – they only involve geometry on Z.

9.4. Local Relative Morse Data is Morse Data

The local relative Morse data for $f\pi$ at p is independent of the choice of ε or δ, or the Riemannian metric. Furthermore, local relative Morse data is Morse data for the map π, i.e., there is a ($\mathscr{S}$-decomposition preserving) homeomorphism

$$X_{\leq v+\varepsilon} \cong (X_{\leq v-\varepsilon}) \cup_{B^\pi} (A^\pi).$$

Proof. The proof is the same as the proofs of Proposition 3.5.3. In Sect. 7.4 a map $F: Z \to \mathbb{R}^2$ is constructed, together with a stratified pair $(Y, W) \subset \mathbb{R}^2$ such that $F^{-1}(Y, W)$ is local Morse data for Z, and such that any two choices $F^{-1}(Y_0, W_0)$ and $F^{-1}(Y_1, W_1)$ for local Morse data are connected by a one-parameter family $F^{-1}(Y_t, W_t)$ of local Morse data. For each $t \in [0, 1]$ the map F is transverse to each stratum of Y_t and W_t. Thus, the map $(F \circ \pi): X \to \mathbb{R}^2$ is also transverse to each stratum of Y_t and W_t. Thus, Lemma 4.3 (Moving the wall) applies to this map as well, giving an $\mathscr{S}$-decomposition preserving homeomorphism between the local relative Morse data,

$$(F \circ \pi)^{-1}(Y_0, W_0) \quad \text{and} \quad (F \circ \pi)^{-1}(Y_1, W_1).$$

Similarly, the proof (Sect. 7.6) that local Morse data is Morse data works also in the relative case: we only observe that the map $F: Z \to \mathbb{R}^2$ is transverse to each stratum of a certain one-parameter family of Whitney stratified sets Y_t, and so the map $(F \circ \pi): X \to \mathbb{R}^2$ is also transverse to each stratum of Y_t. Thus, Lemma 4.3 applies to the composition $F \circ \pi$, giving a homeomorphism between

$$(F \circ \pi)^{-1}(Y_0) = (X_{\leq v-\varepsilon}) \cup_{B^\pi} (A^\pi) \quad \text{and} \quad (F \circ \pi)^{-1}(Y_1) = X_{\leq v+\varepsilon}. \quad \square$$

9.5. The Main Theorem in the Relative Case

If $f: Z \to \mathbb{R}$ is a Morse function, or a function with a nondepraved critical point $p \in Z$, and if $\pi: X \to Z$ is a surjective stratified map, then the local relative Morse data for the composition $(f\pi): X \to \mathbb{R}$ is homeomorphic (by an $\mathcal{S}$-decomposition preserving homeomorphism of pairs) to the product

(tangential Morse data of f) $\times$ (relative normal Morse data for $f\pi$)

where the first factor is trivially decomposed as a single piece and the second factor is $\mathcal{S}$-decomposed by its intersection with the strata of X.

Proof. The proof is a relative version of Sect. 8. The characteristic covectors of a map $F: Z \to \mathbb{R}^2$ are the same as the characteristic covectors of the map $(F \circ \pi): X \to \mathbb{R}^2$, so each of the moving wall arguments in Sect. 8 may be applied directly to X. $\square$

9.6. Halflinks

Relative normal Morse data and relative upper and lower halflinks are well-defined (i.e., are independent of the choice of Riemannian metric on M, normal slice through the critical point $p \in Z$, and of the choice of ε and δ), and depend only on the differential $df(p)$ of the Morse function f. The same holds when p is a nondepraved critical point of f.

Theorem. *The boundaries of the relative halflinks, $\partial \ell^{\pi\pm}$ are homeomorphic by some stratum preserving homeomorphism. The union*

$$(\ell^{\pi+}) \cup_{\partial \ell^\pi} (\ell^{\pi-})$$

is homeomorphic (by an $\mathcal{S}$-decomposition preserving homeomorphism) to the relative link $L^\pi = \pi^{-1}(L)$ of the stratum containing the point p.

Proofs. The proofs of these theorems are simply relative versions of the proofs found in Sect. 7. As in the preceding sections, we need only observe that if $F: Z \to \mathbb{R}^2$ is transverse to some Whitney stratified subset $Y \subset \mathbb{R}^2$, then so is the map $(F \circ \pi): X \to \mathbb{R}^2$. $\square$

9.7. Normal Morse Data and the Halflink

Define the *special fibre* $p^\pi = \pi^{-1}(p)$. Let N^π denote the (relative) normal slice at p. Choose $\varepsilon > 0$ and $\delta > 0$ so that the point (δ, ε) is an element of the fringed set A_2 which was defined in Sect. 6.5. For notational simplicity we suppose $v = f(p) = 0$. Let

$$U = \pi^{-1}(N \cap B_\delta(p) \cap Z) \qquad \partial U = \pi^{-1}(N \cap \partial B_\delta(p) \cap Z).$$

By [G2] there is a (weak) deformation retraction $\phi: U \to p^\pi$. Construct the relative halflink with respect to this choice of ε and δ, i.e.,

$$\ell^{\pi\pm} = \pi^{-1}(Z \cap B_\delta(p) \cap f^{-1}(\pm\varepsilon)).$$

Definition. The specialization map $\phi: \ell^{\pi\pm} \to p^\pi$ is the restriction of ϕ to the halflink.

Proposition. *The relative normal Morse data for $f\pi$ has the homotopy type of the pair*

$$(\mathrm{cyl}(\ell^{\pi-} \to p^\pi), \ell^{\pi-})$$

where cyl *denotes the mapping cylinder of the specialization map ϕ.*

Proof. It is easy to see that $U - p^\pi$ is homeomorphic to $\partial U \times (0, 1]$. The map $\phi | \partial U: \partial U \to p^\pi$ is homotopic to the inclusion $\partial U \to U$. Therefore, the following triples are homotopy equivalent:

$$(\mathrm{cyl}(\partial U \to p^\pi), \partial U, \partial U \cap \pi^{-1} f^{-1}(-\infty, -\varepsilon))$$

and (*)

$$(U, \partial U, \partial U \cap \pi^{-1} f^{-1}(-\infty, -\varepsilon)).$$

Now use the same moving wall argument (and notation) as in Sects. 7.8 and 7.9 to find homeomorphisms as follows:

$$\begin{aligned}
\text{Normal Morse data} &= \pi^{-1} F^{-1}(\mathrm{Box}(\delta, \varepsilon), \mathrm{Bottom}(\delta, \varepsilon))\\
&\cong \pi^{-1} F^{-1}(\mathrm{Halfspace}(\delta), \mathrm{RS}^-(\delta, \varepsilon))\\
&\cong (U, \partial U \cap \pi^{-1} f^{-1}(-\infty, -\varepsilon))\\
&\sim (\mathrm{cyl}(\partial U \to p^\pi), \partial U \cap \pi^{-1} f^{-1}(-\infty, -\varepsilon))\\
&\cong (\mathrm{cyl}(\pi^{-1} F^{-1}(\mathrm{RS}^+ \cup \mathrm{RS} \cup \mathrm{RS}^-) \to p^\pi), \pi^{-1} F^{-1}(\mathrm{RS}^-))\\
&\cong (\mathrm{cyl}(\pi^{-1} F^{-1}(\mathrm{Top} \cup \mathrm{RS} \cup \mathrm{Bottom}) \to p^\pi), \pi^{-1} F^{-1}(\mathrm{Bottom}))\\
&\cong (\mathrm{cyl}(\pi^{-1}(\ell^+ \cup_{\partial\ell} \ell^-) \to \pi^{-1}(p)), \pi^{-1}(\ell^-)). \qquad (**)
\end{aligned}$$

However, the same homotopy argument as in Corollary 3.11 shows that this pair has the same homotopy type as the pair

$$(\mathrm{cyl}(\pi^{-1}(\ell^-) \to \pi^{-1}(p)), \pi^{-1}(\ell^-)). \quad \square$$

Remark. It can be shown that the homotopy equivalence (*) is a homeomorphism; however, it does not preserve the $\mathcal{S}$-decomposition of the spaces involved unless the map π is finite (i.e., the fibres of π consist of finitely many points).

9.8. Summary of Homotopy Consequences

Suppose Z is a Whitney stratified space, $\pi: X \to Z$ is a proper surjective stratified map, $f: Z \to \mathbb{R}$ is a proper *Morse* function, and $[a, b] \subset \mathbb{R}$ is an interval which contains no critical values except for a single isolated critical value $v \in (a, b)$ which corresponds to a critical point p which lies in some stratum S of Z. Let λ be the Morse index of $f \mid S$ at the point p.

Theorem. *The space $X_{\leq b}$ has the homotopy type of a space which is obtained from $X_{\leq a}$ by attaching the pair*

$$(D^\lambda, \partial D^\lambda) \times (\mathrm{cyl}(\ell^{\pi -} \to \pi^{-1}(p)), \ell^{\pi -}).$$

Proof. By Sect. 9.4, the local Morse data is Morse data. By Sect. 9.5, this is a product of normal and tangential Morse data. By remark 3.5.4, the tangential Morse data has the homotopy type of the pair $(D^\lambda, \partial D^\lambda)$. By Sect. 9.7, the normal Morse data has the homotopy type of the pair $(\mathrm{cyl}(\ell^{\pi -} \to \pi^{-1}(p)), \ell^{\pi -})$. □

Chapter 10. Nonproper Morse Functions

It is often necessary to consider a "Morse function" $f: X \to \mathbb{R}$ which is not proper, but which can be extended to a proper function $\bar{f}: Z \to \mathbb{R}$ where Z contains X as a dense open subset. For example, Z may be a compactification of some noncompact algebraic variety $X \subset \mathbb{CP}^n$, and f may be a smooth function defined on the ambient $\mathbb{CP}^n$. We shall assume that it is possible to find a stratification of Z so that $X \subset Z$ is a union of strata. Thus, X is obtained from Z by removing certain strata. The main theorems (Sects. 3.7, 3.10, 3.11) continue to apply to X, because we simply remove the same strata from both sides of the homeomorphisms. Since these homeomorphisms were originally proven to be decomposition preserving, it is a triviality that they induce homeomorphisms on unions of pieces in the decomposition. However, Proposition 3.2 is no longer strictly true in this context unless we also consider the effect on $X_{\leq v}$ of critical values v which correspond to critical points $p \in Z$ which do not lie in X. Our main theorems also apply to these "critical points at infinity". For all the applications which we consider, X will be an open dense subset of Z. However, the results of this chapter apply to any union of strata $X \subset Z$, and so we will not even assume that X is locally closed in Z.

10.1. Definitions

Throughout this chapter, Z will denote a Whitney stratified (closed) subset of some smooth manifold M. The strata of Z will be indexed by a partially ordered set $\mathcal{S}$. We will be interested in applying Morse theory to a subset $X \subset Z$ which is a union of strata corresponding to a partially ordered subset $\mathcal{T} \subset \mathcal{S}$, i.e.,

$$X = Z \cap |\mathcal{T}| = \bigcup_{i \in \mathcal{T}} S_i$$

where $\{S_i\}$ are the strata of Z (i.e., the pieces of the $\mathcal{S}$-decomposition: see Sect. 1.1).

For each stratum S of Z we define the link of S *in* X to be the intersection of X with the link of S in Z. This is well-defined whether or not S is a stratum of X.

We fix a smooth function $f: M \to \mathbb{R}$ whose restriction to Z is proper and has a nondepraved critical point $p \in Z$ with critical value $f(p) = 0$. Such a critical point $p \in Z$ may be one of two types: an "ordinary" critical point ($p \in X$) or a "critical point at infinity" ($p \in Z - X$).

As in Sect. 3, we use the symbol $X_{\leq a}$ to denote the space $X \cap f^{-1}(-\infty, a]$, together with its decomposition into strata, indexed by $\mathscr{T}$.

10.2. Regular Values

Suppose $[a, b] \subset \mathbb{R}$ contains no critical values of $f: Z \to \mathbb{R}$.

Theorem. *There is a $\mathscr{T}$-decomposition preserving homeomorphism*

$$X_{\leq a} \cong X_{\leq b}.$$

Proof. By Proposition 3.2 there is an $\mathscr{S}$-decomposition preserving homeomorphism $Z_{\leq a} \cong Z_{\leq b}$. This restricts to a homeomorphism

$$X_{\leq a} = |\mathscr{T}| \cap Z_{\leq a} \cong |\mathscr{T}| \cap Z_{\leq b} = X_{\leq b}. \quad \square$$

10.3. Morse Data in the Nonproper Case

Let S denote the stratum of Z which contains the critical point p. Let (J_Z, K_Z) denote the local Morse data (resp. normal Morse data, resp. normal slice, resp. upper or lower halflink) for $f: Z \to \mathbb{R}$ at the point p. (This involves a choice of a distance function r, an $\varepsilon > 0$, a $\delta > 0$ and a normal slice N. See Sects. 3.4, 3.5, 3.6, 3.9, 7.3, 7.4.) Define the local Morse data (resp. normal Morse data, resp. normal slice, resp. upper or lower halflink) for $f|X: X \to \mathbb{R}$ to be the intersection

$$(J_X, K_X) = (J_Z, K_Z) \cap |\mathscr{T}| = (X \cap J_Z, X \cap K_Z).$$

Since (J_Z, K_Z) is an $\mathscr{S}$-decomposed space, the pair (J_X, K_X) is canonically $\mathscr{T}$-decomposed.

We define the *tangential* Morse data (A, B) for $f: X \to \mathbb{R}$ to be the tangential Morse data

$$(A, B) = S \cap B_\delta(p) \cap (f^{-1}[v-\varepsilon, v+\varepsilon], f^{-1}(v-\varepsilon))$$

for the function $f: Z \to \mathbb{R}$ (where $\varepsilon > 0$ and $\delta > 0$ are chosen as in Sects. 3.3 or 7.3). Tangential Morse data is a subset of a single stratum S, so it has a trivial decomposition.

10.4. Local Morse Data is Morse Data

Let (J_X, K_X) denote the local Morse data for X at the critical point p.

Theorem. *For $\varepsilon > 0$ sufficiently small, there is a $\mathscr{T}$-decomposition preserving homeomorphism*

$$X_{\leq \varepsilon} \cong X_{\leq -\varepsilon} \cup_{K_X} J_X.$$

Proof. By Sect. 7.6 there is an $\mathscr{S}$-decomposition preserving homeomorphism

$$Z_{\leq \varepsilon} \cong Z_{\leq -\varepsilon} \cup_{K_Z} J_Z$$

and this restricts to the desired homeomorphism on those pieces of the decomposition which are indexed by the subset $\mathcal{T}$. □

10.5. The Main Theorem in the Nonproper Case

Let (J_X, K_X) denote the local Morse data for the map $f: X \to \mathbb{R}$ at the critical point p. Let (A, B) denote the tangential Morse data at p, and let (P_X, Q_X) denote the normal Morse data for X at p. Then there is a $\mathcal{T}$-decomposition preserving homeomorphism

$$(J_X, K_X) \cong (A, B) \times (P_X, Q_X).$$

Proof. By Sects. 3.7 and 8 there is an $\mathcal{S}$-decomposition preserving homeomorphism

$$(J_Z, K_Z) \cong (A, B) \times (P_Z, Q_Z)$$

where (J_Z, K_Z) (resp. (P_Z, Q_Z)) denotes the local (resp. normal) Morse data for $f: Z \to \mathbb{R}$ at the point p. This homeomorphism restricts to a homeomorphism of those pieces of the decomposition which are indexed by $\mathcal{T}$. □

10.6. Halflinks

Let ℓ_X^+, ℓ_X^-, $\partial \ell_X$ denote the upper halflink, lower halflink, and boundary of the halflink in X of the point p and let L_X denote the link of the stratum S in X at the point p (these are the intersections with X of the upper halflink, lower halflink, boundary of the halflink, and link of S in Z).

Theorem. *There is a $\mathcal{T}$-decomposition preserving homeomorphism*

$$L_X \cong \ell_X^+ \cup_{\partial \ell_X} \ell_X^- .$$

Proof. By Sect. 7.7 there is an $\mathcal{S}$-decomposition preserving homeomorphism

$$L_Z \cong \ell_Z^+ \cup_{\partial \ell_Z} \ell_Z^-$$

which therefore restricts to the desired homeomorphism on L_X. □

10.7. Normal Morse Data and the Halflink

Throughout this section N will denote a normal slice through the critical point p, and $\delta > 0$ will be fixed so that $B_\delta(p) \cap N$ is compact and $\partial B_\delta(p)$ is transverse to each stratum of N. Let $\ell_X^\pm = \ell_Z^\pm \cap X$ denote the halflink for X at the point p.

Proposition 1. *If $p \in X$, then the normal Morse data has the homotopy of the pair*

$$(\operatorname{cone}(\ell_X^-), \ell_X^-).$$

Proof. This follows from Sects. 3.11.2 and 3.11.3. The deformation retraction in Sect. 3.11.3 is $\mathcal{S}$-decomposition preserving, so it restricts to a deformation retraction on those pieces of the $\mathcal{S}$-decomposition which are indexed by elements of $\mathcal{T} \subset \mathcal{S}$. □

Now suppose that $p \in Z - X$ is a critical point at infinity. Since the homeomorphism in Sect. 3.11 preserves the $\mathcal{S}$-decomposition, we see by removing the cone point that the normal Morse data for $f|X$ at p is homeomorphic to the pair

$$(\ell_X^+ \cup_{\partial \ell_X} \ell_X^-, \ell_X^-) \times (0, 1].$$

It follows from the excision property for Morse data (Sect. 3.3) that the pair $(\ell_X^+, \partial \ell_X^+)$ is homotopy Morse data (see Sect. 3.3) for the normal slice $N \cap B_\delta(p)$, even though it is *not* homotopy equivalent to the normal Morse data. Nevertheless, for the purpose of understanding homotopy type the normal Morse data can be replaced by the pair $(\ell_X^+, \partial \ell_X^+)$, and this pair is more convenient to use. We make this explicit in the following definition: Any pair (P, Q) which is homotopy equivalent to the pair $(\ell_X^+, \partial \ell_X^+)$ will be called *homotopy normal Morse data*.

Proposition 2. *If $p \in Z - X$ and if (P, Q) is homotopy normal Morse data and (A, B) is tangential Morse data for f at p, then the pair*

$$(A, B) \times (P, Q)$$

is homotopy Morse data for $f|X$ at p.

Proof. By Sects. 3.11 and 10.5, the local Morse data for $f|X$ at p is homeomorphic to the pair

$$(A, B) \times (\ell_X^+ \cup_{\partial \ell_X^+} \ell_X^-, \ell_X^-) \times (0, 1].$$

Thus, (by the excision property for Morse data and the fact that $\partial \ell_X^+$ is collared in ℓ_X^+) the pair

$$(A, B) \times (\ell_X^+, \partial \ell_X^+) \times (0, 1]$$

is Morse data for f at p. Thus, $(A, B) \times (\ell_X^+, \partial \ell_X^+)$ is homotopy Morse data for $f|X$ at p.

10.8. Summary of Homotopy Consequences

Suppose Z is Whitney stratified space, and $X \subset Z$ is a union of strata of Z. Suppose $f: Z \to \mathbb{R}$ is a proper *Morse* function, and $[a, b] \subset \mathbb{R}$ is an interval which contains no critical values except for a single isolated critical value $v \in (a, b)$ which corresponds to a critical point p which lies in some stratum S of Z. Let λ be the Morse index of $f|S$ at the point p.

Theorem. *If $p \in X$, then the space $X_{\leq b}$ has the homotopy type of a space which is obtained from $X_{\leq a}$ by attaching the pair*

$$(D^\lambda, \partial D^\lambda) \times (\operatorname{cone}(\ell_X^-), \ell_X^-).$$

If $p \notin X$ then the space $X_{\leq b}$ has the homotopy type of a space obtained from $X_{\leq a}$ by attaching the pair

$$(D^\lambda, \partial D^\lambda) \times (\ell_X^+, \partial \ell_X^+).$$

Proof. If $p \in X$ then, by Sect. 10.4, the local Morse data is Morse data. By Sect. 10.5, this is a product of normal and tangential Morse data. By Remark 3.5.4, the tangential Morse data has the homotopy type of the pair $(D^\lambda, \partial D^\lambda)$. By Sect. 10.7, the normal Morse data has the homotopy type of $(\text{cone}(\ell^-), \ell^-)$. Thus, the pair $(D^\lambda, \partial D_\lambda) \times (\text{cone}(\ell^-), \ell^-)$ is homotopy Morse data.

If $p \notin X$, then by Sect. 10.7, the pair $(D^\lambda, \partial D^\lambda) \times (\ell^+, \partial \ell^+)$ is homotopy Morse data.

Thus, the space $X_{\leq b}$ has the homotopy type of the adjunction space $X_{\leq a} \cup$ (homotopy Morse data) (for some choice of attaching map). □

Chapter 11. Relative Morse Theory of Nonproper Functions

This section is the common generalization of Chapters 9 and 10.

11.1. Definitions

We shall assume Z is a closed Whitney stratified subset of some smooth manifold M, and that $f: M \to \mathbb{R}$ is a smooth function such that $f|Z$ is proper and has a nondepraved critical point $p \in Z$ with isolated critical value $f(p)=0$. Let S denote the stratum of Z which contains the point p.

Let $\bar{X}$ be a Whitney stratified set and let $\bar{\pi}: \bar{X} \to Z$ be a stratified surjective proper map (see Sect. 1.6), i.e., $\bar{\pi}$ takes each stratum of $\bar{X}$ submersively to a stratum of Z. Let us suppose that the strata in $\bar{X}$ are indexed by some partially ordered set $\mathcal{S}$ and that $X \subset \bar{X}$ is a union of strata of $\bar{X}$, corresponding to a partially ordered subset $\mathcal{T} \subset \mathcal{S}$. We wish to consider the Morse theory of the composition $(f \circ \pi): X \to \mathbb{R}$ (where $\pi = \bar{\pi}|X$).

$$\begin{array}{ccc} X \subset \bar{X} & & \\ {\scriptstyle \pi}\searrow \quad \downarrow{\scriptstyle \bar{\pi}} & & \\ Z & \xrightarrow[f]{} & \mathbb{R} \end{array}$$

11.2. Regular Values

Suppose the closed interval $[a, b]$ contains no critical values of the map $f: Z \to \mathbb{R}$.

Theorem. *There is a $\mathcal{T}$-decomposition preserving homeomorphism $X_{\leq a} \cong X_{\leq b}$.*

Proof. By Sect. 9.2, the homeomorphism $\bar{X}_{\leq a} \cong \bar{X}_{\leq b}$ preserves the $\mathcal{S}$-decomposition. □

11.3. Morse Data in the Relative Nonproper Case

Let (J_Z, K_Z) denote the local Morse data for the map $f: Z \to \mathbb{R}$ at the point p. Let (P_Z, Q_Z) denote the normal Morse data, (A, B) denote the tangential Morse data, and $\ell_Z^{\pm}$ denote the halflink for Z at the point p. By Sect. 9.3 each of the preimages

$$\bar{\pi}^{-1}(J_Z, K_Z), \quad \bar{\pi}^{-1}(P_Z, Q_Z), \quad \bar{\pi}^{-1}(\ell_Z^+, \partial\ell_Z^+), \quad \bar{\pi}^{-1}(\ell_Z^-, \partial\ell_Z^-)$$

are canonically $\mathscr{S}$-decomposed, and the homeomorphisms (Sects. 9.4, 9.5, 9.6) are $\mathscr{S}$-decomposition preserving.

Definition. *The relative local Morse data* (J_X^π, K_X^π) *(resp. the relative normal Morse data* (P_X^π, Q_X^π)*, resp. the relative upper halflink* $(\ell_X^{\pi+}, \partial\ell_X^{\pi+})$*) is defined by:*

$$(J_X^\pi, K_X^\pi) = (\pi^{-1}(J_Z), \pi^{-1}(K_Z)) = (X \cap \bar{\pi}^{-1}(J_Z), X \cap \bar{\pi}^{-1}(K_Z))$$
$$(P_X^\pi, Q_X^\pi) = (\pi^{-1}(P_Z), \pi^{-1}(Q_Z)) = (X \cap \bar{\pi}^{-1}(P_Z), X \cap \bar{\pi}^{-1}(Q_Z))$$
$$(\ell_X^{\pi+}, \partial\ell_X^{\pi+}) = (\pi^{-1}(\ell_Z^+), \pi^{-1}(\partial\ell_Z^+)) = (X \cap \bar{\pi}^{-1}(\ell_Z^+), X \cap \bar{\pi}^{-1}(\partial\ell_Z^+)).$$

11.4. Local Morse Data is Morse Data

Let (J_X^π, K_X^π) denote the local Morse data for X at the critical point p.

Theorem. *For* $\varepsilon > 0$ *sufficiently small, there is a* $\mathscr{T}$*-decomposition preserving homeomorphism*

$$X_{\leq \varepsilon} \cong (X_{\leq -\varepsilon}) \cup_{K_X^\pi} (J_X^\pi).$$

Proof. This follows immediately from Sect. 9.4 and the fact the homeomorphism preserves the $\mathscr{S}$-decomposition. □

11.5. The Main Theorem in the Relative Nonproper Case

Let (J_X^π, K_X^π) denote the local Morse data for the map $f\pi\colon X \to \mathbb{R}$ at the critical point p. Let (A, B) denote the tangential Morse data at p, and let (P_X^π, Q_X^π) denote the normal Morse data for X at p. Then there is a $\mathscr{T}$-decomposition preserving homeomorphism

$$(J_X^\pi, K_X^\pi) \cong (A, B) \times (P_X^\pi, Q_X^\pi).$$

Proof. This follows immediately from Sect. 9.5 and the fact the homeomorphism preserves the $\mathscr{S}$-decomposition. □

11.6. Halflinks

Let $\ell_X^{\pi+}$, $\ell_X^{\pi-}$, $\partial\ell_X^\pi$ denote the upper halflink, lower halflink, and boundary of the halflink in X of the point p and let L_X^π denote the link of the stratum S in X at the point p (these are the intersections with X of the relative upper halflink, relative lower halflink, boundary of the relative halflink, and relative link of S in $\bar{X}$).

Theorem. *There is a* $\mathscr{T}$*-decomposition preserving homeomorphism*

$$L_X^\pi \cong (\ell_X^{\pi+}) \cup_{\partial\ell_X^\pi} (\ell_X^{\pi-}).$$

Proof. This follows immediately from Sect. 9.6 and the fact that the homeomorphism preserves the $\mathcal{S}$-decomposition. □

11.7. Normal Morse Data and the Halflink

The statements in Sect. 9.7 do not immediately generalize to the case of a nonproper Morse function because the homotopy equivalence in Sect. 9.7(∗) does not preserve strata near $\bar{\pi}^{-1}(p)$. However, if the fibre $\bar{\pi}^{-1}(p)$ is contained in either X or $\bar{X}-X$, then we obtain results analogous to those of Sect. 10.7.

Theorem 1. *If X is locally closed in $\bar{X}$ and if the whole fibre $\bar{\pi}^{-1}(p)$ is contained in X, then the normal relative Morse data at p has the homotopy type of the pair*

$$(\operatorname{cone}(\ell_X^{\pi+}), \ell_X^{\pi+}).$$

Proof. The projection $\pi: X \to Z$ is proper near the fibre $\bar{\pi}^{-1}(p)$, so the results of Sect. 9.7 apply. □

If $\bar{\pi}^{-1}(p) \subset \bar{X}-X$ (i.e., if p is a critical point at infinity), then the normal Morse data can be replaced by the pair $(\ell_X^{\pi+}, \partial \ell_X^{\pi})$. We make this explicit in the following definition and proposition: Any pair (P, Q) which is homotopy equivalent to the pair $(\ell_X^{\pi+}, \partial \ell_X^{\pi})$ will be called *homotopy relative normal Morse data.*

Theorem 2. *Suppose $\bar{\pi}^{-1}(p) \subset \bar{X}-X$, that (P, Q) is homotopy relative normal Morse data at p and (A, B) is tangential Morse data for f at p. Then the pair $(A, B) \times (P, Q)$ is homotopy Morse data for f at p.*

Proof. If we remove the fibre over p in the argument of Sect. 9.7, we obtain a *stratum preserving* homeomorphism between the normal Morse data for $f \circ \bar{\pi}: \bar{X} \to \mathbb{R}$, and the pair

$$(\ell^{\pi+} \cup_{\partial \ell^{\pi}} \ell^{\pi-}, \ell^{\pi-}) \times (0, 1]$$

which therefore restricts to a homeomorphism between the normal Morse data for $f \circ \bar{\pi} | X: X \to \mathbb{R}$ at the point p, and the pair

$$(\ell_X^{\pi+} \cup_{\partial \ell_X^{\pi}} \ell_X^{\pi-}, \ell_X^{\pi-}) \times (0, 1].$$

Thus (by the excision property for Morse data, Sect. 3.3), the pair

$$(A, B) \times (\ell_X^{\pi+}, \partial \ell_X^{\pi}) \times (0, 1]$$

is Morse data for $f \circ \bar{\pi} | X: X \to \mathbb{R}$ at p, so the pair

$$(A, B) \times (\ell_X^{\pi+}, \partial \ell_X^{\pi})$$

is homotopy Morse data at p. □

11.8. Summary of Homotopy Consequences

Suppose Z is a Whitney stratified space, and $\bar{\pi}: \bar{X} \to Z$ is a proper surjective stratified map. Let $X \subset \bar{X}$ be an open dense subset which is a union of strata of $\bar{X}$. Suppose $f: Z \to \mathbb{R}$ is a proper *Morse* function, and $[a, b] \subset \mathbb{R}$ is an interval which contains no critical values except for a single isolated critical value $v \in (a, b)$ which corresponds to a critical point p which lies in some stratum S of Z. Let λ be the Morse index of $f \mid S$ at the point p.

Theorem. *If $\bar{\pi}^{-1}(p) \subset X$ then the space $X_{\leq b}$ has the homotopy type of a space which is obtained from $X_{\leq a}$ by attaching the pair*

$$(D^\lambda, \partial D^\lambda) \times (\operatorname{cyl}(\ell^{\pi-} \to \bar{\pi}^{-1}(p)), \ell^{\pi-}).$$

If $\bar{\pi}^{-1}(p) \subset \bar{X} - X$, then the space $X_{\leq b}$ has the homotopy type of a space obtained from $X_{\leq a}$ by attaching the pair

$$(D^\lambda, \partial D^\lambda) \times (\ell_X^{\pi+}, \partial \ell_X^{\pi+}).$$

Proof. By Sect. 11.4, the local Morse data is Morse data. By Sect. 11.5, this is a product of normal and tangential Morse data. By Remark 3.5.4, the tangential Morse data has the homotopy type of the pair $(D^\lambda, \partial D^\lambda)$. By Sect. 11.7, the normal Morse data has the homotopy type of either $(\operatorname{cyl}(\ell^{\pi-} \to \bar{\pi}^{-1}(p)), \ell^{\pi-})$ or $(\ell_X^{\pi+}, \partial \ell_X^{\pi+})$, depending on whether $\bar{\pi}^{-1}(p) \subset X$ or $\bar{\pi}^{-1}(p) \subset \bar{X} - X$. □

Chapter 12. Normal Morse Data of Two Morse Functions

In this chapter we analyze the normal Morse data at a critical point $p \in Z$ of a function $f_1: Z \to \mathbb{R}$ under the assumption that there exists a second function $f_2: Z \to \mathbb{R}$ such that the map $(f_1, f_2): Z \to \mathbb{R}^2$ has a nondegenerate critical point at p (see below).

Our main application of this section is to the situation where Z is a complex analytic variety and f_1 is the real part of a complex analytic function $f = f_1 + if_2: Z \to \mathbb{C}$. However, the same situation occurs when Z is a real analytic variety such that the set of degenerate characteristic covectors at the point p has codimension >1 in the space of characteristic covectors (see Sect. 1.8). We find that in this case the halflink ℓ is a product,

$$\ell = \ell' \times [0, 1]$$

that there is a "monodromy" homeomorphism $\mu: \ell' \to \ell'$, and that the link L of p and the normal Morse data can be completely described in terms of this auxiliary space ℓ', using a stratified generalization of the Milnor fibration theorem [Mi2]. If f_1 is the real part of a complex analytic function f, then the space ℓ' is called the *complex link*, and will be the main object of study in Part II.

12.1. Definitions

Throughout this chapter we will fix a smooth manifold M, a closed Whitney stratified subset $Z \subset M$, a point $p \in Z$ which lies in some stratum S of Z, a normal slice $N = N' \cap Z$ (where N' is a smooth submanifold of M which meets the stratum S transversally in a single point $\{p\}$). We suppose the strata of Z are indexed by some partially ordered set $\mathcal{S}$. We also fix smooth functions f_1 and f_2 which are defined on M, and whose restriction to Z is proper. We shall use the notation $f = (f_1, f_2): Z \to \mathbb{R}^2$. We suppose that p is a critical point of f, i.e., $df(p)(T_p S) = 0$, and that this critical point is nondegenerate in the following sense: for every generalized tangent space Q at the point p, we have

$$(df_1(p), df_2(p))(Q) = \mathbb{R}^2$$

except for the single case $Q = T_p S$. We fix a Riemannian metric on M, and let $r(z)$ denote the square of the distance between the points p and z.

The following two sections (12.2 and 12.3) contain technical lemmas on choices of ε and δ, which will be used in the definition of the complex link. The main definitions and results continue in Sects. 12.4 and 12.5.

12.2. Characteristic Covectors of the Normal Slice for a Pair of Functions

Throughout this section, $N=N'\cap Z$ will denote a normal slice through p.

Definition. If $A\subset\mathbb{R}^+\times\mathbb{R}^+$ is a fringed set of type $0<v\ll u\ll 1$, we shall use the symbol $A^\#$ to denote the subset of $\mathbb{R}^+\times\mathbb{R}\times\mathbb{R}$ which is given by

$$A^\#=\{(u,v_1,v_2)\in\mathbb{R}^+\times\mathbb{R}^2\,|\,(u,\sqrt{(v_1^2+v_2^2)})\in\bar{A}\}.$$

Lemma. (1) *There exists a fringed set $A\subset\mathbb{R}^+\times\mathbb{R}^+$ such that the map $(r,f_1,f_2)\colon N\to\mathbb{R}\times\mathbb{R}^2$ has no characteristic covectors in the region $A^\#$.*

(2) *There exists $r_0>0$ such that if $\lambda=(u,v_1,v_2,\ a\,du+b\,dv_1+c\,dv_2)$ is a nonzero characteristic covector of the map $(r,f_1,f_2)|N$, and if $0<u\le r_0$ then $-\infty<a/(bv_1+cv_2)<0$.*

Remark. Property (2) says that if a plane P is the kernel of the characteristic covector λ, then P intersects the r axis at some point $(w,0,0)$ such that $w<u$.

Proof of 1. The proof is similar to that in Sect. 6.2. Since the set of characteristic covectors is closed in $T^*\mathbb{R}^3$, the existence of a set $A^\#$ with no characteristic covectors is equivalent to the statement that there are no characteristic covectors over any point $(u,0)\in\mathbb{R}\times\mathbb{R}^2$ for u sufficiently small. Suppose this were false, i.e., that there is a sequence of points $p_i\in N\cap f^{-1}(0)$ which converge to p, such that $dr(p_i)$, $df_1(p_i)$, and $df_2(p_i)$ are linearly dependent when restricted to the stratum Y of N which contains the point p_i. This means that $df_1(p_i)$ and $df_2(p_i)$ are linearly dependent when restricted to $T_{p_i}Y\cap\ker dr(p_i)$. By restricting to a subsequence if necessary, we may suppose that the points p_i are all contained in the same stratum Y of N, that the tangent planes at p_i converge to some limiting plane τ, and that the secant lines $\ell_i=\overline{pp_i}$ converge to some limiting line ℓ.

We may assume that $Y>S$. By Whitney's condition B,

$$\tau=\lim_{i\to\infty}\,[\ell_i\oplus T_{p_i}Y\cap\ker dr(p_i)]$$

and this limit is a perpendicular direct sum. It follows that the rank of $(df_1(p), df_2(p))|\tau$ is less than or equal to one, because

(a) $df_1(p)(\ell)=df_2(p)(\ell)=0$ since $f(p_i)=f(p)$ and

(b) $df_1(p)$ and $df_2(p)$ are linearly dependent when restricted to $\lim(T_{p_i}Y\cap\ker dr(p_i))$.

This contradicts the assumption that $(df_1(p), df_2(p))$ is nondegenerate.

Proof of 2. (This proof follows the method of Step 3 in Sect. 6.2.) Assume that the second part of the proposition is false, i.e., that there is a sequence of points $p_i\in N$ converging to p and a sequence of numbers a_i, b_i, $c_i\in\mathbb{R}$ such that the covector

$$a_i\, dr(p_i) + b_i\, df_1(p_i) + c_i\, df_2(p_i)$$

vanish on the tangent spaces $T_{p_i} Y_i$, (where Y_i is the stratum of N with contains the point p_i) such that

$$a_i/(b_i\, f_1(p_i) + c_i\, f_2(p_i)) \geq 0.$$

Replace the sequence $\{p_i\}$ by a subsequence so that all the points p_i are contained in the same stratum Y of N, so that the secant lines $\overline{pp_i}$ converge to a limiting line ℓ, so that the tangent planes $T_{p_i} Y$ converge to a limiting plane τ, and so that the angle $\theta_i = \arctan(b_i/c_i)$ converges to some limiting angle θ. Multiplying the numbers (a_i, b_i, c_i) by a factor, we may also assume that $a_i \geq 0$ for all i. Let $v_i = (p_i - p)/|p_i - p|$. These converge to a limiting unit vector $v \in \tau$. Consider the equation

$$a_i\, dr(p_i)(v_i) + b_i\, df_1(p_i)(v_i) + c_i\, df_2(p_i)(v_i) = 0.$$

The first term is clearly positive (since $p_i \neq p$), so

$$(\sin \theta_i)\, df_1(p_i)(v_i) + (\cos \theta_i)\, df_2(p_i)(v_i) < 0.$$

Taking the limit,

$$(\sin \theta)\, df_1(p)(v) + (\cos \theta)\, df_2(p)(v) \leq 0.$$

Since (f_1, f_2) is nondegenerate, this quantity does not vanish, i.e.,

$$0 > (\sin \theta)\, df_1(p)(v) + (\cos \theta)\, df_2(p)(v) = \lim_{i \to \infty} (b_i\, f_1(p_i) + c_i\, f_2(p_i))/D$$

where

$$D = |p_i - p| \sqrt{(b_i^2 + c_i^2)}.$$

Thus, for sufficiently large i, we have $b_i\, f_1(p_i) + c_i\, f_2(p_i) < 0$, which contradicts the assumption. □

12.3. Characteristic Covectors of a Level

We would like to have an understanding (similar to that in Sect. 12.2) of the characteristic covectors of the map

$$(r, f_2) \,|\, N \cap f_1^{-1}(\eta) \colon N \cap f_1^{-1}(\eta) \to \mathbb{R}^2.$$

Unfortunately, estimates as in Sect. 12.2 do not hold for such a map unless we choose $|\eta|$ to be extremely small, and unless we also exclude a neighborhood of the origin in $\mathbb{R}^2$.

Lemma. *Fix $(\delta, \varepsilon) \in A$ such that $\delta \leq r_0$ (where A and r_0 were found in Lemma 12.2). There exists $\eta > 0$ with the following property: If $\lambda = (u, v_1, v_2, a\,du + b\,dv_1 + c\,dv_2) \in T^*\mathbb{R}^3$ is a characteristic covector of the map $(r, f_1, f_2) \colon N \to \mathbb{R}^3$ and if $0 < |u| \leq \delta$, $|v_1| \leq \eta$, and $v_1^2 + v_2^2 \geq \varepsilon^2$ then $a/c\,v_2 < 0$.*

Proof. Step 1. We consider the special case $v_1 = 0$. If λ is a characteristic covector, then there is a point $q \in N$ such that $a\,dr(q) + b\,df_1(q) + c\,df_2(q)$ vanishes on $T_q Y$. By Sect. 12.2 this implies (since $u \neq 0$)

$$a/(bf_1(q)+cf_2(q))<0.$$

But $f_1(q)=0$, so $a/cf_2(q)<0$, as desired.

Step 2. Sublemma. Let $g\colon \mathbb{R}^3\to\mathbb{R}$ denote the projection $g(u, v_1, v_2)=v_1$. Let $\pi\colon T^*\mathbb{R}^3\to\mathbb{R}^3$ be the projection. Suppose K is a compact subset of $\mathbb{R}^3$, V is a closed conical subset of unit vectors in $T^*\mathbb{R}^3$, and that for any

$$\lambda=(u, 0, v_2, a\,du+b\,dv_1+c\,dv_2)\in\pi^{-1}(g^{-1}(0)\cap K)\cap V$$

we have

$$a/cv_2<0.$$

Then there exists $\eta_0>0$ so that for any $\eta\in[-\eta_0, \eta_0]$ and for any

$$\lambda=(u, \eta, v_2, a\,du+b\,dv_1+c\,dv_2)\in\pi^{-1}(g^{-1}(\eta)\cap K)\cap V$$

we have $a/cv_2<0$.

Proof of sublemma. The sublemma is easily proven by contradicton, using the fact that K is compact and V is closed.

Step 3. Let V denote the closed subset of $T^*\mathbb{R}^3$ consisting of all unit characteristic covectors of the map

$$(r, f_1, f_2)\,|\,N\colon N\to\mathbb{R}^3.$$

Let $K'\subset N$ denote the compact set

$$K'=\{z\in N\,|\,r(z)\le\delta \text{ and } f_1(z)^2+f_2(z)^2\ge\varepsilon^2\}.$$

Let K denote the compact subset of $\mathbb{R}^3$, $K=(r, f_1, f_2)(K')$. Note: $p\notin K'$.

In step 1 we verified the hypotheses of the sublemma, so we conclude that there is an $\eta_0>0$ so that for any $\eta\in[-\eta_0, \eta_0]$, every characteristic covector

$$\lambda=(u, \eta, v_2, a\,du+b\,dv_1+c\,dv_2)$$

of the map $(r, f_1, f_2)\colon N\to\mathbb{R}^3$ satisfies $a/cv_2<0$, provided that $(u, \eta, v_2)\in K$. Thus, the number η_0 satisfies the requirements of the lemma. □

12.4. The Quarterlink and Related Spaces

Suppose as above that $p\in Z$ is a nondepraved critical point of a pair of functions $(f_1, f_2)\colon Z\to\mathbb{R}^2$. By Sect. 12.2, Lemma 1 there is a fringed set $A\subset\mathbb{R}^+\times\mathbb{R}^+$ of type $0<\varepsilon\ll\delta\ll1$ such that the map $(r, f_1, f_2)\,|\,N\colon N\to\mathbb{R}^3$ has no characteristic covectors in the region $A^\#$. Now fix $(\delta, \varepsilon)\in A$.

Definition. 0. The disk and ball,

$$D_\varepsilon=\{\zeta\in\mathbb{R}^2\,|\,|\zeta|\le\varepsilon\} \qquad \partial D_\varepsilon=\{\zeta\in\mathbb{R}^2\,|\,|\zeta|=\varepsilon\}$$
$$N_\delta=N\cap B_\delta(p) \qquad \partial N_\delta=N\cap\partial B_\delta(p)$$
$$D_\varepsilon^0=D_\varepsilon-\partial D_\varepsilon \qquad N_\delta^0=N_\delta-\partial N_\delta.$$

1. The quarterlink and its boundary:

$$\ell'=f^{-1}(\varepsilon, 0)\cap N_\delta \qquad \partial\ell'=f^{-1}(\varepsilon, 0)\cap\partial N_\delta.$$

2. The cylindrical neighborhood of p, its interior and boundary (called the "particular neighborhood" in [GM3]):

$$C = f^{-1}(D_\varepsilon) \cap N_\delta \qquad C^0 = f^{-1}(D_\varepsilon^0) \cap N_\delta^0$$

$$L = \partial C = C - C^0 = L_h \cup L_v \qquad \text{(see below).}$$

3. The horizontal and vertical parts of the link

$$L_h = f^{-1}(D_\varepsilon) \cap \partial N_\delta \qquad L_h^0 = f^{-1}(D_\varepsilon^0) \cap \partial N_\delta$$

$$L_v = f^{-1}(\partial D_\varepsilon) \cap N_\delta \qquad L_v^0 = f^{-1}(\partial D_\varepsilon) \cap N_\delta^0$$

$$\partial L_h = \partial L_v = f^{-1}(\partial D_\varepsilon) \cap \partial N_\delta.$$

Each of these spaces is $\mathscr{S}$-decomposed by its intersection with the strata of N (with are also the strata of Z) and are canonically Whitney stratified since each is a transversal intersection of Whitney stratified spaces. The following schematic diagram illustrates the cylindrical neighborhood for the pair of functions $f\colon \mathbb{R}^3 \to \mathbb{R}^2$ which is given by linear projection:

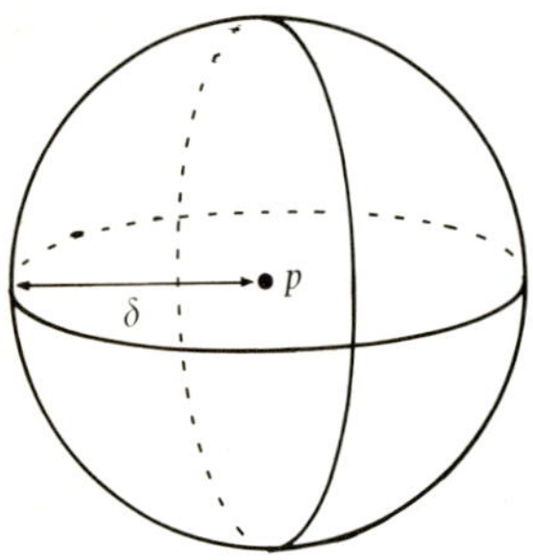

The ball $B_\delta(p)$ in the normal slice

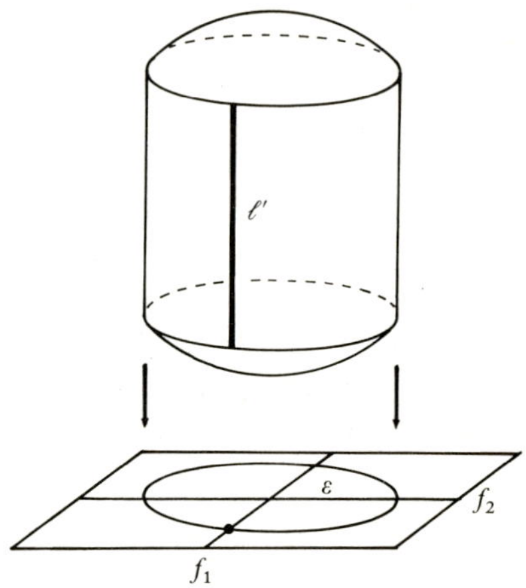

The cylindrical neighborhood C

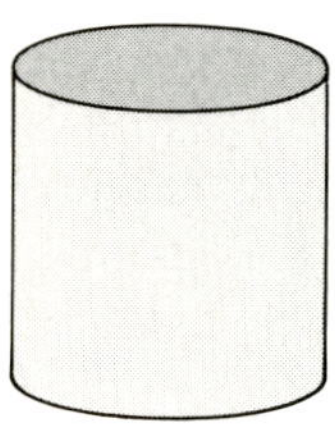

The vertical part L_v of the link

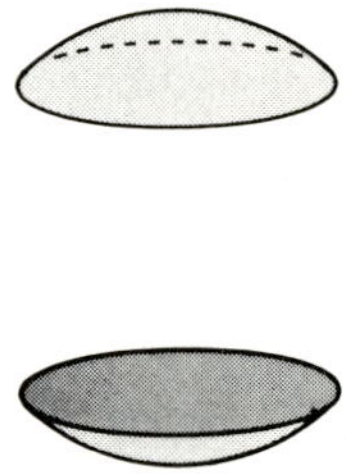

The horizontal part L_h of the link

Theorem. *The homeomorphism type of each of the above spaces is independent of the choices of the normal slice N, the metric r, or the choice of ε and δ. The homeomorphisms may be taken to preserve the $\mathscr{S}$-decompositions and the*

Whitney stratifications. Furthermore, there is a natural choice for such a homeomorphism, which is well defined up to isotopy.

Proof. The proof is similar to that in I Sects. 7.4 and 7.5, so we will only sketch it here. First we show that between any two allowable choices for N, r, ε and δ, there is a smooth one-parameter family of such choices. Let $N_0 = N'_0 \cap Z$ and $N_1 = N'_1 \cap Z$ be two normal slices at p, where N'_0 and N'_1 are smooth submanifolds of M which meet S transversally at the point p. Then, there is a smooth one-parameter family of submanifolds N'_t which connects them (i.e., there is a smooth submanifold $N' \subset M \times \mathbb{R}$ whose projection to $\mathbb{R}$ has no critical points, such that $N'_0 = N' \cap M \times \{0\}$ and $N'_1 = N' \cap M \times \{1\}$).

If r_0 and r_1 are two distance functions corresponding to two choices of Riemannian metrics g_0 and g_1, then $g_t = t g_1 + (1-t) g_0$ is a one-parameter family of metrics connecting them, which gives rise to a one-parameter family of distance functions from p, r_t.

We now have a one-parameter family of spaces N_t and maps $(r_t, f)\colon N_t \to \mathbb{R}^3$ to a wall space. Consider the set $\tilde{A} \subset [0,1] \times \mathbb{R}^+ \times \mathbb{R}^+$ consisting of all points (t, δ, ε) such that the map

$$(r_t, f_1, f_2) \,|\, N_t\colon N_t \to \mathbb{R}^3$$

has no characteristic covectors at point in the set

$$\{(u, v_1, v_2) \in \mathbb{R}^+ \times \mathbb{R}^2 \,|\, u = \delta, \sqrt{(v_1^2 + v_2^2)} \le \varepsilon\}.$$

It is easy to see that this set $\tilde{A}$ is open in $[0,1] \times \mathbb{R}^+ \times \mathbb{R}^+$ (since $(df_1(p), df_2(p))$ is nondegenerate). Thus, $\tilde{A}$ contains a one-parameter family of fringed sets, $A_t \subset \mathbb{R}^+ \times \mathbb{R}^+$, whose union is also an open subset of $[0,1] \times \mathbb{R}^+ \times \mathbb{R}^+$, such that $A_t^{\#}$ contains no characteristic covectors of the map (r_t, f_t). By I Sect. 5.4, there is a smooth one-parameter family of choices $(\delta_t, \varepsilon_t) \in A_t$ connecting the two original choices of ε and δ.

Each of the spaces ℓ', $\partial\ell'$, L, C, L_h, L_v, etc., can be written as the preimage under (r, f) of some stratified subset T of the wall space $\mathbb{R}^3$. Between any two choices of the data $(N, r, \varepsilon, \delta)$ there is a one-parameter family of the corresponding subsets of $\mathbb{R}^3$. The lemma on moving the wall then gives a (stratum preserving) homeomorphism between the corresponding preimages. For example, $\ell' = (r, f)^{-1}(T)$, where

$$T = \{(u, \varepsilon, 0) \in \mathbb{R}^3 \,|\, u \le \delta\}.$$

Given a one-parameter family as constructed above, let

$$T_t = \{(u, \varepsilon_t, 0) \in \mathbb{R}^3 \,|\, u \le \delta_t\}.$$

Then moving the wall defines a homeomorphism between

$$\ell'_0 = (r_0, f_0)^{-1}(T_0) \quad \text{and} \quad \ell'_1 = (r_1, f_1)^{-1}(T_1).$$

Proof of "furthermore". Suppose we have two different one-parameter families of normal slices connecting N'_0 and N'_1, and two different one-parameter families of distance functions connecting r_0 and r_1, and two different one-parameter families of choices for (δ, ε). These give rise to two different homeomorphisms G and H. It is easy to find a two-parameter family of normal slices $N_{(s,t)}$ and distance functions $r_{(s,t)}$ which connect these one-parameter families. Proposition I Sect. 5.5 now provides the appropriate two-parameter family of choices $(\delta_{(s,t)}, \varepsilon_{(s,t)})$, so that for each value of s and t the corresponding map

$$(r_{(s,t)}, f_1, f_2) \mid N_{(s,t)}: N_{(s,t)} \to \mathbb{R}^3$$

has no characteristic covectors at points in the set

$$\{(u, v_1, v_2) \mid u = \delta_{(s,t)}, \sqrt{(v_1^2 + v_2^2)} \leq \varepsilon_{(s,t)}\}.$$

This gives rise (by moving the wall as above) to a one-parameter family of homeomorphisms between G and H. $\square$

12.5. Local Structure of the Normal Slice: The Milnor Fibration

The following statements (a), (b), (c), and (d) are the key technical lemmas which allow us to analyze the normal Morse data for complex analytic varieties. We use the notation established above, i.e., $f: Z \to \mathbb{R}^2$ has a nondegenerate critical point at p, and N is a normal slice through p. Choose numbers $0 < \eta \ll \varepsilon \ll \delta$ so that (δ, ε) lies in the fringed set A, and η satisfies the conclusion of Lemma 12.3.

We show in the proposition below that the vertical part of the link is a fibre bundle over the circle, with fibre homeomorphic to the quarterlink (see [Mi2] or [Lê3]). In Part (e) we show that the horizontal part of the link is a product of a two-disk with the boundary of the quarterlink. The intersection of these two pieces of the link is collared in both pieces and the collared neighborhood is a trivial bundle over the circle. This collaring restricts in each fibre to a collaring of $\partial \ell'$ in ℓ'. We also show (in parts (b) and (c)) that although the cylindrical neighborhood C is a different shape from the usual "conical neighborhood" N_δ, it is also conical and, when cut off by the values $f_1^{-1}(\pm\eta)$ it gives the normal Morse data for the function f_1. However, it may be necessary to choose this η very much smaller than ε and δ. Finally in part (d) we show that the halflink is topologically a product of the quarterlink with an interval, and that this homeomorphism preserves boundaries.

Proposition. (a) *Milnor fibration.* The restriction $f \mid L_v: L_v \to \partial D_\varepsilon$ is a topological fibre bundle with fibre ℓ'. In fact, the restriction $f: (C - f^{-1}(0)) \to D_\varepsilon - \{0\}$ is a fibre bundle with fibre ℓ'.

(b) *The cylindrical neighborhood is conical.* For any $\eta > 0$ sufficiently small, there are homeomorphisms

$$(C, L) \xrightarrow{G} (N_\delta, \partial N_\delta) \xrightarrow{F} (cone(\partial N_\delta), \partial N_\delta)$$

with the following properties:

(i) G preserves the following levels: $f_1 = -\eta$, $f_1 = 0$, $f_1 = \eta$. In other words, $f_1(z) \in K \Leftrightarrow f_1 G(z) \in K$, where K is any of the following sets:

$$(-\infty, -\eta),\ \{-\eta\},\ (-\eta, 0),\ \{0\},\ (0, \eta),\ \{\eta\},\ (\eta, \infty).$$

(ii) F preserves the level $f_1=0$. In other words, F takes the sets
$N_\delta \cap f_1^{-1}(-\infty, 0)$ homeomorphically to cone $(\partial N_\delta \cap f_1^{-1}(-\infty, 0))$ − conepoint
$N_\delta \cap f_1^{-1}(0)$ homeomorphically to cone $(\partial N_\delta \cap f_1^{-1}(0))$
$N_\delta \cap f_1^{-1}(0, \infty)$ homeomorphically to cone $(\partial N_\delta \cap f_1^{-1}(0, \infty))$ − conepoint.

(c) *The cylindrical neighborhood gives normal Morse data.* The pair $(C \cap f_1^{-1}[-\eta, \eta], C \cap f_1^{-1}(-\eta))$ is normal Morse data, i.e., is homeomorphic to the pair

$$(N_\delta \cap f_1^{-1}[-\varepsilon, \varepsilon], N_\delta \cap f_1^{-1}(-\varepsilon)).$$

(d) *The halflink is the quarterlink times an interval.* There are homeomorphisms of pairs,

$$(\ell^+, \partial \ell^+) \cong (\ell', \partial \ell') \times (I, \partial I)$$
$$(\ell^-, \partial \ell^-) \cong (\ell', \partial \ell') \times (I, \partial I)$$

where ℓ^+ (resp. ℓ^-) denotes the upper (resp. lower) halflink, I denotes the unit interval $[0, 1]$, and $\partial I = \{0, 1\}$.

(e) *Collarings of the boundary of the quarterlink.* There is a number $\omega > 0$ and a neighborhood U of L_h in C and a homeomorphism

$$H: \partial \ell' \times [\delta - \omega, \delta] \times D_\varepsilon \to U$$

which commutes with the projection $(r, f): U \to \mathbb{R} \times \mathbb{R}^2$, i.e.,

$$r(H(z, u, v)) = u \quad \text{and} \quad f(H(z, u, v)) = v$$

and which restricts to homeomorphisms,

(i) $H_1: \partial \ell' \times D_\varepsilon \to L_h$ which commutes with $f: L_h \to D_\varepsilon$.
(ii) $H_2: \partial \ell' \times \partial D_\varepsilon \to \partial L_h = \partial L_v$ which commutes with $f: \partial L_h \to \partial D_\varepsilon$.
(iii) $H_3: \partial \ell' \times [\delta - \omega, \delta] \times \partial D_\varepsilon \to U \cap L_v$ which commutes with

$$(r, f): U \cap L_v \to [\delta - \omega, \delta] \times \partial D_\varepsilon.$$

Each of the spaces in the preceding proof has a canonical $\mathscr{S}$-decomposition, and the homeomorphisms may be chosen so as to preserve these $\mathscr{S}$-decompositions.

Corollary. *The normal Morse data for the covector $df_1(p)$ has the homotopy type of the pair*

$$(\mathrm{cone}(\ell'), \ell').$$

Proof. By Corollary Sect. 3.11, the normal Morse data is homotopy equivalent to

$$(\mathrm{cone}(\ell^-), \ell^-).$$

Furthermore (by part (d) above), $\ell^- \cong \ell' \times [0, 1]$, so this pair is homotopy equivalent to the pair $(\mathrm{cone}(\ell'), \ell')$. □

12.6. Proof of Proposition 12.5

Proof of (a). By Sect. 12.2 (Lemma 1), the map $f: C \to D_\varepsilon$ has no characteristic covectors except at the origin. By Thom's isotopy lemma (Sect. 1.5) it is a locally trivial fibre bundle.

Remark. This generalization of [Mi2] appears in [Lê3] and [Du], for example. More generally, if M_1 and M_2 are smooth manifolds containing Whitney stratified subsets $W_1 \subset M_1$ and $W_2 \subset M_2$, and if $f\colon M_1 \to M_2$ is a smooth map such that $f(W_1)=W_2$, $f \mid W_1$ is proper, and such that f is stratified by the given stratifications of W_1 and W_2, and if $p \in W_1$ and if $q=f(p) \in W_2$, then one may ask whether (for $\varepsilon \ll \delta$ sufficiently small), the map

$$W_1 \cap B_\delta(p) \cap f^{-1}(\partial B_\varepsilon(q)) \to W_2 \cap \partial B_\varepsilon(q)$$

is a fibre bundle over each stratum $S_2 \cap \partial B_\varepsilon(q)$ of $W_2 \cap \partial B_\varepsilon(q)$. In [Lê4], it is shown that in order to obtain such a generalized Milnor fibration theorem, it is sufficient to assume that the map f is a *Thom mapping*, i.e., that it satisfies condition A_f of [T5]. Our assumption that the function f has a "nondegenerate" critical point at p implies Thom's condition A_f.

Proof of (b). The pair (C, L) is the preimage of the box

$$\{(r, |f|) \in \mathbb{R}^2 \mid r \le \delta \text{ and } |f| \le \varepsilon\}$$

in $\mathbb{R}^2$, modulo the right hand side and top.

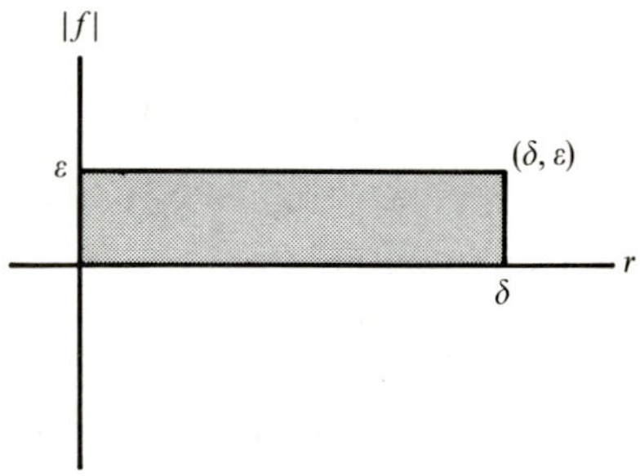

The pair (C, L) in the wall space

The pair $(N_\delta, \partial N_\delta)$ is the preimage of the halfspace

$$\{(r, |f|) \in \mathbb{R}^2 \mid r \le \delta\}$$

modulo the boundary line $r=\delta$.

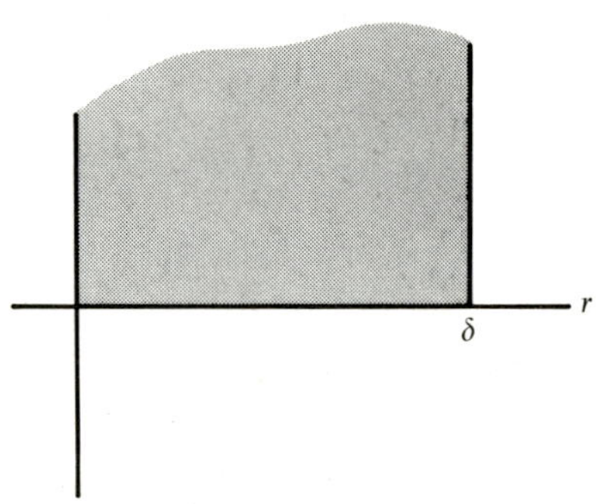

The pair $(N_\delta, \partial N_\delta)$ in the wall space

The idea for constructing G is to move the wall in $(r, |f|)$ space, just as in I Sect. 7.8 (where $|f|=(f_1^2+f_2^2)^{1/2}$). However, we want to guarantee that the levels $f_1=-\eta$, $f_1=0$, and $f_1=\eta$ are preserved by the resulting homeomorphism, so we add new strata to N which correspond to these levels. Since the homeomorphism which results from moving the wall preserves strata, these levels of f_1 will be preserved assuming we have not introduced any bad characteristic covectors in $(r, |f|)$-space by adding these new strata. Here are the details:

Step 1. Divide each stratum R of N into the following seven pieces:

$$R_1=R\cap f_1^{-1}(-\infty,-\eta),\quad R_2=R\cap f_1^{-1}(-\eta),\quad R_3=R\cap f_1^{-1}(-\eta,0),$$
$$R_4=R\cap f_1^{-1}(0),\quad R_5=R\cap f_1^{-1}(0,\eta),\quad R_6=R\cap f_1^{-1}(\eta),$$
$$R_7=R\cap f_1^{-1}(\eta,\infty).$$

Except in the single case $R=\{p\}$, each of these pieces is nonempty. It is easy to see that this refinement is a Whitney stratification of N. We shall denote this space with its *new stratification* by $\tilde{N}$. Let $A\subset\mathbb{R}^+\times\mathbb{R}^+$ denote the fringed set of Sect. 12.2. Consider the map

$$(r, |f|)\colon \tilde{N}\to\mathbb{R}\times\mathbb{R}.$$

Claim. This map has no characteristic points in the region

$$A^\flat=\{(u,v)\in\mathbb{R}^+\times\mathbb{R}\mid(u,|v|)\in\bar{A}\}.$$

Furthermore, if $\lambda\in T^*\mathbb{R}^2$ is a nontrivial characteristic covector of this map, then the slope of $\ker(\lambda)$ is positive.

Proof of claim. The map $(r,f)\colon N\to\mathbb{R}\times\mathbb{R}^2$ has no characteristic covectors in the region $A^\#$ (by Sect. 12.2). Thus, the map $(r,|f|)\colon N\to\mathbb{R}\times\mathbb{R}$ has no characteristic covectors in the region $A^\flat$, and the slope of the kernel of any characteristic covector (which lies outside $A^\flat$) is positive. However, we must consider separately the new strata in $\tilde{N}$ which are of the form $R\cap f_1^{-1}(t)$ where $t=-\eta$, 0, or η. Suppose $\lambda=(u,v,a\,du+b\,dv)\in T^*\mathbb{R}^2$ is a characteristic covector of the map $(r,|f|)|\tilde{N}$, corresponding to a point $q\in\tilde{N}$. Let us assume q lies in some stratum $\tilde{S}=S\cap f_1^{-1}(t)$. Then $a\,dr(q)+b\,d|f|(q)$ vanishes on $T_q\tilde{S}=T_qS\cap\ker df_1(q)$. Thus,

$$a\,dr(q)+\frac{b}{|f|}\,df_1(q)+\frac{b}{|f|}\,df_2(q)=0.$$

However, Lemma 12.3 applies (since $|f_1(q)|\le\eta$) and gives $a/b\,|f(q)|<0$. Therefore, $a/b<0$, as desired.

Step 2. Define

$$\mathrm{Box}(\delta,\varepsilon)=\{(u,v)\in\mathbb{R}^2\mid 0\le u\le\delta \text{ and } 0\le v\le\varepsilon\}$$
$$\mathrm{HS}(\delta)=\{(u,v)\in\mathbb{R}^2\mid u\le\delta\}.$$

We want to find a homeomorphism G between

$$(\tilde{C},\tilde{L})=(r,|f|)^{-1}(\mathrm{Box}(\delta,\varepsilon))\cap\tilde{N}$$

and

$$(\tilde{N}_\varepsilon,\partial\tilde{N}_\varepsilon)=(r,|f|)^{-1}(\mathrm{HS}(\delta))\cap\tilde{N}$$

which can be achieved by moving the wall as follows:

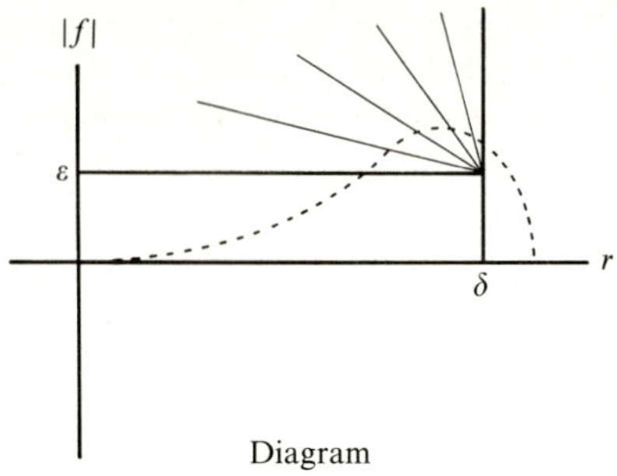

Diagram

The preceding claim guarantees that there are no characteristic covectors to prevent this motion of the wall, so we obtain a stratum preserving homeomorphism between $(\tilde{C}, \tilde{L})$ and $(\tilde{N}_\varepsilon, \partial\tilde{N}_\varepsilon)$ which therefore preserves the levels $f_1 = -\eta$, $f_1 = 0$, and $f_1 = \eta$.

The second homeomorphism F is obtained in a similar manner: refine the stratification of the pair $(N_\varepsilon, \partial N_\varepsilon)$ by adding strata from $f_1^{-1}(0)$. The conical structure of N_ε (which is obtained by integrating a controlled lift of the vectorfield d/dr, i.e., by moving the wall in r-space, from $r=\delta$ to $r=0$) will respect this stratification and will therefore preserve the sign of f.

Proof of (c). part (b) of the lemma gives us a homeomorphism of pairs

$$(C \cap f_1^{-1}[-\eta, \eta], C \cap f_1^{-1}(-\eta)) \cong (N_\delta \cap f_1^{-1}[-\eta, \eta], N_\delta \cap f_1^{-1}(-\eta)).$$

However, the second pair is homeomorphic to the pair

$$(N_\delta \cap f_1^{-1}[-\varepsilon, \varepsilon], N_\delta \cap f_1^{-1}(-\varepsilon))$$

by moving the wall as follows:

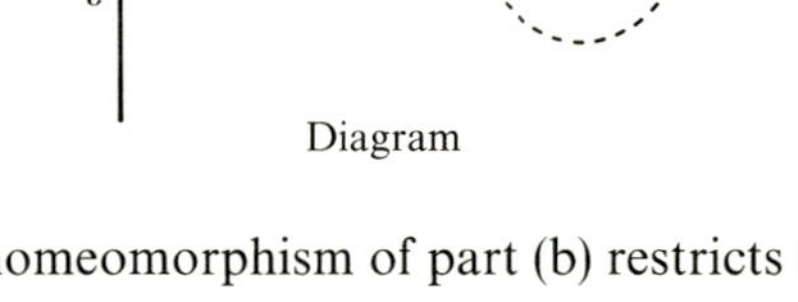

Diagram

Proof of (d). The homeomorphism of part (b) restricts to a homeomorphism

$$\begin{aligned}(\ell, \partial\ell) &= (N_\delta \cap f_1^{-1}(\eta), \partial N_\delta \cap f_1^{-1}(\eta))\\ &\cong (C \cap f_1^{-1}(\eta), L \cap f_1^{-1}(\eta))\\ &= (C \cap f_1^{-1}(\eta), f_1^{-1}(\eta) \cap (L_v \cup L_h))\\ &= (f^{-1}(\eta, 0) \cap C) \times [-\omega, \omega], f^{-1}(\eta, -\omega)\end{aligned}$$

$$\cup f^{-1}(\eta, \omega) \cup (f^{-1}(\eta, 0) \cap \partial N_\delta) \times [-\omega, \omega]))$$
$$= (\ell' \times [-\omega, \omega], (\ell' \times \partial[-\omega, \omega]) \cup (\partial\ell' \times [-\omega, \omega]))$$

where $\omega = \sqrt{(\varepsilon^2 - \eta^2)}$ and where we have used part (a) of the proposition to realize C and L_h as fibre bundles over the punctured disk. □

Proof of (e). (This is the only part of the argument which uses the fact that there are no characteristic covectors on the r axis, $f=0$.) Choose ω so that the rectangle $[\delta-\omega, \delta] \times (0, \varepsilon]$ is contained in the fringed set A.

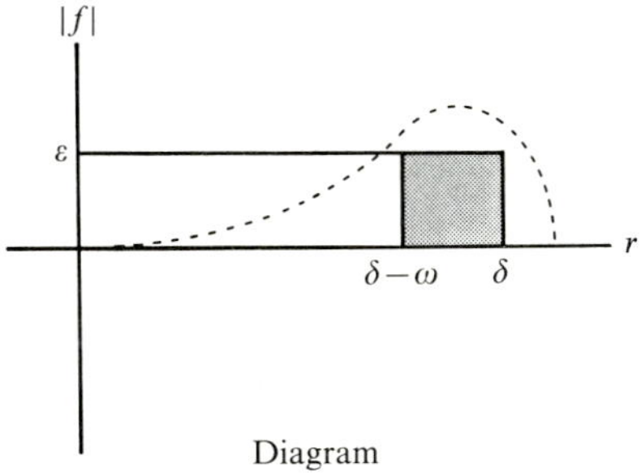

Diagram

Thus, $[\delta-\omega, \delta] \times D_\varepsilon$ is contained in the region $A^\#$, where there are no (nontrivial) characteristic covectors of the map $(r, f)\colon N \to \mathbb{R} \times \mathbb{R}^2$. It follows from the first isotopy lemma that

$$U = (r, f)^{-1}([\delta-\omega, \delta] \times D_\varepsilon)$$

is a fibre bundle over $[\delta-\omega, \delta] \times D_\varepsilon$, with fibre

$$\partial\ell' = (r, f)^{-1}(\delta, \varepsilon, 0).$$

This fibre bundle is trivial since $[\delta-\omega, \delta] \times D_\varepsilon$ is contractible, and the isotopy lemma can be used to find a global trivialization

$$H\colon \partial\ell' \times [\delta-\eta, \delta] \times D_\varepsilon \to U$$

which preserves strata. Thus, H restricts to a homeomorphism on any closed union of strata of U, for example in the following three cases:

(i) $(r, f)^{-1}(\delta \times D_\varepsilon) = L_h$

(ii) $(r, f)^{-1}(\delta \times \partial D_\varepsilon) = \partial L_h = \partial L_v$

(iii) $(r, f)^{-1}([\delta-\omega, \delta] \times \partial D_\varepsilon) = U \cap L_v$. □

12.7. Monodromy

In this section we define a stratum preserving homeomorphism $\mu\colon \ell' \to \ell'$ which fixes a neighborhood of $\partial\ell'$ such that the space

$$\ell' \times [0, 1]/((z, 0) \sim (\mu(z), 1))$$

is homeomorphic (by a stratum preserving homeomorphism) to L_v.

Since the map $f: L_v \to \partial D_\varepsilon$ is a stratified fibre bundle, it is possible to choose control data ([Ma1], [Ma2]) on L_v which is compatible with f. Choose an orientation of the circle ∂D_ε and let $d/d\tau$ denote a unit vectorfield on ∂D_ε. By Thom's first isotopy lemma, it has a controlled lift to a vectorfield F_1 on L_v, whose time 2π flow defines a homeomorphism $\ell' \to \ell'$. This homeomorphism must be modified so as to restrict to the identity on $\partial \ell'$.

By Sect. 12.5(e iii) there is a neighborhood $W = U \cap L_v$ of ∂L_v in L_v and a stratum preserving homeomorphism

$$H_3: \partial \ell' \times [\delta - \eta, \delta] \times \partial D_\varepsilon \to W$$

which commutes with the projection to ∂D_ε. Let $V_2 = (H_3)_*(0 \times 0 \times d/d\tau)$. This is a controlled lift of $d/d\tau$ to this neighborhood, and its time 2π flow is the identity on $\partial \ell' \times [\delta - \eta, \delta]$. Now patch the controlled vectorfields V_1 and V_2 together, using a partition of unity. The result is a controlled vectorfield whose time 2π flow gives a homeomorphism $\ell' \to \ell'$, and which restricts to the identity in a neighborhood of $\partial \ell'$.

12.8. Monodromy is Independent of Choices

The isotopy class of the monodromy $\mu: \ell' \to \ell'$ is independent of the choice of control data, the normal slice N, the metric r, and the allowable choices of ε and δ in the following sense:

Proposition. *Suppose we are given two choices N_0 and N_1 of normal slice through the point p, two allowable choices $(\delta_0, \varepsilon_0)$ and $(\delta_1, \varepsilon_1)$ of the parameters δ and ε, choices r_0 and r_1 of distance functions from the point p (which are associated to two choices of Riemannian metrics on the ambient manifold M), and two choices $f_0: N_0 \to \mathbb{R}^2$ and $f_1: N_1 \to \mathbb{R}^2$ of functions whose differentials are nondegenerate at p (as in Sect. 12.1), which are connected by a smooth one-parameter family of functions $f_t: N_t \to \mathbb{R}^2$, whose differentials are nondegenerate at p. From this data we construct two representatives ℓ_0' and ℓ_1' of the quarterlink. Suppose we make choices of control data on the associated fibre bundles $(L_v)_0$ and $(L_v)_1$ and construct the corresponding monodromy homeomorphisms $\mu_0: \ell_0' \to \ell_0'$ and $\mu_1: \ell_1' \to \ell_1'$ (with respect to the same orientation of the circle ∂D_ε). Then there is a homeomorphism $f: \ell_0' \to \ell_1'$ such that the composition $f \circ \mu_1 \circ f^{-1}$ is isotopic to μ_0, and the isotopy may be chosen to be the identity in a neighborhood of $\partial \ell_0'$ and to preserve the $\mathcal{S}$-decomposition of ℓ_0'.*

Sketch of Proof. Using the techniques of Sect. 12.4, it is possible to find a one-parameter family of data $(N_t, \delta_t, \varepsilon_t, f_t, (L_v)_t, \ell_t')$ connecting these choices. The one-parameter family f_t maps the union of the $(L_v)_t$ to a cylinder, and this map is a stratified fibre bundle. It is now possible using standard controlled vectorfield techniques to construct the homeomorphism f and the isotopy. □

12.9. Relative Normal Morse Data for Two Nonproper Functions

This section is a common generalization of Sects. 11 and 12.4–12.6. Suppose (as in Sect. 12.1) that Z is a closed Whitney stratified subset of a smooth manifold,

$p \in Z$, N is a normal slice at p to the stratum of Z which contains the point p, and that f_1 and f_2 are smooth functions defined on M whose restriction to Z is proper and which satisfy the nondegeneracy condition of Sect. 12.1, i.e.,

$$(df_1(p), df_2(p))(Q) = \mathbb{R}^2$$

for every generalized tangent space Q at the point p, except for the single case Q equals the tangent space to the stratum which contains p.

Suppose also (as in Sect. 11.1) that $\bar{\pi}: \bar{X} \to Z$ is a Whitney stratified surjective proper map, that the strata of $\bar{X}$ are indexed by some partially ordered set $\mathcal{S}$ and that $X \subset \bar{X}$ is a dense open subset which is a union of strata of $\bar{X}$ corresponding to a partially ordered subset $\mathcal{T} \subset \mathcal{S}$. We wish to consider the normal Morse data of the composition $(f_1 \circ \bar{\pi}) \mid X: X \to \mathbb{R}$.

Let N denote a normal slice (in Z) through the point p, and let A denote the fringed set of Sect. 12.2. A choice of $(\delta, \varepsilon) \in A$ determines a quarterlink and its boundary $(\ell', \partial \ell')$ (for Z), a cylindrical neighborhood C and its boundary L, the horizontal and vertical parts of the link L_h and L_v, and their common boundary $\partial L_h = \partial L_v$.

Definition. The relative quarterlink (resp. relative normal slice, resp. relative cylindrical neighborhood, resp. relative horizontal and vertical parts of the link) is the intersection with X of the preimage under $\bar{\pi}$ of the quarterlink (resp. normal slice, resp. cylindrical neighborhood, resp. horizontal and vertical parts of the link)

$$(\ell_X'^{\pi}, \partial \ell_X'^{\pi}) = \bar{\pi}^{-1}(\ell', \partial \ell') \cap X.$$

Proposition. The results of Proposition 12.5 remain true when the spaces ℓ', $\partial \ell'$, L, C, ∂C, L_v, L_h, N_δ, ∂N_δ are replaced with the corresponding relative spaces, except for the existence of the homeomorphism F in Proposition 12.5(b).

Proof (see Sect. 11.5). Except for the statements concerning the homeomorphism F in (Sect. 12.5(b)), each of the results is obtained by moving the wall with respect to some map $H: N \to \mathbb{R}^k$ (for some k) and checking that this map has no characteristic covectors $\lambda \in T^* \mathbb{R}^k$ such that $\ker(\lambda)$ contains the tangent space to a stratum of the wall in $\mathbb{R}^k$. However the composition $H \circ \bar{\pi}: \bar{\pi}^{-1}(N) \to \mathbb{R}^k$ has precisely the same characteristic covectors as the map H, so the same arguments can be applied to $\bar{\pi}^{-1}(N)$. Furthermore, the resulting homeomorphisms preserve the $\mathcal{S}$-decomposition of the spaces (which are induced by their intersection with the strata of N). Therefore they restrict to homeomorphisms on any union of pieces of the decomposition. □

Corollary 1. *If $\bar{\pi}^{-1}(p) \subset X$, i.e., if π is proper near the fibre over p, then the relative normal Morse data for f_1 at the point p has the homotopy type of the pair*

$$(\mathrm{cyl}(\ell'^{\pi} \to \bar{\pi}^{-1}(p)), \ell'^{\pi})$$

where cyl *denotes the mapping cylinder of the specialization map (I Sect. 9.7),*

$$\phi: \pi^{-1}(\ell') \to \pi^{-1}(p).$$

Proof. By Part I, Sect. 9.7, the relative normal Morse data has the homotopy type of the pair

$$(\mathrm{cyl}(\ell^{\pi-} \to \pi^{-1}(p), \ell^{\pi-}).$$

However, by Proposition 12.9 we have a homeomorphism

$$\ell^{\pi-} \cong \ell'^{\pi} \times [0, 1].$$

Corollary 2. *Suppose $\bar{X}$ equals Z and π equals the identity. If the critical point p is a "critical point at infinity", $p \in Z - X$, then the following pair is homotopy normal Morse data (see Sect. 10):*

$$(\ell' \cap X, \partial \ell' \cap X) \times (I, \partial I)$$

where I denotes the unit interval $[0, 1]$. *More generally, the same result holds for arbitrary π provided $\bar{\pi}^{-1}(p) \subset \bar{X} - X$, and provided we replace $(\ell', \partial \ell')$ by $((\ell')^{\pi}_{X}, (\partial \ell')^{\pi}_{X})$.*

Proof. By Sect. 10.6.2, homotopy normal Morse data is the pair $(\ell^+, \partial \ell^+)$. The relative version of Sect. 12.5(d) gives a homeomorphism between $(\ell^+, \partial \ell^+)$ and $(\ell', \partial \ell') \times (I, \partial I)$. □

12.10. Normal Morse Data for Many Morse Functions

The preceding results readily generalize to the case of many normal Morse functions.

In this section we assume $f = (f_1, f_2, \ldots, f_{k+1}) \colon Z \to \mathbb{R}^{k+1}$ is a collection of smooth functions and that p is a critical point of f, i.e., $df(p)(T_p S) = 0$, where S is the stratum of Z which contains the point p. We also assume this critical point is nondegenerate, i.e., for every generalized tangent space Q at the point p, we have $df(p)(Q) = \mathbb{R}^{k+1}$, except for the single case $Q = T_p S$. Fix a normal slice $N = N' \cap Z$ through the point p, where N' is a smooth submanifold which is transverse to S and $N' \cap S = \{p\}$.

Proposition. *There exists a fringed set $A \subset \mathbb{R}^+ \times \mathbb{R}^+$ such that the map*

$$(r, f) \mid N \colon N \to \mathbb{R} \times \mathbb{R}^{k+1}$$

has no characteristic covectors in the region

$$A^{\#\#} = \{(u, v_1, v_2, \ldots, v_{k+1}) \in \mathbb{R}^+ \times \mathbb{R}^{k+1} \mid (u, |v|) \in \bar{A}\}$$

where $|v| = \sqrt{v_1^2 + v_2^2 + \ldots + v_{k+1}^2}$.

Proof. The proof is identical to that in Sect. 12.2.

Definition. Fix $(\delta, \varepsilon) \in A$. The disk and ball:

$$D_\varepsilon^{k+1} = \{\zeta \in \mathbb{R}^{k+1} \mid |\zeta| \le \varepsilon\} \qquad \partial D_\varepsilon^{k+1} = \{\zeta \in \mathbb{R}^{k+1} \mid |\zeta| = \varepsilon\}$$

$$N_\delta = N' \cap Z \cap B_\delta(p) \qquad \partial N_\delta = N' \cap Z \cap \partial B_\delta(p).$$

(1) The littlelink and its boundary:

$$\ell^{(k)} = f^{-1}(\varepsilon, 0, \ldots, 0) \cap N_\delta \qquad \partial \ell^{(k)} = f^{-1}(\varepsilon, 0, \ldots, 0) \cap \partial N_\delta.$$

(2) The cylindrical neighborhood of p, and its boundary:

$$C^{(k)} = f^{-1}(D_\varepsilon^{k+1}) \cap N_\delta \qquad L^{(k)} = \partial C^{(k)} = L_h^{(k)} \cup L_v^{(k)}.$$

(3) The horizontal and vertical parts of the link:

$$L_h^{(k)} = f^{-1}(D_\varepsilon^{k+1}) \cap \partial N_\delta \qquad L_v^{(k)} = f^{-1}(\partial D_\varepsilon^{k+1}) \cap N_\delta$$

$$\partial L_h^{(k)} = \partial L_v^{(k)} = f^{-1}(\partial D_\varepsilon^{k+1}) \cap \partial N_\delta.$$

Each of these spaces is $\mathscr{S}$-decomposed by its intersection with the strata of N (which are also the strata of Z) and are canonically Whitney stratified, since each is a transversal intersection of Whitney stratified spaces.

Theorem. *The homeomorphism type of each of the above spaces is independent of the choices of the normal slice N', the metric r, or the choice of ε and δ. The homeomorphisms may be taken to preserve the $\mathscr{S}$-decompositions and the Whitney stratifications.*

Proof. The proof is the same as Sect. 12.4. □

Proposition.

(a) *The restriction $f \mid L_v^{(k)}: L_v^{(k)} \to \partial D_\varepsilon^{k+1}$ is a topological fibre bundle with fibre $\ell^{(k)}$.*

(b) *There are homeomorphisms of pairs,*

$$(\ell^{(m-1)}, \partial \ell^{(m-1)}) \cong (\ell^{(m)}, \partial \ell^{(m)}) \times (I, \partial I)$$

where I denotes the unit interval $[0, 1]$. Therefore

$$(\ell^+, \partial \ell^+) \cong (\ell^{(k)}, \partial \ell^{(k)}) \times (I, \partial I)^k$$

$$(\ell^-, \partial \ell^-) \cong (\ell^{(k)}, \partial \ell^{(k)}) \times (I, \partial I)^k.$$

Each of the spaces has a canonical $\mathscr{S}$-decomposition, and the homeomorphisms may be chosen so as to preserve these $\mathscr{S}$-decompositions.

Proof. The proof is identical to Sect. 12.6. □

Corollary. *The normal Morse data, halflink and quarterlink for the function f_1 are homeomorphic to the normal Morse data, halflink, and quarterlink for the function f_j $(1 \leq j \leq k)$. If $k \geq 2$, then the monodromy (defined on the quarterlink of f_1) is isotopic to the identity. The normal Morse data for f_1 has the homotopy type of the pair*

$$(\text{cone}\,(\ell^{(k)}), \ell^{(k)}).$$

Analogous results hold in the relative and noncompact case.

Part II. Morse Theory of Complex Analytic Varieties

Chapter 0. Introduction

The main technical results of complex Morse theory are the following: Let W be a closed subvariety of a Whitney stratified complex analytic variety Z, and suppose that W is a union of strata in Z. Let $f\colon Z \to \mathbb{R}$ be a proper Morse function, which has a nondegenerate critical point $p \in Z$ which lies in some stratum S of complex codimension c in Z. Let λ denote the Morse index of $f|S$ at the point p, and let $v = f(p)$ be the associated critical value. We will consider the Morse theory of $f|X$, where $X = Z - W$. Suppose the interval $[a, b]$ contains no critical values of $f|Z$ other than v, and that $v \in (a, b)$.

(1) *If* $p \in X$, then (Sect. 3.3) the space $X_{\leq b}$ has the homotopy type of a space obtained from $X_{\leq a}$ by attaching the pair of spaces

$$(D^\lambda, \partial D^\lambda) \times (\operatorname{cone}(\mathscr{L}), \mathscr{L})$$

where $\mathscr{L} = \mathscr{L}_Z = \mathscr{L}_X$ is the complex link of the stratum S, and D^λ is a cell of dimension λ.

(2) If $p \notin X$, then (Sect. 3.3) the space $X_{\leq b}$ has the homotopy type of a space obtained from $X_{\leq a}$ by attaching the pair

$$(D^{\lambda+1}, \partial D^{\lambda+1}) \times (\mathscr{L}_X, \partial \mathscr{L}_X)$$

where $\mathscr{L}_X = X \cap \mathscr{L}_Z$ is the complex link for X (see Sect. 2.6).

(3) If $p \in X$, or if $W = \phi$, then (Sect. 4.5.2*) $\mathscr{L}_X$ has the homotopy type of a CW complex of dimension $\leq c-1$. Therefore, $H_i(X_{\leq b}, X_{\leq a}; \mathbb{Z}) = 0$ for all $i > \lambda + c$, and $H_{\lambda+c}(X_{\leq b}, X_{\leq a}; \mathbb{Z})$ is torsion free.

(4) If X is a local complete intersection, then (Sect. 4.6.2) the pair $(\mathscr{L}_X, \partial \mathscr{L}_X)$ is $c-2$ connected. It follows from this and Sects. 4.6.1, 4.6.2, that $\pi_i(X_{\leq b}, X_{\leq a}) = 0$ for all $i < \lambda + c$ and $H_i(X_{\leq b}, X_{\leq a}; \mathbb{Z}) = 0$ for all $i < \lambda + c$.

(5) *If* $p \in X$, then (Sect. 6.4) $IH_i(X_{\leq b}, X_{\leq a}; \mathbb{Z}) = 0$ unless $i = \lambda + c$, and $IH_{\lambda+c}(X_{\leq b}, X_{\leq a}; \mathbb{Z})$ is the image of the variation map,

$$(I-\mu)_*\colon IH_{c-1}(\mathscr{L}, \partial \mathscr{L}; \mathbb{Z}) \to IH_{c-1}(\mathscr{L}; \mathbb{Z})$$

and is torsion free; where IH denotes the "middle intersection homology" groups [GM3] (with compact supports) and μ denotes the monodromy homomorphism.

(6) If $p \notin X$, then (Sect. 6.4)

$$IH_i(X_{\leq b}, X_{\leq a}; \mathbb{Z}) = IH_{i-\lambda-1}(\mathscr{L}_X, \partial \mathscr{L}_X; \mathbb{Z})$$

and this vanishes for all $i \geq \lambda + c + 1$.

Various extensions of these results also appear in Sects. 4 and 6.4.

These technical results are applied (in Chapter 5) to very general Lefschetz theorems and to theorems on the homotopy dimension of certain varieties. However, the method which we use is very simple and tends to be obscured by technical details having to do with the generality of the situations. For this reason we will give an outline of the proof of a special case of Theorem 1.1*: an affine n-dimensional complex analytic variety Z has the homotopy type of a CW complex of dimension n. (This is due to [Kr2] in the complex algebraic case, and [H3], corrected in [H4], in the case of a Stein space.)

We consider the function $f: Z \to \mathbb{R}$ which is the distance from a generic point $p \notin Z$. By I Sect. 2.2, a point p can be found so that f is a Morse function. By calculating the Levi form (Sect. 4.A.5) of f we see that the Morse index λ (at any critical point) of the restriction $f|S$ to any stratum S of Z is bounded as follows: $\lambda \leq \dim_{\mathbb{C}}(S)$. If Z is nonsingular, we conclude that at each critical value v, the set $Z_{\leq v+\varepsilon}$ has the homotopy type of a space obtained from $Z_{\leq v-\varepsilon}$ by attaching a cell of dimension $\leq n$, as in [AF]. However, if Z is singular we can only conclude (I Sect. 3.7) that the space $Z_{\leq v+\varepsilon}$ has the homotopy type of a space obtained from $Z_{\leq v-\varepsilon}$ by attaching the product (tangential Morse data) $\times$ (normal Morse data).

Let us say $p \in S$ is the critical point and the stratum containing the critical point. Then the tangential Morse data has the homotopy type of the pair $(D^k, \partial D^k)$ where D^k denotes the k-dimensional disk, and $k \leq \dim_{\mathbb{C}}(S)$. The normal Morse data (Sects. 3.2, 4.5) has the homotopy type of the pair $(c(\mathscr{L}), \mathscr{L})$ where $\mathscr{L}$ denotes the "complex link" of the stratum S at the point p. This is an $n-\dim(S)-1$ dimensional complex analytic space with boundary. The function $g: \mathscr{L} \to \mathbb{R}$, which is given by the distance from a generic point near p, is a Morse function (I Sect. 2.2), and the Morse index of any critical point of g on any stratum S' of $\mathscr{L}$ is bounded by the same estimate: $\lambda \leq \dim_{\mathbb{C}}(S')$. Since $\mathscr{L}$ is a Stein space, we conclude by induction that the space $\mathscr{L}$ has the homotopy type of a CW complex of dimension $\leq \dim_{\mathbb{C}}(\mathscr{L}) = n-k-1$. It follows that the normal Morse data at p has the homotopy type of a CW complex of dimension $\leq n-k$. Thus, $Z_{\leq v+\varepsilon}$ is obtained from the space $Z_{\leq v-\varepsilon}$ (up to homotopy) by attaching a CW complex of dimension $\leq n$. Passing each critical value in this manner, we conclude that Z has the homotopy type of a CW complex of dimension n.

This argument generalizes in several directions:

(1) We can remove a subvariety W from Z (see Sect. 1.2*). In this case, two kinds of critical points must be considered: those in W and those in $Z-W$. The normal Morse data is analyzed in Sects. 4.6* and 5.2*. The "homotopy dimension" of Z may rise, depending on how many equations were used in defining W.

(2) We can replace Z with a "relative space" $\pi: Y \to Z \subset \mathbb{C}^n$ with some estimate on the dimension of the fibres of π (see Sect. 1.1*). In this case the study of the normal Morse data is carried out in I Sect. 12 and II Sect. 3.4. The "homotopy dimension" of Z may rise, depending on the dimension of the fibres of π.

(3) We can put Z in $\mathbb{CP}^n$ and study the distance from a hyperplane, or from a codimension c linear subspace. This gives lower bounds on the Morse index λ (instead of upper bounds) so we obtain Lefschetz type theorems instead of "homotopy dimension" theorems (see Sects. 1.1 and 1.2).

(4) We can replace the pair $(\mathbb{CP}^n, H)$ by an arbitrary "q-defective" pair (Y, A) on which there exists a Morse function which takes a minimum on A and has a Levi form with certain positivity properties (Sect. 7). This gives Lefschetz theorems for subvarieties of (Griffiths-)"positive" vectorbundles. Sommese [Sm4] shows how such pairs arise as subvarieties of homogeneous spaces.

(5) We can replace "homotopy" by "homology" or by "intersection homology" (Sects. 6.8, 6.9, 6.10, 6.11).

(6) We also consider (in Sects. 1.3 and 1.3*) "local" versions of each of these theorems, replacing the variety Z by the link of a stratum in Z.

Extensions. There are by now a number of "tricks" which can be used to deduce surprising new results from Lefschetz theorems. For example, in [FL1] it is shown how Barth theorems can be derived from Lefschetz theorems by reembeding the given variety Z into a larger projective space so that the subspace $Y \subset Z$ becomes a hyperplane section. See [FL1] Sect. 9, for a catalog of such tricks and applications. It would be interesting to find similar applications along the lines of [FH], [FL2], [Fa], or [Go].

Chapter 1. Statement of Results

The proofs of the results stated here will appear in Sect. 5. The four main theorems (Sects. 1.1, 1.1*, 1.2, 1.2*) are followed by four analogous "local" theorems (1.3.1, 1.3.2, 1.3*.1, 1.3*.2) which may be considered as generalizations of the global theorems. These results are further generalized in Sects. 7.2 and 7.3.

1.0. Notational Remarks and Basepoints

If $B \subset A$ are topological spaces, then we shall write $\pi_0(B, A) = 0$ or $\pi_0(B) \twoheadrightarrow \pi_0 A$ to indicate that the set of connected components of B surjects to the set of connected components of A.

If k is a nonnegative integer and if, for any $b \in B$, the relative homotopy group

$$\pi_i(A, B, b) = 0 \qquad \text{for all } i \leq k$$

then we shall say that the pair (A, B) is k-connected. This is equivalent to the following statement: for any $b \in B$, the homomorphism

$$\pi_i(B, b) \to \pi_i(A, b)$$

is an isomorphism for all $i < k$ and is a surjection for $i = k$. In this case we will ignore the basepoints and write either

$$\pi_i(A, B) = 0 \qquad \text{for } i \leq k$$

or $\pi_i(B) \to \pi_i(A)$ is an isomorphism for $i < k$ and is a surjection for $i = k$.

1.1. Relative Lefschetz Theorem with Large Fibres

The following Lefschetz theorem was conjectured by P. Deligne [D1]. An outline of the proof described here was published in [GM1].

Theorem. *Let X be a purely n-dimensional nonsingular connected algebraic variety. Let $\pi: X \to \mathbb{CP}^N$ be an algebraic map and let $H \subset \mathbb{CP}^N$ be a linear subspace of codimension c. Let H_δ be the δ-neighborhood of H with respect to some (real analytic) Riemannian metric. Define $\phi(k)$ to be the dimension of the set of points $z \in \mathbb{CP}^N - H$ such that the fibre $\pi^{-1}(z)$ has dimension k. (If this set is empty, we set $\phi(k) = -\infty$.) If δ is sufficiently small, then the homomorphism induced* *by inclusion,* $\pi_i(\pi^{-1}(H_\delta)) \to \pi_i(X)$ *is an isomorphism for all* $i < \hat{n}$ *and is a surjection for* $i = \hat{n}$, *where*

$$\hat{n} = n - \sup_k (2k - (n - \phi(k)) + \inf(\phi(k), c-1)) - 1.$$

Furthermore: (1) In this theorem, π is not necessarily proper, and $\pi^{-1}(H_\delta)$ may be replaced by $\pi^{-1}(H)$ if H is generic or if π is proper.

(2) The assumption that X is algebraic may be replaced by the assumption that X is the complement of a closed subvariety of a complex analytic variety $\bar{X}$ and that π extends to a proper analytic map $\bar{\pi}: \bar{X} \to \mathbb{CP}^N$.

Proof. The proof will appear in Sect. 5.1.

Remark. The fibre dimension estimates in the above formula are sharp: see Sect. 8.1 for counterexamples.

Special cases. (1) This implies *Bertini's theorem* [Ber], [D1]. Let Y be an irreducible algebraic variety and $\pi: Y \to \mathbb{CP}^N$ be an algebraic morphism. Fix $c < \dim \overline{\pi(Y)}$. Let $H \subset \mathbb{CP}^N$ be a generic linear subspace of codimension c. Then $\pi^{-1}(H)$ is irreducible. If Y is locally irreducible as a complex analytic space, then $\pi_1(\pi^{-1}(H)) \to \pi_1(Y)$ is a surjection.

To see this, apply the theorem to $X = Y - Y'$, where Y' is a subvariety of Y which contains the singularities of Y and also contains the set

$$\{y \in Y \mid \dim \pi^{-1}\pi(y) > \dim(Y) - \dim(\pi(Y))\}.$$

Thus, X and $X \cap \pi^{-1}(H)$ are nonsingular (since H may be chosen transverse to X), and the map $X \to \pi(X)$ has equidimensional fibres of dimension $\alpha = \dim(Y) - \dim(\pi(Y))$. Thus, $\hat{n} \geq 1$, so $\pi_0(X \cap \pi^{-1}(H)) = \pi_0(X)$ and $\pi_1(X \cap \pi^{-1}(H)) \to \pi_1(X)$ is a surjection. The first implies that $X \cap \pi^{-1}(H)$ is connected (i.e., $\pi^{-1}(H)$ is irreducible) while the second implies that the composition

$$\pi_1(X \cap \pi^{-1}(H)) \to \pi_1(X) \to \pi_1(Y)$$

is a surjection, if Y is locally irreducible. However, this homomorphism factors through $\pi_1(\pi^{-1}(H))$, which proves that $\pi_1(\pi^{-1}(H)) \to \pi_1(Y)$ is surjective if Y is locally irreducible. □

(2) Recall that if X is nonsingular, a proper surjective algebraic map $\pi: X \to Y$ is *small* if

$$\operatorname{cod}\{y \in Y \mid \dim f^{-1}(y) \geq r\} > 2r$$

and is *semismall* if

$$\operatorname{cod}\{y \in Y \mid \dim f^{-1}(y) \geq r\} \geq 2r.$$

The Étale cohomology version of Theorem 1.1 was proved by Artin [Art] in the case that π is proper and semismall. The same method of proof works for singular cohomology. By carefully analyzing Artin's method, Deligne (unpublished) arrived at the singular cohomology version of Theorem 1.1 and conjectured the homotopy version. (See also Sect. 1.2 for the case that π is finite.)

(3) We remark that Theorem 1.1 remains valid if we replace the projective space $\mathbb{CP}^N$ by affine space $\mathbb{C}^N$, provided H is generic, i.e., we get a (relative)

Lefschetz theorem for (maps to) an affine algebraic variety because: If $\pi': X \to \mathbb{C}^N$ is an algebraic map and if $i: \mathbb{C}^N \to \mathbb{CP}^N$ denotes the inclusion, then Theorem 1.1 may be applied to the composition $\pi = i \circ \pi': X \to \mathbb{CP}^N$.

Remarks. (1) It is not possible to prove this result by induction on c, the codimension of H.

(2) If $c=1$, then it is possible to replace the linear space H with an arbitrary hypersurface W of (codimension one), because for any such W there exists an embedding $g: \mathbb{CP}^N \to \mathbb{CP}^{N'}$ such that $W = g^{-1}(L)$, where L is a linear subspace of codimension 1.

1.1*. Homotopy Dimension with Large Fibres

Theorem. *Let X be an n-dimensional (possibly singular) complex analytic variety. Let $\pi: X \to \mathbb{CP}^N - H$ be a proper analytic map, where H is a linear subspace of codimension c. Let $\phi(k)$ denote the dimension of the set of points $y \in \pi(X)$ such that the fibre $\pi^{-1}(y)$ has dimension k. (If this set is empty, we set $\phi(k) = -\infty$.) Then X has the homotopy type of a CW complex of (real) dimension less than or equal to*

$$\hat{n}^* = n + \sup_k (2k - (n - \phi(k)) + \inf(\phi(k), c-1)).$$

Proof. The proof will appear in Sect. 5.1*.

Remark. The estimates on the fibre dimension are sharp: see Sect. 8.1* for counterexamples.

Special cases. An affine variety X (or a Stein space X) of complex dimension n has the homotopy type of a CW complex of dimension n. This was conjectured by Kato [Kt1], [Kt2], was proved by Karchyauskas [Kr2], [Kr3] (in the complex algebraic case) and by H. Hamm (in the case of a Stein space) [H3], corrected in [H5], and follows from Theorem 1.1* by setting $\phi(k)=0$ and $c=1$. L. Kaup [Ku1] and Narasimhan [N] had previously shown that the homology groups of X vanished in dimensions greater than n and that $H_n(X; \mathbb{Z})$ was torsion free. (We show in Sect. 6.9 that the same result holds for the intersection homology, i.e., $IH_i(X; \mathbb{Z}) = 0$ for $i < n$ and $IH_n(X); \mathbb{Z})$ is torsion free.) The homology of a nonsingular affine variety X was done by Andreotti and Frankel [AF1], following Thom's suggestion that Morse theory could be used. (See also Milnor [Mi].) There are various refinements of this result which would have implications for intersection homology. For example, we have the following:

Conjecture. A Whitney stratified complex n-dimensional affine variety X deformation retracts (by a stratum preserving retraction) to a (real) n-dimensional Whitney stratified subset of X which intersects each stratum S of X in a subset of dimension $s = \dim_{\mathbb{C}}(S)$. Considerable progress on this conjecture has been made by [Lê2].

1.2. Lefschetz Theorem with Singularities

An outline of proof for the following theorem was published in [GM1]. See also [H3] or [HL2] for a similar result, but where the assumptions on the number of equations has been replaced with assumptions on the rectified homotopical depth (see [Gro] exp. XIII).

Theorem. *Let X be an algebraic subvariety of some algebraic manifold M. Let $\pi: X \to \mathbb{CP}^N$ be a (not necessarily proper) algebraic map with finite fibres. Let H be a linear subspace of codimension c in $\mathbb{CP}^N$, and let H_δ be a δ-neighborhood of H (with respect to some real analytic Riemannian metric, as in Theorem 1.1). Let $\phi(k)$ denote the dimension of the set of points $p \in X - \pi^{-1}(H)$ such that a neighborhood (in X) of p can be defined (in M) by k equations and no fewer. (If this set is empty, we set $\phi(k) = -\infty$.) If $\delta > 0$ is sufficiently small, then the homomorphism*

$$\pi_i(\pi^{-1}(H_\delta)) \to \pi_i(X)$$

is an isomorphism for all $i < \hat{n}$ and is a surjection for $i = \hat{n}$, where

$$\hat{n} = \inf_k (\dim_{\mathbb{C}}(M) - k - \inf(\phi(k), c-1)) - 1.$$

Furthermore: (1) It is possible to replace $\pi^{-1}(H_\delta)$ by $\pi^{-1}(H)$ if H is generic or if π is proper.

(2) The assumption that X is algebraic may be replaced by the assumption that X is the complement of a closed subvariety of an analytic variety $\bar{X}$ and that π is an analytic map which extends to a proper finite analytic map $\bar{\pi}: \bar{X} \to \mathbb{CP}^N$.

(3) If X is purely n-dimensional (and has arbitrary singularities) and H is generic, then the homeomorphism

$$IH_i(\pi^{-1}(H); \mathbb{Z}) \to IH_i(X; \mathbb{Z})$$

is an isomorphism for all $i < n - c$ and is a surjection for $i = n - c$. (Here, IH_* denotes the middle intersection homology with compact supports [GM3], [GM4].)

Proof. The proof will appear in Sect. 5.2, and the intersection homology part will appear in Sect. 6.10.

Remarks. The numerical estimates above are sharp: see Sect. 8.2 for counterexamples. The Lefschetz theorem is false for constructible sets: see Sect. 8.3. It is not possible to find a simple common generization of Theorems 1.1 and 1.2 by adding fibre and singularity defects: see Sect. 8.4.

Special cases. The following special case of Theorem 1.2 appears in Hamm [H4]: Let X be an n-dimensional projective algebraic variety and suppose that $Z \subset X$ is a subvariety and H is a linear hyperplane in the ambient projective space. If $X - (Z \cup H)$ is a local complete intersection, then the homomorphism

$$\pi_i((X - Z) \cap H) \to \pi_i(X - Z)$$

is an isomorphism for all $i<n-1$ and is a surjection for $i=n-1$. This generalizes the papers of Kaup [Ku3], [KW], [GK], who proved analogous results with $Z=\phi$ and for homology instead of homotopy. The case of nonsingular X was proven by Hamm and Lê [HL1], following Zariski's theorem [Z] for surfaces. See also Oka [Ok] and Kato [Kt1], [Kt2]. Similar results appear in Ogus [Og] for varieties defined over fields of positive characteristic. The case $Z=\phi$ and X nonsingular is the "classical" Lefschetz theorem, the nicest proof of which is due to Thom (see Andreotti and Frankel [AF1] and Milnor [Mi]).

We remark that Theorem 1.2 remains valid if we replace the projective space $\mathbb{CP}^N$ by affine space, $\mathbb{C}^N$, provided H is generic. (However, H must be generic, or at least transversal to the strata at infinity of the closure $\bar{X}$; see Sect. 8.2 for counterexamples.) Thus, we obtain Lefschetz theorems for the homotopy groups, homology, and intersection homology groups of affine algebraic varieties (see [H4]). (If $\pi'\colon X\to\mathbb{C}^N$ is an algebraic map and $i\colon \mathbb{C}^N\to\mathbb{CP}^N$ is the inclusion, then Theorem 1.2 may be applied to the composition $\pi=i\circ\pi'\colon X\to\mathbb{CP}^N$).

1.2*. Homotopy Dimension of Nonproper Varieties

Theorem. *Let X be a complex n-dimensional analytic variety and let $\pi\colon X\to\mathbb{CP}^N-H$ be a proper finite analytic map, where H is a linear subspace of codimension c. Let W be an analytic subvariety of X. We consider the extent to which the inclusion $W\subset X$ fails to be a local complete intersection morphism by defining for each k the number $\phi(k)$ to be the dimension of the set of all points $p\in W$ such that a neighborhood of p (in W) can be defined (as a subset of X) by $n-\dim_p(W)+k$ equations, and no fewer. (If this set is empty, we set $\phi(k)=-\infty$.) Then the space $X-W$ has the homotopy type of a CW complex of dimension $\leq\hat{n}^*$, where*

$$\hat{n}^*=\sup_{k\geq 1}(n+k-1+\inf(\phi(k),c-1)).$$

Proof. The proof will appear in Sect. 5.2*.

Remark. These estimates are sharp: see Sect. 8.2*.

Examples. If X is an affine n-dimensional variety and $W\subset X$ is a local complete intersection morphism of codimension k, then the complement $X-W$ has the homotopy type of a CW complex of dimension $n+k-1$. (If W is a hypersurface, then this is obvious since in this case, $X-W$ is also an affine algebraic variety.)

1.3. Local Lefschetz Theorems

The Lefschetz theorems of Sects. 1.1 and 1.2 are special cases of a more general result: the local Lefschetz theorems (see [Gro] or [HL3]). In its simplest form this theorem applies to an isolated singularity p of an n-dimensional complex algebraic subvariety X of some algebraic manifold P. If $L=X\cap\partial B_\delta(p)$ denotes the link (in X) of the point p and if H denotes a hyperplane section of X (with respect to some local coordinate system on P) which contains the singular point p, and is generic among all hyperplane sections which contain the point p, then the homomorphism

$$\pi_i(L \cap H) \to \pi_i(L)$$

is an isomorphism for all $i < n-2$ and is a surjection for $i = n-2$. This result is local near the point p, and there are no projectivity assumptions on X. (Thus, the manifold P may be replaced with an open subset of $\mathbb{C}^N$, with no loss in generality.) However, the usual Lefschetz theorem on hyperplane sections follows from this local result: suppose Y is a nonsingular projective algebraic variety and $\tilde{H}$ is a generic hyperplane in projective space. Let X denote the (complex) cone on Y, with conepoint p, and let H denote the cone on $\tilde{H}$. It follows that p is an isolated singular point of X, and that the link L of p in X is a circle bundle over Y. Similarly $H \cap L$ is a circle bundle over $\tilde{H}$. Consider the following diagram of exact sequences in homotopy for these fibre bundles:

$$\begin{array}{ccccccccccccc}
\longrightarrow & \pi_{i+1}(L \cap H) & \longrightarrow & \pi_{i+1}(Y \cap \tilde{H}) & \longrightarrow & \pi_i(S^1) & \longrightarrow & \pi_i(L \cap H) & \longrightarrow & \pi_i(Y \cap \tilde{H}) & \longrightarrow & \pi_i(S^1) & \longrightarrow \\
& \downarrow & & \downarrow & & \downarrow & & \downarrow & & \downarrow & & \downarrow & \\
\longrightarrow & \pi_{i+1}(L) & \longrightarrow & \pi_{i+1}(Y) & \longrightarrow & \pi_i(S^1) & \longrightarrow & \pi_i(L) & \longrightarrow & \pi_i(Y) & \longrightarrow & \pi_i(S^1) & \longrightarrow
\end{array}$$

Applying the five lemma together with the local Lefschetz theorem for the homomorphism $\pi_i(L \cap H) \to \pi_i(L)$ gives the (usual) Lefschetz theorem, i.e., the homomorphism $\pi_i(Y \cap H) \to \pi_i(Y)$ is an isomorphism for all $i < \dim(Y) - 1$ and is a surjection for $i = \dim(Y) - 1$.

In this section X will denote a complex algebraic subvariety of some nonsingular variety M, and $\pi: X \to P$ will be a complex algebraic map, where P is a nonsingular algebraic variety. Fix $p \in \overline{\pi(X)}$ and let $\partial B_\delta(p)$ denote the boundary of a ball of radius δ about the point p (with respect to some Riemannian metric on P). Let H be an affine linear subspace of codimension c in P (with respect to some local coordinate system about p) which passes through the point p, and let H_ε denote an ε-neighborhood of H, with respect to some real analytic Riemannian metric on P.

Theorem 1. *Suppose X is nonsingular, connected, and purely n-dimensional. Let $\phi(k)$ denote the dimension of the set of points $z \in P - H$ such that the fibre $\pi^{-1}(z)$ has dimension k. (If this set is empty, we set $\phi(k) = -\infty$.) If δ is sufficiently small, then for any $\varepsilon > 0$ sufficiently small, the homomorphism induced by inclusion,*

$$\pi_i(X \cap \pi^{-1}(\partial B_\delta(p) \cap H_\varepsilon)) \to \pi_i(X \cap \pi^{-1}(\partial B_\delta(p)))$$

is an isomorphism for all $i < \hat{n}$ and is a surjection for $i = \hat{n}$, where

$$\hat{n} = n - \sup_k (2k - (n - \phi(k)) + \inf(\phi(k), c-1)) - 2.$$

Furthermore: (1) If H is a generic affine subspace or if π is proper, then the neighborhood H_ε may be replaced by H in the above formula.

(2) The assumption that X is algebraic may be replaced by the assumption that P is a nonsingular analytic variety, X is the complement of a closed subvariety of a complex analytic variety $\bar{X}$, and that π extends to a proper analytic map $\bar{\pi}: \bar{X} \to P$.

Proof. The proof will appear in Sect. 5.3.

Remark. The subspace H may be replaced by an arbitrary complete intersection of codimension c, because such a complete intersection may be locally realized as a linear subspace H' by composing with an embedding $g: P \to P'$. In this case, if π is proper or if H is generic (with respect to the local general linear group action at the point p, in some local coordinate system), then the neighborhood H_ε may be replaced by H. Otherwise, the neighborhood H_ε must be taken to be a neighborhood of the form $g^{-1}(H'_\varepsilon)$, where H'_ε is an ε-neighborhood of the linear subspace $H' \subset P'$.

Theorem 2. *Suppose π is finite (but not necessarily proper). Let $\phi(k)$ denote the dimension of the set of points $x \in X - \pi^{-1}(H)$ such that a neighborhood of x (in X) can be defined by k equations, and no fewer. (If this set is empty, we set $\phi(k) = -\infty$.) If δ is sufficiently small, then for all $\varepsilon > 0$ sufficiently small, the homomorphism induced by inclusion*

$$\pi_i(X \cap \pi^{-1}(\partial B_\delta(p) \cap H_\varepsilon)) \to \pi_i(X \cap \pi^{-1}(\partial B_\delta(p)))$$

is an isomorphism for all $i < \hat{n}$ and is a surjection for $i = \hat{n}$, where

$$\hat{n} = \inf_k (\dim_{\mathbb{C}}(M) - k - \inf(\phi(k), c-1))) - 2.$$

Furthermore: (1) If H is a generic subspace or if π is proper, then the neighborhood H_ε may be replaced by H in the above formula.

(2) The assumption that X is algebraic may be replaced by the assumption that P is an analytic variety, X is the complement of a closed subvariety of a complex analytic variety $\bar{X}$, and that π extends to a proper analytic map $\bar{\pi}: \bar{X} \to P$.

(3) *If* X is purely n-dimensional (with arbitrary singularities) and H is generic, then the homomorphism

$$IH_i(X \cap \pi^{-1}(\partial B_\delta(p) \cap H); \mathbb{Z}) \to IH_i(X \cap \pi^{-1}(\partial B_\delta(p)); \mathbb{Z})$$

is an isomorphism for all $i < n - c - 1$ and is a surjection for $i = n - c - 1$. (Here, IH_i denotes the middle intersection homology with compact supports.)

Proof. The proof will appear in Sect. 5.3, and the intersection homology part of the proof will appear in Sect. 6.

Special cases. An outline of a proof for the following special case of Theorem 2 appears in Hamm [H4]: Suppose that X is an n-dimensional local complete intersection properly embedded in $\mathbb{C}^N$ or $\mathbb{CP}^N$. Let H be a linear hyperplane, $p \in X \cap H$, and let $L_p(X) = X \cap \partial B_\delta(p)$ denote the intersection of X with the boundary of a ball $B_\delta(p)$ of sufficiently small radius δ which is centered at p. Then the homomorphism

$$\pi_i(L_p(X) \cap H) \to \pi_i(L_p(X))$$

is an isomorphism for all $i<n-2$ and is a surjection for $i=n-2$.

Similar results for homology (in place of homotopy) are proven in Kaup [Ku3]. See also Lê [Lê1]. An important related result is the theorem of Hamm [H1], [H2] (see Sects. 4.6 and 4.6*) on the local connectivity of local complete intersections.

Remark. The subspace H may be replaced by an arbitrary complete intersection of codimension c, because such a complete intersection may be locally realized as a linear subspace H' by composing with an embedding $g: P \to P'$. In this case, if π is proper or if H is generic (with respect to the local general linear group action at the point p, in some local coordinate system), then the neighborhood H_ε may be replaced by H. Otherwise, the neighborhood H_ε must be taken to be a neighborhood of the form $g^{-1}(H'_\varepsilon)$, where H'_ε is an ε-neighborhood of the subspace $H' \subset P'$.

1.3*. Local Homotopy Dimension

In this section we suppose that X is an n-dimensional connected analytic subvariety of some nonsingular analytic variety M, and that $\pi: X \to P$ is an analytic map to some nonsingular variety P. Fix $p \in Z = \pi(X)$ and let ∂B_δ denote the boundary of a ball of radius δ about the point p (with respect to some Riemannian metric on P). Let H be an affine linear subspace of codimension c in P (with respect to some local coordinate system about p) which passes through the point p.

Theorem 1. *Suppose π is proper. Let $\phi(k)$ denote the dimension of the set of all points $z \in Z$ such that the fibre $\pi^{-1}(z)$ has dimension k. (If this set is empty, we set $\phi(k) = -\infty$.) If δ is chosen sufficiently small, then the space*

$$\pi^{-1}(Z \cap \partial B_\delta(p) - H)$$

has the homotopy type of a CW complex of dimension less than or equal to

$$\hat{n} = n + \sup_k (2k - (n - \phi(k)) + \inf(\phi(k), c-1)).$$

Proof. The proof will appear in Sect. 5.3*.

Theorem 2. *Suppose that X is an n-dimensional complex analytic subvariety of some complex analytic manifold M. Fix a subvariety $W \subset X$ which is locally determined in X by k equations. Fix a point $p \in W$ and let $\partial B_\delta(p)$ denote the boundary of a small of radius δ which is centered at the point p. Define*

$$L_p(X) = \partial B_\delta(p) \cap X \qquad L_p(W) = \partial B_\delta(p) \cap W.$$

If δ is sufficiently small, then the space $L_p(X) - L_p(W)$ has the homotopy type of a CW complex of dimension $\leq n+k-1$.

Proof. The proof will appear in Sect. 4.6*.

Special cases. Related results appear in Hamm [H1], [H2] and Hamm [H4]. A slightly more refined version of Theorem 2 may be true, which would have consequences for intersection homology:

Conjecture. The space $L_p(X)-L_p(W)$ deformation retracts (by a stratum preserving deformation) to a Whitney stratified subset of $L_p(X)$ which intersects each stratum S of X in a subset of dimension $\leq s+k-1$, where $s=\dim_{\mathbb{C}}(S)$. See also [Lê2].

Chapter 2. Normal Morse Data for Complex Analytic Varieties

2.0. Introduction

In this chapter we describe the local topological structure of a complex analytic variety and a generic complex analytic function on that variety. Most of the material described in this section is fairly well known, see for example [Mi2], [Du], [H1], [H2], [HL3], [LK], [Kp4], [Lê3], [Lê4], [LT1]. However the proofs we give here are rigorous and are easy, given the technique of "moving the wall" which was developed in Part I.

The main application of this section is to the analysis of *normal* Morse data of a real valued function defined on a complex analytic variety. We will show that the homeomorphism type of the link of a singular point and of the normal Morse data at that point are determined by the complex link $\mathscr{L}$ and the monodromy homeomorphism $\mu: \mathscr{L} \to \mathscr{L}$. In Part I it was shown that the normal Morse data of a nondepraved critical point depends only on the differential of the Morse function, so throughout this chapter we will fix a particular nondegenerate covector ω and study the normal Morse data associated to it. The key trick (Sect. 2.1.4) in complex Morse theory is to realize that this covector ω is also the differential of the real part of a complex analytic function. This allows us to replace any smooth Morse function (locally) with a complex analytic function when analyzing the normal Morse data.

This local analysis is made in terms of the *complex link* $\mathscr{L}$ of an i-dimensional stratum of a Whitney stratified complex analytic subset of n-space. This is (roughly) the intersection of a small tubular neighborhood of that stratum with a nearby generic plane of dimension $n-i-1$.

The complex link of a stratum S of a singular variety X is obtained from a generic projection $f: X \to \mathbb{C}$ in the same way that the Milnor fibre [Mi2] of a hypersurface $Y = g^{-1}(0)$ is obtained from a singular projection $g: X' \to \mathbb{C}$ of a nonsingular variety X'. In particular, there is a related Milnor fibration and monodromy homeomorphism $\mu: \mathscr{L} \to \mathscr{L}$. The complex link together with this monodromy homeomorphism determine the link of the stratum S (Sect. 2.4; the technicalities in the proof are actually carried out in Part I, Chap. 12)), but the complex link has the added interesting property that it is a complex analytic space, which admits a canonical Morse function, so it can be studied using complex Morse theory. Such a study is carried out in Chap. 4. We believe that the complex link, together with its complex structure and monodromy homeomorphism constitute the local "complex" nature of the singularities of X.

2.1. Nondegenerate Covectors

In this chapter we will study a complex analytic subvariety Z of some complex analytic manifold M. We fix a (complex analytic) Whitney stratification of Z, whose strata are indexed by some partially ordered set $\mathscr{S}$. We fix a stratum S of Z and let $\pi: T_S^* M \to S$ denote the conormal bundle of S in M whose fibre $\pi^{-1}(p) = T_{S,p}^* M$ consists of all covectors $\omega \in T_p^* M$ such that $\omega(T_p S) = 0$. (Recall from Part I, Sect. 1.8 that

$$T_p^* M = \operatorname{Hom}_{\mathbb{R}}(T_p M, \mathbb{R}) = \operatorname{Hom}_{\mathbb{C}}(T_p M, \mathbb{C})$$

and in particular that $T_S^* M \to S$ is a complex vectorbundle.)

2.1.1. Definition. The set of *nondegenerate conormal vectors*, is the set

$$C_S = \{\omega \in T_S^* M \mid \omega(Q) \neq 0 \text{ for any generalized tangent space } Q \neq T_p S\}$$

where a generalized tangent space is any limit of tangent planes from any stratum $R > S$ in Z (see Part I, Sect. 1.8).

2.1.2. Remark. By Part I, Proposition 1.8, the set C_S is the complement of the complex codimension one subvariety $V \subset T_S^* M$ of degenerate covectors. The zero section is contained in V. By [Te1], or [HM] and [NT], for any point $p \in S$, the intersection $V \cap T_{S,p}^* M$ of V with the conormal space of S at p consists of a subvariety of the conormal space, which has codimension ≥ 1, i.e., there are no "exceptional points" (see Part I, Sect. 1.8).

Fix a nondegenerate covector ω at a point $p = \pi(\omega)$.

2.1.3. Definition. A quintuple $(N', f, r, \delta, \varepsilon)$ is *normal projection data* for ω at p if:

(a) N' is a complex analytic submanifold of M which meets the stratum S transversally at a single point, p. We define the *normal slice* $N = N' \cap Z$ and we canonically identify the spaces $T_{S,p}^* M \to T^* N'$.

(b) $f: N' \to D^0 \subset \mathbb{C}$ is a proper complex analytic map to the unit disk such that $f(p) = 0$ and $d(\operatorname{Re}(f))(p) = \omega$ under the above canonical identification.

(c) r is a Riemannian metric on M. By abuse of notation we will also denote by $r(z)$ the square of the distance (p, z) in this metric.

(d) $\delta > 0$ is so small that:

(i) $N \cap B_\delta(p)$ is compact, where

$$B_\delta(p) = \{q \in M \mid r(q) \leq \delta\}$$

(ii) $\partial B_\delta(p) = \{q \in M \mid r(q) = \delta\}$ is transverse to each stratum of N.

(iii) The same holds for every $\delta' \leq \delta$, i.e., for each stratum A of Z, the restriction $r \mid (A \cap N')$ has no critical values in the interval $(0, \delta]$.

(e) $\varepsilon > 0$ is so small that

(i) For every stratum A of Z, the restriction $f \mid A \cap N'$ has no critical values in the disk $D_\varepsilon \subset \mathbb{C}$, except for the isolated critical value 0.

(ii) For every stratum A of Z and for any point $z \in A \cap N' \cap \partial B_\delta(p)$, if $|f(z)| \leq \varepsilon$ then the complex linear map

$$(dr(z), df(z)) \mid T_z(A \cap N'): T_z(A \cap N') \to \mathbb{C}^2$$

has rank 2.

(iii) For any $\delta' \leq \delta$ there exists $\varepsilon' \leq \varepsilon$ such that for any stratum A of Z and for any point $z \in A \cap N' \cap \partial B_{\delta'}(p) \cap f^{-1}(D_\varepsilon)$, the map

$$(dr(z), df(z)) \mid T_z(A \cap N'): T_z(A \cap N') \to \mathbb{C}^2$$

has rank 2.

2.1.4. Lemma. *Normal projection data exists for any $\omega \in C_S$.*

Proof. By choosing local coordinates on N' in some neighborhood of the point p, we may replace the manifold N' by Euclidean space $\mathbb{C}^m$. The covector ω gives rise to a complex analytic function f by

$$f(z) = \omega(z) - i\omega(iz)$$

such that $d(\mathrm{Re}(f))(p) = \omega$. Choices for ε and δ exist by Part I, Sect. 12.2 provided $\varepsilon \ll \delta$. In fact, given N', f, and r as above, there is a fringed set $A \subset \mathbb{R}^+ \times \mathbb{R}^+$ of type $0 < \delta \ll \varepsilon \ll 1$ so that the map

$$(r, f) \mid (Z \cap N'): Z \cap N' \to \mathbb{R} \times \mathbb{C}$$

has no characteristic covectors in the region

$$A^\# = \{(u, v) \in \mathbb{R}^+ \times \mathbb{C} \mid (u, |v|) \in \bar{A}\}.$$

2.2. The Complex Link and Related Spaces

The complex link is the central object of study in complex Morse theory. It is a well-known object which is analogous to the "Milnor fibre", and was studied in depth in [LK]. See also [LT1], [H4], [Db1]. The complex link is closely related to the *polar varieties* of [LT1]. See [LT2] for formulas relating the Euler characteristic of the complex links to the multiplicities of the polar varieties. The Euler characteristic of the complex link is also related to the *local Euler obstruction* of [MP1] (see [LT2], or [Db1], [Db2]).

Choose a nondegenerate covector $\omega \in C_S$ at in Sect. 2.1, and choose a set $\{N', f, r, \delta, \varepsilon\}$ of normal projection data. Define $N = N' \cap Z$ and let $\xi = \varepsilon + 0i \in \mathbb{C}$.

Definition. The complex link $\mathcal{L}$ and its boundary $\partial\mathcal{L}$ are the spaces

$$\mathcal{L} = f^{-1}(\xi) \cap N \cap B_\delta(p) \qquad \partial\mathcal{L} = f^{-1}(\xi) \cap N \cap \partial B_\delta(p).$$

We also define the following related spaces:

(0) The disk and normal ball:

$$\begin{aligned} D_\varepsilon &= \{\zeta \in \mathbb{C} \mid |\zeta| \leq \varepsilon\} & \partial D_\varepsilon &= \{\zeta \in \mathbb{C} \mid |\zeta| = \varepsilon\} \\ N_\delta &= B_\delta(p) \cap N & \partial N_\delta &= \partial B_\delta(p) \cap N \\ D_\varepsilon^0 &= D_\varepsilon - \partial D_\varepsilon & N_\delta^0 &= N_\delta - \partial N_\delta. \end{aligned}$$

(1) The cylindrical neighborhood of p, its interior, and boundary:

$$C = f^{-1}(D_\varepsilon) \cap N_\delta \qquad C^0 = f^{-1}(D_\varepsilon^0) \cap N_\delta^0$$
$$L = \partial C = C - C^0.$$

(2) The horizontal and vertical parts of the link:

$$L_h = f^{-1}(D_\varepsilon) \cap \partial N_\delta \qquad L_h^0 = f^{-1}(D_\varepsilon^0) \cap \partial N_\delta$$
$$L_v = f^{-1}(\partial D_\varepsilon) \cap N_\delta \qquad L_v^0 = f^{-1}(\partial D_\varepsilon) \cap N_\delta^0$$
$$\partial L_h = \partial L_v = f^{-1}(\partial D_\varepsilon) \cap \partial N_\delta.$$

(3) The cut off spaces:

$$C_{<0} = f^{-1}\{\zeta \mid \mathrm{Re}(\zeta) < 0\} \cap C \qquad C_{<0}^0 = C_{<0} \cap B_\delta^0(p)$$
$$L_{<0} = C_{<0} \cap L.$$

Each of these spaces is $\mathscr{S}$-decomposed by its intersection with the strata of N (which are also the strata of Z) and is canonically Whitney stratified since each is a transversal intersection of Whitney stratified spaces.

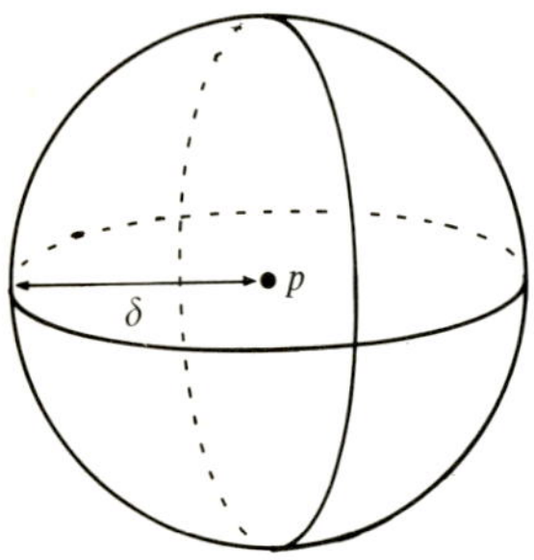

The normal ball $B_\delta(p)$

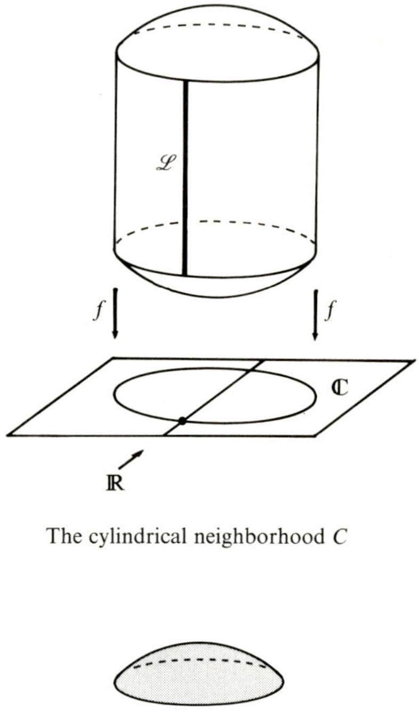

The cylindrical neighborhood C

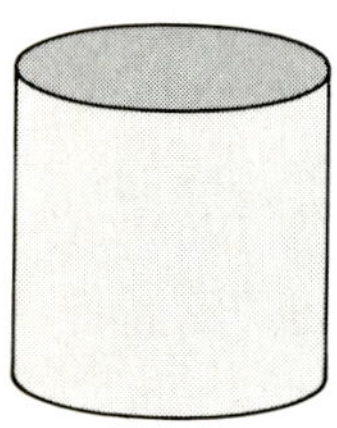

The vertical part L_v of the link

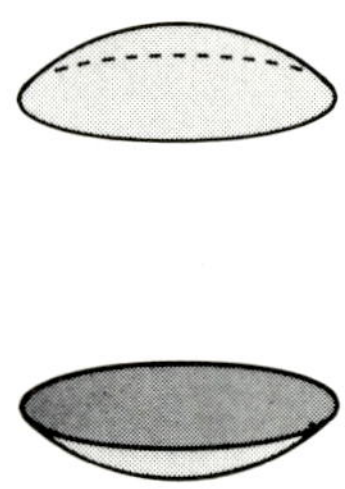

The horizontal part L_h of the link

2.3. The Complex Link is Independent of Choices

Suppose ω_0 and ω_1 are nondegenerate covectors in $T_S^* M$ and suppose that $p_0 = \pi(\omega_0)$ and $p_1 = \pi(\omega_1)$ lie in the same connected component of the stratum S. (By abuse of notation we now denote this connected component by S.) Let $(N_0', f_0, r_0, \delta_0, \varepsilon_0)$ be normal projection data for ω_0 and let $(N_1', f_1, r_1, \delta_1, \varepsilon_1)$ be normal projection data for ω_1. Let T_0 denote any one of the twenty spaces defined in Sect. 2.2, as constructed with respect to the choices $(\omega_0, N_0', f_0, r_0, \delta_0, \varepsilon_0)$ and let T_1 denote the corresponding space as defined with respect to the choices $(\omega_1, N_1', f_1, r_1, \delta_1, \varepsilon_1)$.

Theorem. There exists a (nonnatural) $\mathscr{S}$-decomposition preserving homeomorphism $H: T_0 \to T_1$. If $p_0 = p_1$ and if $\omega_0 = \omega_1$, then this homeomorphism H has a natural choice up to isotopy.

This theorem follows immediately from three lemmas which we now state:

2.3.1. Lemma. *The space $C_S \subset T_S^* M$ of nondegenerate conormal vectors is smoothly path connected.*

2.3.2. Lemma. *It is possible to associate to each $\omega \in C_S$ a set of normal projection data $\{N_\omega', f_\omega, r_\omega, \delta_\omega, \varepsilon_\omega\}$ by an association which is "smooth" in the following sense: Let $T_\omega \subset Z$ denote the space corresponding to T which is determined by the choices $\{N_\omega', f_\omega, r_\omega, \delta_\omega, \varepsilon_\omega\}$. Let*

$$\tilde{T} = \{(q, \omega) \in M \times C_S \mid q \in T_\omega\}$$

be the family of these T_ω. Then the projection to the second factor, $\tilde{T} \to C_S$ is a (locally trivial) stratified submersion.

2.3.3. Lemma. *Fix $\omega \in C_S$ and $p = \pi(\omega)$. If $(N_0', f_0, r_0, \delta_0, \varepsilon_0)$ and $(N_1', f_1, r_1, \delta_1, \varepsilon_1)$ are two choices of normal projection data for ω, then the corresponding spaces T_0 and T_1 are homeomorphic by a homeomorphism H which has a canonical choice up to isotopy.*

Proof of Lemma 2.3.1. This is just proposition Part I, Sect. 1.8 and the fact that S is connected. □

Proof of Lemma 2.3.3. This is just Part I, Sect. 12.4 with the slight difference that here we are allowing the function to vary, but we are fixing its differential at p, whereas in Part I, Sect. 12.4 the function was fixed once and for all. But, the required modification is trivial. □

Proof of Lemma 2.3.2. First we recall the complex analytic tubular neighborhood theorem: there is a neighborhood U of the zero section of the normal bundle $\theta: TM/TS \to S$ and a smooth embedding $\phi: U \to M$ so that $\phi | S =$ identity, and so that for each $p \in S$ the fibre $\theta^{-1}(p) \cap U$ is embedded by ϕ as a complex analytic submanifold of M which is transverse to S at the point p. This allows us to define the normal slices N_ω and the local normal functions f_ω as follows:

$$N_\omega' = \phi(\theta^{-1}(\theta(\omega))$$
$$f_\omega = \omega \circ (\phi^{-1} | N_\omega').$$

Take the Riemannian metric r_ω to be any fixed metric (independent of ω). It remains to find choices for δ_ω and ε_ω.

Consider the set

$$B \subset C_S \times \mathbb{R}^+ \times \mathbb{R}^+$$

consisting of triples $(\omega, \delta, \varepsilon)$ such that the quintuple $(N'_\omega, f_\omega, r_\omega, \delta, \varepsilon)$ is normal projection data for ω. (This is a condition on δ and ε: see Sect. 2.1 (d) and (e).) It is easy to see that B is an open set (since Sect. 2.1 (d) and (e) are open conditions and ω is nondegenerate). Thus, B is a fringed set parameterized by the (noncompact) manifold C_S. By Part I, Proposition Sect. 5.5 there is a section $s: C_S \to B$ of the projection $\pi_1: B \to C_S$, i.e., a way of associating to each $\omega \in C_S$ a pair $(\delta_\omega, \varepsilon_\omega) \in \mathbb{R}^+ \times \mathbb{R}^+$ with the required properties.

The local triviality of the map $\tilde{T} \to C_S$ now follows from Thom's isotopy lemma (Part I, Sect. 1.5). In fact, it is a (fibrewise) transversal intersection of Z with various smooth manifolds and manifolds with boundaries. □

Remark. The space C_S is very much noncompact. As one approaches a point $\omega \in C_S$ near the "edge" of C_S, the values of δ_ω and ε_ω may shrink rapidly.

2.4. Local Structure of Analytic Varieties

The following statements, (a), (b), (c), and (d), were announced without proof in [GM3]. They are the key technical lemmas which allow us to analyze the normal Morse data for complex analytic varieties. Similar results have appeared in the literature in varying degrees of generality and detail, beginning with Milnor [Mi2] (in the case of isolated singularities), and followed by [Lê3] (for general singularities). See also [Du], [H1], [H4], [Lê1], [Lê4], [LT2]. We refer the reader to Sect. 8.5 for Hironaka's counterexamples to similar sounding statements.

For the purpose of applications to Lefschetz theorems and estimates on homotopy dimension (Sects. 5.1 and 5.2), we will need only part (d) of the following proposition, and its Corollaries 1 and 1* (in the relative case, Sect. 2.6). The local Lefschetz theorems and connectivity theorems (Sect. 5.3) will also use part (a) and part (c). The rest of the proposition is needed for the Lefschetz theorems in intersection homology (Sect. 6) [GM3]. Parts (a), (b), (c), and (d) of this proposition have analogous statements for an arbitrary proper complex analytic map to $\mathbb{C}$. This is discussed in the appendix, Sect. 2.A.

We use the notation established above, i.e., S is a stratum of Z, $\omega \in T_S^* M$ is a nondegenerate covector, $p = \pi(\omega)$ is a point in S, N' is a transverse submanifold through S at the point p and $N = N' \cap Z$ is a "normal slice", $f: N \to D^0 \subset \mathbb{C}$ has $d(\mathrm{Re}(f)(p) = \omega$, and $\varepsilon > 0$ and $\delta > 0$ are chosen so that (ε, δ) satisfy the conditions Sect. 2.1 (d) and (e). Set $f = f_1 + i f_2$.

We show in part (a) that the vertical part of the link is a fibre bundle over the circle, with fibre homeomorphic to the complex link. In part (e) we show that the horizontal part of the link is a product of a two-disk with the boundary of the complex link. The intersection of these two pieces of the link is collared in both pieces and the collared neighborhood is a trivial bundle over the circle. This collaring restricts in each fibre to a collaring of $\partial\mathscr{L}$ in $\mathscr{L}$. We also show (in parts (b) and (c)) that although the cylindrical neighborhood C is a different shape from the usual "conical neighborhood" N_δ, it is also conical, and, when cut off by the values $f_1^{-1}(\pm\eta)$ it gives the normal Morse data for the function f_1. However, it may be necessary to choose this η very much smaller than ε and δ. Finally, in part (d) we show that the halflink is topologically a product of the complex link with an interval, and this homeomorphism preserves boundaries.

Proposition. (a) *Milnor fibration* ([Mi2], [Lê3], [LT2]). The restriction $f\,|\,L_v\colon L_v \to \partial D_\varepsilon$ is a topological fibre bundle with fibre $\mathscr{L}$. In fact, the restriction $f\colon (C-f^{-1}(0)) \to D_\varepsilon - \{0\}$ is a fibre bundle with fibre $\mathscr{L}$.

(b) *The cylindrical neighborhood is conical.* For any $\eta>0$ sufficiently small, there are ($\mathscr{S}$-decomposition preserving) homeomorphisms

$$(C, L) \xrightarrow{G} (N_\delta, \partial N_\delta) \xrightarrow{F} (\operatorname{cone}(\partial N_\delta), \partial N_\delta)$$

with the following properties:

(i) G preserves the levels $f_1=-\eta$, $f_1=0$, $f_1=\eta$. In other words, $f_1(z)\in K \Leftrightarrow f_1\,G(z)\in K$, where K is any of the following sets:

$$(-\infty, -\eta), \quad \{-\eta\}, \ (-\eta, 0), \quad \{0\}, \ (0, \eta), \quad \{\eta\}, \ (\eta, \infty).$$

(ii) F preserves the level $f_1=0$. In other words, F takes the sets

$N_\delta \cap f_1^{-1}(-\infty, 0)$ homeomorphically to $\operatorname{cone}(\partial N_\delta \cap f_1^{-1}(-\infty, 0))$ − conepoint
$N_\delta \cap f_1^{-1}(0)$ homeomorphically to $\operatorname{cone}(\partial N_\delta \cap f_1^{-1}(0))$
$N_\delta \cap f_1^{-1}(0, \infty)$ homeomorphically to $\operatorname{cone}(\partial N_\delta \cap f_1^{-1}(0, \infty))$ − conepoint.

(c) *The cylindrical neighborhood gives Morse data.* For any $\eta>0$ sufficiently small, the pair $(C\cap f_1^{-1}[-\eta, \eta],\ C\cap f_1^{-1}(-\eta))$ is normal Morse data, i.e., is homeomorphic (by an $\mathscr{S}$-decomposition preserving homeomorphism) to the pair

$$(N_\delta \cap f_1^{-1}[-\varepsilon, \varepsilon],\ N_\delta \cap f_1^{-1}(-\varepsilon)).$$

(d) *The halflink is the complex link times an interval.* There are ($\mathscr{S}$-decomposition preserving) homeomorphisms of pairs,

$$(\ell^+, \partial\ell^+) \cong (\mathscr{L}, \partial\mathscr{L}) \times (I, \partial I)$$
$$(\ell^-, \partial\ell^-) \cong (\mathscr{L}, \partial\mathscr{L}) \times (I, \partial I)$$

where ℓ^+ (resp. ℓ^-) denotes the upper (resp. lower) halflink, I denotes the unit interval $[0, 1]$, and $\partial I = \{0, 1\}$.

(e) *Collarings of $\partial\mathscr{L}$.* There is a number $\omega>0$ and a neighborhood U of L_h in C and a homeomorphism

$$H\colon \partial\mathscr{L} \times [\delta-\omega, \delta] \times D_\varepsilon \to U$$

which commutes with the projection $(r, f)\colon U \to \mathbb{R} \times \mathbb{C}^2$, i.e.,

$$r(H(z, u, v)) = u \quad \text{and} \quad f(H(z, u, v)) = v$$

and which restricts to homeomorphisms

(i) $H_1: \partial\mathscr{L} \times D_\varepsilon \to L_h$ which commutes with $f: L_h \to D_\varepsilon$

(ii) $H_2: \partial\mathscr{L} \times \partial D_\varepsilon \to \partial L_h = \partial L_v$ which commutes with $f: \partial L_h \to \partial D_\varepsilon$

(iii) $H_3: \partial\mathscr{L} \times [\delta-\omega, \delta] \times \partial D_\varepsilon \to U \cap L_v$ which commutes with

$$(r, f): U \cap L_v \to [\delta-\omega, \delta] \times \partial D_\varepsilon.$$

Each of the spaces in the preceding proof has a canonical $\mathscr{S}$-decomposition, and the homeomorphisms may be chosen so as to preserve these $\mathscr{S}$-decompositions.

Proof. See Part I, Sect. 12.5. □

Corollary 1. *If $g: Z \to \mathbb{R}$ is any Morse function with a nondegenerate critical point at p, then the normal Morse data of g has the homotopy type of the pair*

$$(cone(\mathscr{L}), \mathscr{L}).$$

Proof. By Theorem 2.3, the normal Morse data is independent of the function, so we may take (for example) g to be any Morse function so that $dg(p) = \omega$. The result now follows from Part I, Sect. 12.5. □

The following diagram illustrates the fibres of the projection of the cylindrical neighborhood, $f: C \to \mathbb{C}$ for the case of an isolated quadratic surface singularity.

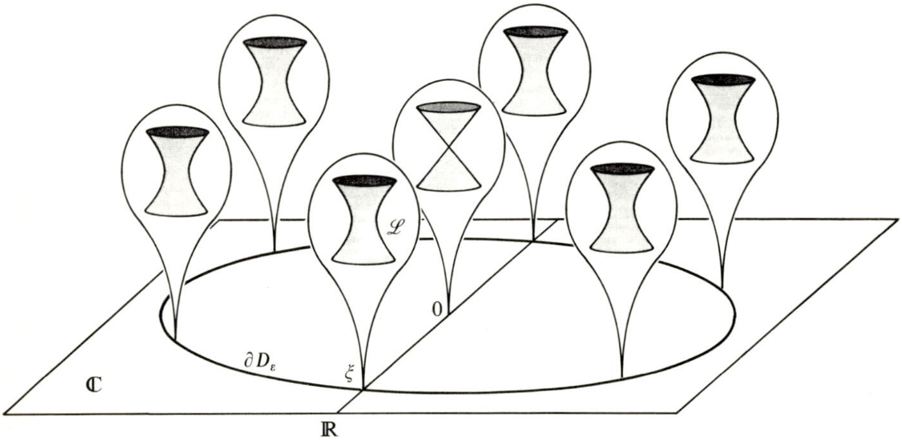

Fibres of the projection of the cylindrical neighborhood

If $\xi \in \partial D_\varepsilon$, then $C \cap f^{-1}(\xi)$ is the complex link $\mathscr{L}$. These form the fibres of a locally trivial bundle over $D_\varepsilon - \{0\}$, which does not extend in a locally trivial way over the origin because the central fibre $f^{-1}(0)$ is singular. However, the horizontal part L_h of the link is a trival bundle over D_ε whose fibre $\partial\mathscr{L}$ consists of two circles.

2.5. Monodromy, the Structure of the Link, and the Normal Morse Data

In Sect. 2.4, Corollary 1, we saw that the homotopy normal Morse data could be described completely in terms of the complex link $\mathscr{L}$. In this section we show that the homeomorphism type of the normal Morse data and the homeomorphism type of the link can be described in terms of the complex link and the monodromy homeomorphism $\mu\colon \mathscr{L} \to \mathscr{L}$.

By Sect. 2.4(a), the projection $L_v \to \partial D_\varepsilon$ is a fibre bundle with fibre $\mathscr{L}$, and is therefore classified by an orientation of the circle ∂D_ε and an isotopy class of homeomorphism $\mu\colon \mathscr{L} \to \mathscr{L}$ (which is called the monodromy). In fact (by Part I, Sect. 12.7), the monodromy homeomorphism may be chosen to preserve strata of $\mathscr{L}$ and to be the identity in a neighborhood of $\partial \mathscr{L}$. By Sect. 2.4(b), the link ∂N_δ is homeomorphic to the boundary $L = \partial C$ of the cylindrical neighborhood. Since L has been decomposed into two pieces,

$$L = L_v \cup_{\partial L_v} L_h$$

and by Sect. 2.4(e), L_h is homeomorphic to $\partial \mathscr{L} \times D^2$ and $\partial L_v = \partial L_h$ is homeomorphic to $\partial \mathscr{L} \times S^1$, we have the following:

Corollary 2. *The link ∂N_δ is homeomorphic to the space*

$$[(\mathscr{L} \times [0, 2\pi])/(\ell, 0) \sim (\mu(\ell), 2\pi)] \cup_{\partial \mathscr{L} \times S^1} [\partial \mathscr{L} \times D^2]$$

which is obtained from the complex link by the following procedure:

(a) *form the product of $\mathscr{L}$ with the interval $[0, 2\pi]$ and attach the ends together using the monodromy μ,*

(b) *attach to this the product $\partial \mathscr{L} \times D^2$ of the boundary of the complex link with the two-disk, along the subspace $\partial \mathscr{L} \times S^1$.* □

The following three diagrams illustrate the reconstruction of the link $L \cong \partial N_\delta$ from the complex link $\mathscr{L}$.

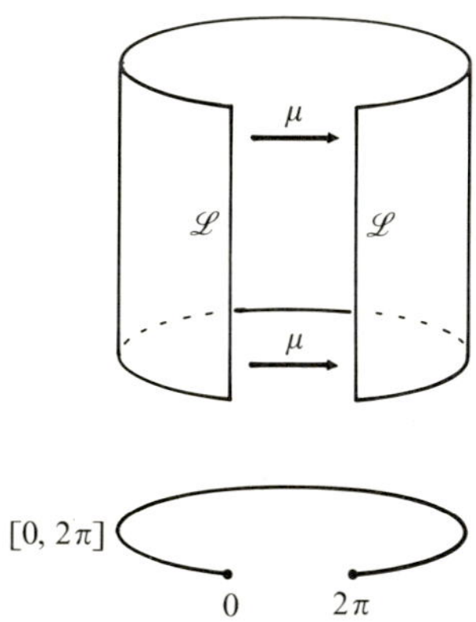

Glue by monodromy $\mathscr{L} \times [0, 2\pi]$

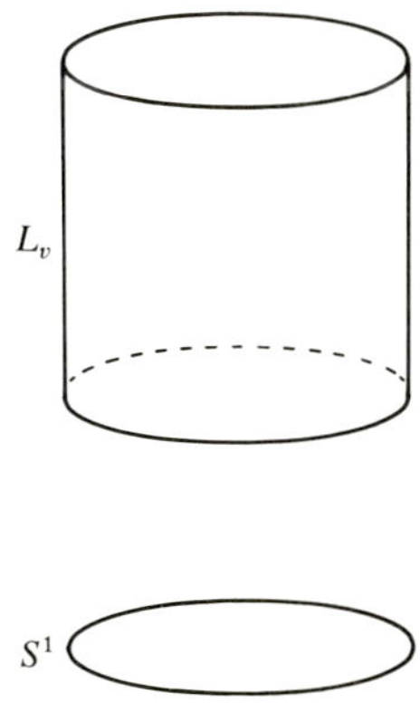

This gives the vertical part L_v of the link

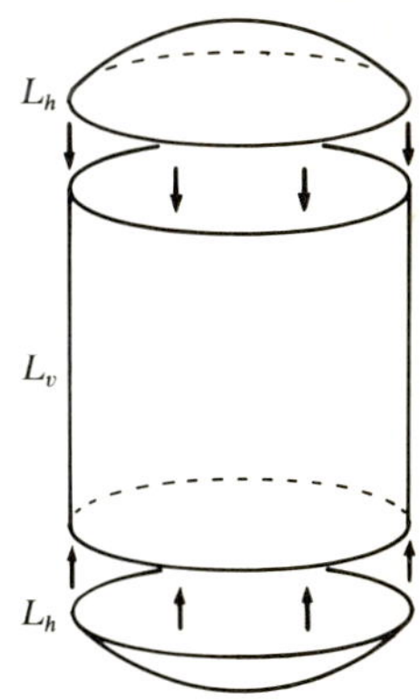

Attach $\partial \mathscr{L} \times D^2$

If we combine this description of the link together with the description Sect. 2.4(c) of the normal Morse data and use the collarings of Sect. 2.4(d), then it is easy to show (see, e.g. [GM3]) that

Corollary 3. *The normal Morse data is homeomorphic to the pair (J, K) where*

$$J = \operatorname{cone}([(\mathscr{L} \times [0, 2\pi])/(\ell, 0) \sim (\mu(\ell), 2\pi)] \cup_{\partial\mathscr{L} \times S^1} [\partial\mathscr{L} \times D^2])$$
$$K = \mathscr{L} \times [0, \pi]$$

and where $K \subset J$ is embedded in the base of the cone. In other words,

$$(J, K) \cong (\operatorname{cone}(L), L_v^-)$$

where $L_v^- = L_v \cap f^{-1}(-\infty, 0]$.

For completeness, we also recall (from Part I, Sects. 12.7 and 12.8) the following facts about the monodromy homeomorphism:

Theorem. *The monodromy is the identity on some neighborhood of $\partial\mathscr{L}$. The isotopy class (modulo some neighborhood of the boundary) of the monodromy is independent of all choices (i.e., the control data, normal slice, Riemannian metric, allowable choices of ε and δ, and the function f).*

Proof. This is proven in Part I, Sects. 12.7 and 12.8, except for the independence under a change of the function f, which follows (in this complex analytic context) from Proposition 2.1. □

2.6. Relative Normal Morse Data for Nonproper Functions

This section is a complex analytic version of Part I, Sects. 9, 10, and 11. We assume as above that $\bar{Z}$ is a complex analytic Whitney stratified subset of some complex analytic manifold M. As in Part I, Sect. 11 we also fix a Whitney stratified subset $\bar{X}$ of some smooth manifold M', and a proper surjective map $\bar{\pi}: \bar{X} \to \bar{Z}$, which is the restriction of a smooth map from M' to M. We suppose that $\bar{\pi}$ is stratified (Part I, Sect. 1.6) by the given stratifications of $\bar{X}$ and $\bar{Z}$, i.e., the restriction of $\bar{\pi}$ to each stratum of $\bar{X}$ is a smooth proper submersion (and hence is a fibre bundle) over a stratum of $\bar{Z}$. We fix an open subset $X \subset \bar{X}$ which is a union of strata, and define $Z = \pi(X) \subset \bar{Z}$.

$$\begin{array}{ccccc} X & \subset & \bar{X} & & \\ \downarrow{\scriptstyle \pi} & & \downarrow{\scriptstyle \bar{\pi}} & & \\ Z & \subset & \bar{Z} & \xrightarrow[f]{} & \mathbb{R} \end{array}$$

We also fix a stratum S of $\bar{Z}$, a point $p \in S$, and a nondegenerate conormal vector $\omega \in T^*_{S,p}M$. We make a choice $(N', f, r, \delta, \varepsilon)$ of normal projection data (Sect. 2.1) corresponding to ω at the point p, and we identify $T^*_p N'$ with $T^*_{S,p}M$. These choices determine a complex link $\mathscr{L}$ of $\bar{Z}$ at the point p.

Definition. The relative complex link $\mathscr{L}_X^\pi$ of f at p (resp. the relative normal slice N_X^π, ball $(N_\delta)_X^\pi$, cylindrical neighborhood (C_X^π, L_X^π), horizontal and vertical parts of the link $(L_h)_X^\pi$, $(L_v)_X^\pi$) is defined by

$$\mathscr{L}_X^\pi = X \cap \bar{\pi}^{-1}(\mathscr{L})$$

i.e., it is the intersection with X of the preimage of the complex link (resp. the normal slice, ball, cylindrical neighborhood, horizontal and vertical parts of the link).

Proposition. *The results of Proposition 2.4 remain valid when each of the above spaces is replaced by the corresponding relative space, with the single exception of the existence of the homeomorphism F in part (b).*

Proof. See Part I, Sect. 12.9. □

Remark. It is not necessary to assume that $\bar{X}$ is a complex analytic variety or that $\bar{\pi}$ is a complex analytic map. However, this will usually be the case because stratified maps occur most naturally as a result of the following fact: If $\phi: A \to B$ is a proper complex analytic map between complex analytic spaces, then there are complex analytic Whitney stratifications of A and B such that ϕ becomes a stratified map (see Sect. 2.7).

Corollary 1. *If $g: \bar{Z} \to \mathbb{R}$ is any Morse function with a nondegenerate critical point at p, and if $p \in \bar{Z} - Z$, i.e., if $\bar{\pi}^{-1}(p) \subset \bar{X} - X$ (i.e., p is a "critical point at infinity"), then the following pair is homotopy normal Morse data for $g \circ \pi$:*

$$(\mathcal{L}_X^\pi, \partial \mathcal{L}_X^\pi) \times (I, \partial I).$$

Proof. By Theorem 2.3, the normal Morse data is independent of the function, so we may take g to be (for example) any Morse function such that $dg(p) = \omega$. The result now follows from Part I, Sect. 12.9. □

Corollary 1*. *If $g: \bar{Z} \to \mathbb{R}$ is any Morse function with a nondegenerate critical point at p and if $\bar{\pi}^{-1}(p) \subset X$ (i.e., π is locally proper near the fibre over p), then the relative normal Morse data for $g \circ \pi$ at p has the homotopy type of the pair*

$$(\mathrm{cyl}(\mathcal{L}^\pi \to \pi^{-1}(p)), \mathcal{L}^\pi)$$

where cyl *denotes the mapping cylinder of the specialization map (Part I, Sect. 9.7), $\phi: \mathcal{L}^\pi \to \pi^{-1}(p)$.*

Proof. By Theorem 2.3, the normal Morse data is independent of the function, so we may take g to be (for example) any Morse function such that $dg(p) = \omega$. The result now follows from Part I, Sect. 12.9. □

2.7. Normal Morse Data for Two Complex Morse Functions

In this section we prove a result of K. Vilonen [MV]. Suppose S is a stratum of Z, $p \in S$, N' is a submanifold of M which meets S transversally at the point p, and suppose there are two conormal vectors,

$$\omega, \eta \in T^*_{S,p} M \cong T^*_p N'$$

which are jointly nondegenerate at p, i.e.,

$$(\omega, \eta)(Q/T_p S) = \mathbb{C}^2$$

for every generalized tangent space Q at the point p (except for the single case $Q=T_pS$). A choice of normal projection data is $(N', f, r, \delta, \varepsilon)$, which is consistent with ω and N' determines a complex link $\mathscr{L}$.

Theorem. (a) *The monodromy* μ: $\mathscr{L}\to\mathscr{L}$ *is isotopic to the identity.*

(b) *There is a homeomorphism*

$$\mathscr{L}\cong\mathscr{L}'\times(D^2, \partial D^2)$$

where D^2 *denotes the two-disk and*

$$\mathscr{L}'=N_\delta\cap f^{-1}(\varepsilon)\cap g^{-1}(0).$$

Here δ *and* ε *must be chosen in accordance with Part I, Sect. 12.10 so as to satisfy* $0<\varepsilon\ll\delta$.

Remark. A similar result holds in the relative and nonproper case.

Proof. See Part I, Sect. 12.10. □

2.A. Appendix: Local Structure of Complex Valued Functions

This section is parallel to Sects. 2.2–2.4 but differs from these sections in two ways:

(a) We consider an arbitrary proper complex analytic map f: $Z\to D^0$, rather than a nondegenerate map defined on the normal slice.

(b) We choose a stratification of f so that $Z_0=f^{-1}(0)$ is a union of strata. In particular, it is possible for Z_0 to contain a whole component of Z.

We then show that parts (a), (b), (c), and (d) of Proposition 2.4 continue to hold in this context. These often quoted results ([Du], [H1], [H4], [HL1], [KT2]) are not central to our development (which is concerned with nondegenerate functions f which arise by complexifying a Morse function), but are included here because the proof is virtually the same as that in Sect. 2.4 and we will use the results in Sect. 4.6*.

These results are surprisingly delicate and rely on the paper [Hi2] which uses the generic wing lemma, the curve selection lemma, etc. We refer the reader to Sect. 8 for Hironaka's counterexamples to similar sounding statements.

2.A.1. The setup. In this section we suppose that Z is a complex analytic subvariety of some complex analytic manifold M; that f: $Z\to D^0\subset\mathbb{C}$ is a proper complex analytic map to the open unit disk, which can be Whitney stratified so that $0\in D^0$ is the only stratum of dimension zero. It follows ([Hi2] Corollary 1 to Theorem 2, p. 248) that the central fibre $Z_0=f^{-1}(0)$ can be refined so that f satisfies Thom's condition A_f, i.e.,

If A is a stratum of Z_0 and if B is a stratum of $Z-Z_0$, and if $b_i\in B$ is a sequence of points converging to some point $a\in A$, and if the planes $T_{b_i}B$ converge to some plane τ and if the kernels $\ker(df(b_i)|T_{b_i}B)$ converge to some limiting plane K, then

$$K\supset\ker(df(a)|\tau).$$

We shall use the notation f_1 and f_2 to denote the real and complex parts of f, i.e., $f = f_1 + if_2$. We also fix a point $p \in Z_0$ and a Riemannian metric on M, and we define the function $r(x)$ to be the square of the distance between p and x.

2.A.2. Lemma. *There exists a fringed set* $D \subset \mathbb{R}^+ \times \mathbb{R}^+$ *of type* $0 \ll \varepsilon \ll \delta$, *so that the map*

$$(r, f): Z \to \mathbb{R}^+ \times \mathbb{C}$$

has no characteristic covectors in the set

$$D^{\#} = \{(u, v) \in \mathbb{R}^+ \times \mathbb{C} \mid (u, |v|) \in \bar{D}\}$$

except over points along the r axis (i.e., $f = 0$*).*

Proof. Take r so small that the boundary of the ball

$$\partial B_\delta(p) = \{z \in M \mid r(z) = \delta\}$$

is transverse to each stratum of Z (by Whitney's condition B). Now suppose there is a sequence of points $z_i \in Z \cap \partial B_\delta(p)$ such that $f(z_i) \neq 0$, $f(z_i) \to 0$, and such that $dr(z_i)$, $df_1(z_i)$, and $df_2(z_i)$ are linearly dependent when restricted to the stratum B_i, which contains the point z_i. By choosing a subsequence if necessary, we may suppose the z_i all lie in the same stratum B of Z, that they converge to some point z_0 in a stratum A of Z_0, and that the subspaces

$$K_i = \ker(df(z_i) \mid T_{z_i}(B)) \quad \text{and} \quad \tau_i = T_{z_i}(B)$$

converge to some limiting spaces K and τ respectively. It follows that $dr(z_0)$ vanishes on K. But,

$$K \supset \ker df(z_0) \mid \tau \supset T_{z_0} A$$

by condition A_f and by Whitney's condition A for the pair $A < B$. Thus, $\partial B_\delta(p)$ is not transverse to A at the point z_0. This is a contradiction. □

Now fix a point $(\varepsilon, \delta) \in D$, let $\zeta = \varepsilon + 0i$, and consider the analogous sets to those of Sect. 2.3, i.e.,

Definition.

$$D_\varepsilon = \{\xi \in \mathbb{C} \mid |\xi| \le \varepsilon\} \qquad \partial D_\varepsilon = \{\xi \in \mathbb{C} \mid |\xi| = \varepsilon\}$$
$$B_\delta = \{z \in Z \mid r(z) \le \delta\} \qquad \partial B_\delta(\varepsilon) = \{z \in Z \mid r(z) = \delta\}$$
$$\tilde{\mathscr{L}} = f^{-1}(\zeta) \cap B_\delta \qquad \partial \tilde{\mathscr{L}} = f^{-1}(\zeta) \cap \partial B_\delta$$
$$\tilde{C} = f^{-1}(D_\varepsilon) \cap B_\delta$$
$$\tilde{L}_h = f^{-1}(D_\varepsilon) \cap \partial B_\delta \qquad \partial \tilde{L}_h = f^{-1}(\partial D_\varepsilon) \cap \partial B_\delta$$
$$\tilde{L}_v = f^{-1}(\partial D_\varepsilon) \cap B_\delta \qquad \partial \tilde{L}_v = \partial \tilde{L}_h$$
$$\tilde{L} = \tilde{L}_v \cup \tilde{L}_h = \partial \tilde{C}$$
$$\tilde{\ell}^+ = f_1^{-1}(\varepsilon) \cap B_\delta \qquad \partial \tilde{\ell}^+ = f_1^{-1}(\varepsilon) \cap \partial B_\delta$$
$$\tilde{\ell}^- = f_1^{-1}(-\varepsilon) \cap B_\delta \qquad \partial \tilde{\ell}^- \cong \partial \tilde{\ell}^+.$$

Proposition. *The topological type of the above spaces is independent of the choice of* ε, δ, *and the Riemannian metric. (However the spaces* $\tilde{\mathscr{L}}$, $\tilde{L}_v$ *and* $\tilde{L}_h$ *may depend on the function* f.*)*

Proof. The proof is identical to that in Part I, Sect. 12.4. □

2.A.3. Proposition. (a) *The restriction*

$$f\,|\,\tilde{L}_v\colon \tilde{L}_v \to \partial D_\varepsilon$$

is a topological fibre bundle with fibre $\mathscr{L}$. *The restriction* $f\,|\,\tilde{C}-f^{-1}(0)$ *is a fibre bundle over* $D_\varepsilon-\{0\}$, *with fibre* $\tilde{\mathscr{L}}$.

(b) *There are stratum preserving homeomorphisms*

$$(\tilde{C}, \tilde{L}) \to (B_\delta, \partial B_\delta) \to (\mathit{cone}(\partial B_\delta), \partial B_\delta).$$

If $\eta>0$ *is sufficiently small, then these homeomorphisms may be chosen so as to preserve the levels* $\mathrm{Re}(f)=0$ *and* $\mathrm{Re}(f)=\pm\eta$.

(c) *For any* $\eta>0$ *sufficiently small, the pair* $(C\cap \mathrm{Re}(f)^{-1}[-\eta, \eta], C\cap \mathrm{Re}(f)^{-1}(-\eta))$ *is homeomorphic (by an* $\mathscr{S}$*-decomposition preserving homeomorphism) to the pair*

$$(N_\delta\cap \mathrm{Re}(f)^{-1}[-\varepsilon, \varepsilon], N_\delta\cap \mathrm{Re}(f)^{-1}(-\varepsilon)).$$

(d) *There are homeomorphisms of pairs*

$$(\tilde{\ell}^{+}, \partial\tilde{\ell}) \cong (\tilde{\mathscr{L}}, \partial\tilde{\mathscr{L}}) \times (I, \partial I)$$
$$(\tilde{\ell}^{-}, \partial\tilde{\ell}) \cong (\tilde{\mathscr{L}}, \partial\tilde{\mathscr{L}}) \times (I, \partial I).$$

Remark. Part (e) of Proposition 2.4 is no longer true in this context, i.e., $\tilde{L}_h$ is not necessarily homeomorphic to the product $\partial\tilde{\mathscr{L}} \times D_\varepsilon$, although it is easy to see that

$$\tilde{L}_h - (f^{-1}(0) \cap \tilde{L}_h)$$

is homeomorphic to $\partial\tilde{L}_v \times (0, 1]$. We refer to Sect. 8.5 for counterexamples to other similar sounding statements.

Proof. The proof is exactly the same as in Part I, Sect. 12.5 – just observe that only the existence of the fringed set A with no characteristic covectors was used (and that the nonexistence of characteristic covectors along the r axis $\{f=0\}$ was used only in the proof of part (e)). □

Chapter 3. Homotopy Type of the Morse Data

3.0. Introduction

We are now in a position to identify the homotopy Morse data (Part I, Sect. 3.3) for Morse functions on complex analytic varieties. Recall that homotopy Morse data is a pair (A, B) which is homotopy equivalent to some choice of Morse data. The importance of homotopy Morse data is the following: Suppose the pair (A, B) is homotopy Morse data for the Morse function $f: X \to \mathbb{R}$ at the critical point p, with critical value $v=f(p)$. Suppose $v \in (a, b)$, and that the closed interval $[a, b]$ contains no other critical values of f. Then there exists a continuous map $h: B \to X_{\leq a}$ such that $X_{\leq b}$ is homotopy equivalent to the adjunction space $X_{\leq a} \cup_B A$ (see Part I, Sect. 3.3). The identification of homotopy Morse data uses the deepest results of Part I and of Part II, Sect. 2: Theorem 3.5.4 of Part I says that local Morse data is Morse data and Theorem 3.7 of Part I says that local Morse data is the product of tangential Morse data with normal Morse data. So, homotopy Morse data is the product of the homotopy type of the tangential Morse data with the homotopy type of the normal Morse data. The homotopy type of tangential Morse data was identified classically by Morse, Thom, and Bott. It is the pair $(D^\lambda, \partial D^\lambda)$, where λ denotes the Morse index of the restriction of f to the stratum which contains the critical point p. So, the problem of identifying the homotopy Morse data is reduced to the problem of identifying the homotopy type of the normal Morse data. This was carried out in Sect. 2. In this short chapter we summarize those results.

3.1. Definitions

Throughout this chapter we suppose Z is a Whitney stratified complex analytic subvariety of some complex manifold M, that $f: Z \to \mathbb{R}$ is a proper smooth function with a critical point $p \in Z$, and that S denotes the stratum of Z which contains the critical point p. Fix $a, b \in \mathbb{R}$ and assume that the critical value by $v=f(p) \in (a, b)$. We further assume that p is the only critical point of f with this critical value, and that the closed interval $[a, b]$ contains no other critical values of f. Suppose f is a Morse function, i.e., that p is a nondegenerate critical point of $f|S$, and that $df(p)(Q) \neq 0$ for every generalized tangent space Q, except for the single case $Q=T_pS$. We denote the Morse index of $f|S$ at p by λ. Let $\mathscr{L}$ denote the complex link of the stratum S at the point p.

If A, B, and X are topological spaces with $B \subset A$ and if $h: B \to X$ is a continuous map (which we call the attaching map), then we denote the adjunction space $X \cup_h A$ by $X \cup (A, B)$.

3.2. Proper Morse Functions: The Main Technical Result

Theorem. *The pair $(D^\lambda, \partial D^\lambda) \times (cone(\mathscr{L}), \mathscr{L})$ is homotopy Morse data for f at p. So (for some attaching map), the space $Z_{\leq b}$ has the homotopy type of the space*

$$Z_{\leq a} \cup (D^\lambda, \partial D^\lambda) \times (cone(\mathscr{L}), \mathscr{L}).$$

Proof. By Part I, Sect. 3.2, the pair $(Z_{\leq b}, Z_{\leq a})$ is homeomorphic to the pair $(Z_{\leq \varepsilon}, Z_{\leq -\varepsilon})$ for $\varepsilon > 0$ arbitrarily small. By Part I, Sect. 3.5.4, $Z_{\leq \varepsilon}$ is homeomorphic to the space $Z_{\leq -\varepsilon} \cup (J, K)$ where (J, K) denotes the *local Morse data* for f at p. By Part I, Sect. 3.7, we have a homeomorphism

$$(J, K) \cong (D^{s-\lambda} \times D^\lambda, D^{s-\lambda} \times \partial D^\lambda) \times (\text{normal Morse data})$$

where $s = \dim_{\mathbb{R}}(S)$. We now describe the normal Morse data. By Part I, Sect. 7.5.1, the normal Morse data depends only on the differential $df(p)$ (in fact, by Theorem 2.3 the normal Morse data (at the point p) of any two functions defined on a complex analytic variety is noncanonically homeomorphic). By Part II, Corollary 2.4 the normal Morse data has the homotopy type of the pair $(\text{cone}(\mathscr{L}), \mathscr{L})$. (The attaching map h is implicitly constructed in the proofs, and it turns out to be an embedding.) □

3.3. Nonproper Morse Functions

Suppose $X \subset Z$ is an open subset which is a union of strata of Z. Let $\mathscr{L}_X = \mathscr{L} \cap X$ denote the complex link in X of the critical point p (see Sect. 2.6).

Corollary 1. *If $p \in Z - X$, then the pair $[(D^{\lambda+1}, \partial D^{\lambda+1}) \times (\mathscr{L}_X, \partial \mathscr{L}_X)]$ is homotopy Morse data for f at p. Therefore, the space $X_{\leq b}$ has the homotopy type of the space*

$$X_{\leq a} \cup [(D^{\lambda+1}, \partial D^{\lambda+1}) \times (\mathscr{L}_X, \partial \mathscr{L}_X)].$$

Corollary 1*. *If $p \in X$, then the pair*

$$[(D^\lambda, \partial D^\lambda) \times (cone(\mathscr{L}_X), \mathscr{L}_X)]$$

is homotopy Morse data for f at p. Therefore, the space $X_{\leq b}$ has the homotopy type of the space

$$X_{\leq a} \cup [(D^\lambda, \partial D^\lambda) \times (cone(\mathscr{L}_X), \mathscr{L}_X)].$$

Proof. The proof is parallel to that in Sect. 3.2, but uses Part I, Sect. 10.2 instead of Part I, Sect. 3.2, Part I, Sect. 10.4 instead of Part I, Sect. 3.5.4, Part I, Sect. 10.5 instead of Part I, Sect. 3.7, and Part II, Sect. 2.6, Corollaries 1 and 1* instead of Part II, Corollary 2.4. □

3.4. Relative and Nonproper Morse Functions

Suppose (as in Sect. 2.5) that $\bar{\pi}: \bar{X} \to \bar{Z}$ is a proper surjective stratified map, that $X \subset \bar{X}$ is an open subset which is a union of strata, and that $p \in \bar{Z}$ is a nondegenerate critical point of a proper Morse function $f: \bar{Z} \to \mathbb{R}$. Let $Z = \pi(X)$ and let

$$\mathscr{L}_X^\pi = X \cap \bar{\pi}^{-1}(\mathscr{L})$$

denote the relative complex link.

Corollary 1. *If $p \in \bar{Z} - Z$, i.e., if $\bar{\pi}^{-1}(p) \subset \bar{X} - X$, then the pair*

$$[(D^{\lambda+1}, \partial D^{\lambda+1}) \times (\mathscr{L}_X^\pi, \partial \mathscr{L}_X^\pi)]$$

is homotopy (relative) normal Morse data for f at p, so the space $X_{\le b}$ has the homotopy type of the space

$$X_{\le a} \cup [(D^{\lambda+1}, \partial D^{\lambda+1}) \times (\mathscr{L}_X^\pi, \partial \mathscr{L}_X^\pi)].$$

Corollary 1*. *If $\bar{\pi}^{-1}(p) \subset X$, then π is proper over some neighborhood of p, so the pair*

$$[(D^\lambda, \partial D^\lambda) \times (\mathrm{cyl}(\mathscr{L}_X^\pi \to \pi^{-1}(p)), \mathscr{L}_X^\pi)]$$

is homotopy (relative) normal Morse data for f at p. Therefore, the space $X_{\le b}$ has the homotopy type of the space

$$X_{\le a} \cup [(D^\lambda, \partial D^\lambda) \times (\mathrm{cyl}(\mathscr{L}_X^\pi \to \pi^{-1}(p)), \mathscr{L}_X^\pi)]$$

where cyl *denotes the mapping cylinder of the specialization map $\phi: \mathscr{L}_X^\pi \to \pi^{-1}(p)$ as in Part I, Sect. 9.7 and Part II, Sect. 2.6.*

Proof. The proof is parallel to that in Sect. 3.2 but uses Part I, Sect. 11.2 instead of Part I, Sect. 3.2, uses Part I, Sect. 11.4 instead of Part I, Sect. 3.5.4, uses Part I, Sect. 11.5 instead of Part I, Sect. 3.7, and uses Part II, Sect. 2.6, Corollaries 1 and 1* instead of Part II, Corollary 2.4. □

Chapter 4. Morse Theory of the Complex Link

4.0. Introduction

This chapter contains the estimates on the connectivity of the Morse data which are necessary for the proofs of the main applications. Since the Morse data is the product of the tangential Morse data with the normal Morse data, this requires estimates on both. The estimates on tangential Morse data are made in terms of the remarkable properties of the Levi form of the Morse function (see the appendix, Sect. 4.A). The normal Morse data is analyzed (inductively) by applying the entire apparatus of Morse theory to the complex link, which is a complex analytic space of smaller dimension.

Since all the arguments in this section are developed in maximal generality, the reader should first follow the special case (of the homotopy dimension of a Stein space) which appears in the introduction, and where the essential lines of the argument are clearly presented.

Philosophy of defects. The Lefschetz theorem (Sects. 1.1, 1.2) says that if $\pi: X \to \mathbb{CP}^N$ is an algebraic map, and if H_ε is an appropriate neighborhood of a linear subspace $H \subset \mathbb{CP}^N$ of codimension c, and if K is the first nonvanishing homotopy group of the pair $(X, \pi^{-1}(H_\varepsilon))$, then $K \geq \hat{n}$, where $\hat{n}$ is a certain integer. This theorem is proven by constructing a Morse function on X such that $X_{\leq \varepsilon} = X \cap H_\varepsilon$, with the property that the Morse data for each critical point is $\hat{n}$-connected. In good cases, $\hat{n}$ is expected to equal $n = \dim(X)$. This holds, for example, if π has finite fibres, X is a local complete intersection, and the linear subspace H is a codimension one hyperplane. In these cases, the fact that the local Morse data is n-connected follows from estimates which show that the tangential Morse data is i-connected and the normal Morse data is $n-i$ connected, where i is the complex dimension of the stratum which contains the associated critical point. In the general case, K may be less than n, because one or both of these estimates fail. We call the degree of failure of these estimates the *defect* of the function f at the critical point p. The *tangential defect* is the amount by which the degree of the first nonvanishing homotopy group of the tangential Morse data falls short of i, and the *normal defect* is the amount by which the degree of the first nonvanishing homotopy group of the normal Morse data falls short of $n-i$. Clearly, a bound on these defects gives a bound on $n-K$,

$$n - K \leq \sup(\text{tangential defect} + \text{normal defect})$$

where the sup is taken over all critical points of the Morse function.

Positive normal defect may be caused by singularities which are worse than local complete intersection singularities, or by large fibres of the map π. These contributions are called (respectively) the *singularity defect* (Sect. 4.6) and the *fibre defect* (Sect. 4.5) (compare [Gro] rectified homotopical depth, or [Og] de Rham depth). Positive tangential defect may be caused by higher codimension of the linear subspace H. This contribution is called the *convexity defect* (Sect. 4.4).

In the dual context, an affine n-dimensional algebraic variety X has the homotopy type of a CW complex of dimension no more than n. This is proven by constructing a Morse function on X with the property that, for each critical point, the tangential Morse data has the homotopy type of a CW complex of dimension $\leq i$, and the normal Morse data has the homotopy type of a CW complex of dimension $\leq n-i$, where i is the complex dimension of the stratum which contains the critical point. More generally, we will consider the homotopy dimension of an algebraic variety X which admits an algebraic map $\pi: X \to \mathbb{CP}^N - H$ to the complement of some linear subspace $H \subset \mathbb{CP}^N$. In this case, the tangential Morse data will have the homotopy type of a CW complex of dimension $\leq i + \Gamma^*$, where Γ^* is the *dual tangential defect;* and the normal Morse data will have the homotopy type of a CW complex of dimension $\leq (n-i) + \Delta^*$, where Δ^* is the *dual normal defect* (Sect. 4.2). Positive dual normal defect may be caused by large fibres (and is bounded by the *fibre defect*, Sect. 4.5*) or by failure of π to be proper (Sect. 4.6*). Positive dual tangential defect may be caused by higher codimension of the linear space H, and is bounded by the *dual convexity defect* (Sect. 4.4).

4.1. The Setup

In this chapter we will consider the most general setup (see Part I, Sect. 11 and Part II, Sect. 2.6): $\bar{Z}$ is a Whitney stratified complex analytic subvariety of some analytic manifold M, $f: \bar{Z} \to \mathbb{R}$ is a proper Morse function (Sect. 3.1) with a nondegenerate critical point $p \in \bar{Z}$ which lies in some stratum S of $\bar{Z}$. We assume that $\bar{\pi}: \bar{X} \to \bar{Z}$ is a proper stratified map and that $X \subset \bar{X}$ is an open subset which is a union of strata. Let $\pi = \bar{\pi} | X$ denote the restriction to X, and let $Z = \pi(X)$ denote the image of X under π.

$$\begin{array}{ccccc} X & \subset & \bar{X} & & \\ \downarrow{\scriptstyle \pi} & & \downarrow{\scriptstyle \bar{\pi}} & & \\ Z & \subset & \bar{Z} & \xrightarrow[f]{} & \mathbb{R}. \end{array}$$

Let $\bar{N}$ denote a normal slice (in $\bar{Z}$) through the stratum S at the point p. We define the following integers:

$d(S)$ = the complex dimension of the stratum S
λ = the Morse index of $f | S$ at the critical point p
n = the complex dimension of X
$c(S) = n - d(S)$ = the complex dimension of $\bar{\pi}^{-1}(\bar{N})$.

4.2. Normal and Tangential Defects

In this section we assess blame for the failure of the Lefschetz hyperplane formula, by assigning defects to the normal and tangential Morse data for the Morse function f.

Definition. The *convexity defect* $\Gamma(p)$ of the function f at the point p is the number

$$\Gamma(p) = \Gamma_f(p) = d(S) - \lambda.$$

The *dual convexity defect* $\Gamma^*(p)$ of f at p is the number

$$\Gamma^*(p) = \Gamma_f^*(p) = \lambda - d(S).$$

The *normal defect* $\Delta(p)$ of $f \circ \pi | X$ at p is the number

$$\Delta(p) = c(S) - h$$

where h is the degree of the first nonvanishing homotopy group of the relative normal Morse data for $f \circ \pi | X$ at p, i.e., $\pi_m(J, K) = 0$ for all $m < h$ and $\pi_h(J, K) \neq 0$, where (J, K) denotes the relative normal Morse data (see Part I, Sect. 11). (If $\pi_m(J, K) = 0$ for all m, then we define the normal defect to be $-\infty$.)

The *normal dual defect* $\Delta^*(p)$ of f at p is the number

$$\Delta^*(p) = h^* - c(S)$$

where h^* is the smallest number m such that J/K has the homoptopy type of a CW complex of dimension m, where (J, K) denotes the relative normal Morse data for X at the point p.

4.3. Homotopy Consequences

The definitions of the normal and tangential defects are justified by the two propositions in this section. First, we recall the homotopy excision theorem of Blakers and Massey ([BM], [Sw]):

Homotopy Excision Theorem. *Suppose $W = A \cup B$ are spaces with the homotopy type of CW complexes, and suppose that $A \cap B$ is a neighborhood deformation retract in A (or else suppose that the interiors of A and B cover W). If $(A, A \cap B)$ is n-connected and $(B, A \cap B)$ is m-connected, then the map*

$$\pi_q(B, A \cap B) \to \pi_q(W, A)$$

is an isomorphism for all $q \leq n + m - 1$ and is a surjection for $q = n + m$.

4.3. Proposition. *Suppose that the closed interval $[a, b]$ contains no critical values of f except for the isolated critical value $f(p) \in (a, b)$. Then, $\pi_i(X_{\leq b}, X_{\leq a}) = 0$ for all $i \leq \hat{n}$, where*

$$\hat{n} = n - (\Delta(p) + \Gamma(p)) - 1.$$

Proof. By Sect. 3.4 the pair $(X_{\leq b}, X_{\leq a})$ has the homotopy type of the pair

$$(X_{\leq a} \cup (D^\lambda, \partial D^\lambda) \times (\text{Normal Morse data}), X_{\leq a}).$$

We apply the homotopy excision theorem with $W=X_{\leq b}$, $A=X_{\leq a}$, and $B=(D^\lambda, \partial D^\lambda)\times$(Normal Morse data). Note that B is $h+\lambda=c-\Delta+\lambda$-connected by assumption. Thus, $(X_{\leq b}, X_{\leq a})$ is $c-\Delta+\lambda=n-\Delta-\Gamma$ connected. □

4.3*. Proposition. *If the interval $[a, b]$ contains no critical values of f except for the isolated critical value $f(p)\in(a, b)$, then the space $X_{\leq b}$ has the homotopy type of a space obtained from $X_{\leq a}$ by attaching cells of dimension less than or equal to $\hat{n}^*$, where*

$$\hat{n}^*=n+(\Delta^*(p)+\Gamma^*(p)).$$

Proof. By Sect. 3.4, the space $X_{\leq b}$ is obtained from $X_{\leq a}$ by attaching the pair

$$(D^\lambda, \partial D^\lambda)\times(\text{Normal Morse data})$$

along the subspace. This has the homotopy type of a CW complex of dimension

$$\lambda+h^*=\lambda+\Delta^*+n-d=n+\Delta^*+\Gamma^*. \quad \square$$

4.4. Estimates on Tangential Defects

In this paragraph we estimate the tangential defect of the distance function (in the Fubini Study metric) on projective space. Let $H\subset\mathbb{CP}^M$ be a linear subspace of codimension e and let $\tilde{H}\subset\mathbb{C}^{M+1}$ be the corresponding linear subspace of codimension e. Let $G\subset\mathbb{CP}^M$ be a linear subspace which is complementary to H, i.e., $G\cap H=\phi$ and $\dim(G)=e-1$. Choosing a linear isomorphism

$$\tilde{H}\oplus\tilde{G}\cong\mathbb{C}^{M+1}$$

gives rise to a homogeneous coordinate system $\{[z_0:z_1:\ldots:z_M]\}$ on $\mathbb{CP}^M$ such that

$$H=\{[z_0:z_1:\ldots:z_M]\mid z_{M-e+1}=z_{M-e+2}=\ldots=z_M=0\}$$
$$G=\{[z_0:z_1:\ldots:z_M]\mid z_0=z_1=\ldots=z_{M-e}=0\}.$$

Let $\bar{f}\colon\mathbb{CP}^M\to\mathbb{R}$ be the distance from H (with respect to G), i.e.,

$$\bar{f}([z_0:z_1:\ldots:z_M])=\frac{\sum_{i=M-e+1}^{M} z_i\bar{z}_i}{\sum_{i=0}^{M} z_i\bar{z}_i}.$$

Clearly $\bar{f}^{-1}(0)=H$ and $\bar{f}^{-1}(1)=G$.

Proposition. *Let $A\subset\mathbb{CP}^M$ be a complex analytic submanifold of dimension $d(A)$. Let $f\colon\mathbb{CP}^M\to\mathbb{R}$ be a smooth function and suppose that $p\in A-(G\cup H)$ is a nondegenerate critical point of the restriction $f\mid A$. If f is sufficiently C^k close (for any $k\geq 2$) to the distance function $\bar{f}$, then the Morse index λ of $f\mid A$ at p satisfies the inequalities*

$$d(A)-(e-1)\leq\lambda\leq d(A)+M-e.$$

Proof. The proof of this well known result appears in the appendix to this chapter, Sect. 4.A.4. See also [Bo1], [AF], [Gr1], [Sm1]. □

Corollary. *The convexity defect and dual convexity defect of f at the point p satisfy the a priori inequalities*

$$\Gamma(p) \leq \min(d(A), e-1)$$
$$\Gamma^*(p) \leq \min(d(A), M-e).$$

Proof. From the proposition we have $\Gamma(p)=d(A)-\lambda \leq e-1$. However, we also have $\Gamma(p) \leq d(A)$ since $\lambda \geq 0$. Thus, $\Gamma(p) \leq \min(d(A), e-1)$.

Similarly, $\Gamma^*(p)=\lambda-d(A) \leq M-e$. But, $\lambda \leq 2d(A)$, so $\Gamma^*(p) \leq d(A)$. Thus, $\Gamma^*(p) \leq \min(d(A), M-e)$. □

Remark. The same results hold for any nondegenerate critical point of a composition

$$f \circ \pi \colon A \to \mathbb{R}$$

when $\pi \colon A \to \mathbb{CP}^M$ is a (not necessarily proper) finite complex algebraic map.

4.5. Estimates on the Normal Defect for Nonsingular X

In this section we show that the normal defect is bounded in terms of the dimension of the fibres of the projection π. Our method for estimating this defect is the following: we observe that if X is nonsingular, then we can remove the fibre $\pi^{-1}(p)$ from the normal Morse data without affecting its homotopy groups in low dimensions. By the homotopy excision theorem, this "punctured" normal Morse data has connectivity which is one greater than the connectivity of the relative complex link mod its boundary $(\mathscr{L}_X^\pi, \partial\mathscr{L}_X^\pi)$ (see Sect. 3.4, Corollary 1). The distance from the critical point p is now a Morse function on the relative complex link, so we use induction and Morse theory to estimate its connectivity.

Definition. For each stratum $A \subset \bar{Z}$ define the *fibre defect*

$$\delta_F(A)=\max(0, 2\dim_{\mathbb{C}} \bar{\pi}^{-1}(p)-(n-d(A)))$$

where $p \in A$, and $d(A)$ denote the complex dimension of A.

4.5.1$_\mathbf{m}$. Proposition. *Suppose* $\dim_{\mathbb{C}}(\bar{Z})=m$, *$X$ is nonsingular, and $p \in \bar{Z}$ is a critical point of f. Then, the normal defect of $f \circ \pi | X$ at the point p satisfies*

$$\Delta(p) \leq \sup_A \delta_F(A)$$

where the sup *is taken over all strata $A \subset Z$ such that $p \in \bar{A}$. In particular, if X is nonsingular and π is finite or is semismall (see Sect. 1.1), then the normal defect of every critical point p is zero.*

4.5.2$_\mathbf{m}$. Proposition. *Suppose* $\dim_{\mathbb{C}}(\bar{Z})=m$ *and X is nonsingular. Then, for every stratum $A \subset \bar{Z}$ we have*

$$\pi_i(\mathscr{L}_X^\pi(A), \partial\mathscr{L}_X^\pi(A))=0$$

whenever

$$i < c(A) - 1 - \sup_{B > A} \delta_F(B).$$

Here $\mathscr{L}_X^\pi(A)$ denotes the relative complex link of the stratum A, and $c(A) = n - \dim_{\mathbb{C}}(A)$, and the sup is taken over all strata $B \neq A$ such that $B \subset Z$ and $A \subset \bar{B}$.

Note. In both cases the sup is taken over strata in Z, rather than strata in $\bar{Z}$.

Proof that 4.5.2$_m$ implies 4.5.1$_m$. We must estimate the degree of the first nonvanishing homotopy group of the normal Morse data at the point p,

$$(J, K) = X \cap \bar{\pi}^{-1}(f^{-1}[-\varepsilon, \varepsilon], f^{-1}(-\varepsilon)) \cap B_\delta(p) \cap N'$$

where N' is an analytic manifold which is transverse to the stratum S which contains the critical point p. We will consider two cases: (1) $p \in Z$ and (2) $p \in \bar{Z} - Z$. We will reduce case (1) to case (2) by removing the critical fibre $\pi^{-1}(p)$ from the normal Morse data.

Case 1. Suppose $p \in Z$, i.e., $\bar{\pi}^{-1}(p) \cap X \neq \phi$. We need the following lemma:

Lemma. *The homotopy groups of the normal Morse data $\pi_i(J, K)$ coincide with the homotopy groups of the "punctured" normal Morse data*

$$\pi_i(J - \pi^{-1}(p), K)$$

for all $i < 2n - 2d(S) - 2\dim(\pi^{-1}(p))$.

Proof of lemma. Since X is nonsingular and $\pm\varepsilon$ is not a critical value of $f \mid (\bar{Z} \cap B_\delta(p) \cap N')$, it follows that the pair (J, K) is a smooth (noncompact) manifold with collared boundary. Any representative of the relative homotopy group can be approximated by a smooth function

$$g: (D^i, \partial D^i) \to (J, K)$$

in the same homotopy class. By transversality, there is a slight perturbation of the function g to a function g' which is transverse to each stratum of $\pi^{-1}(p)$. This means that $g'(D^i)$ completely misses the fibre $\pi^{-1}(p)$ provided that $i + \dim_{\mathbb{R}} \pi^{-1}(p) < \dim_{\mathbb{R}}(J)$. □

Proof of Case 1. Following the same method as in Sect. 3.4, we see that the punctured normal Morse data has the same connectivity as the pair

$$(\mathscr{L}_X^\pi, \partial \mathscr{L}_X^\pi) \times (I, \partial I)$$

which is homotopy normal Morse data for the pair $(J - \pi^{-1}(p), K)$. (The argument proceeds as follows: by Part I, Sect. 11, the punctured normal Morse data is homotopy equivalent to the pair $(\ell_X^{\pi+} \cup_{\partial\ell} \ell_X^{\pi-}, \ell_X^{\pi-})$ where $\ell_X^{\pi\pm}$ denotes the relative halflink. Using either the excision property of Morse data (Part I, Sect. 3.2) or the homotopy excision theorem of Blakers and Massey, this pair has the same connectivity as the pair $(\ell_X^{\pi+}, \partial\ell_X^{\pi+})$ which (by Sect. 2.6, part (d)) is homeomorphic to the product

$$(\mathscr{L}_X^\pi, \partial \mathscr{L}_X^\pi) \times (I, \partial I).$$

By Proposition $4.5.2_{\mathrm{m}}$, the first nonvanishing homotopy group of the pair $(\mathscr{L}_X^\pi, \partial\mathscr{L}_X^\pi)$ occurs in dimension $\geq \hat{n}$, where

$$\hat{n} = \dim_{\mathbb{C}}(\mathscr{L}_X^\pi) - \sup_{B>S} \delta_F(B)$$

(the sup being taken over all strata $B \neq S$ such that $B \subset Z$ and $S \subset \bar{B}$). Thus, the first nonvanishing homotopy group of the normal data occurs in degree h, where

$$\begin{aligned} h &\geq \min(\hat{n}+1, 2n - 2d(S) - 2\dim_{\mathbb{C}} \pi^{-1}(p)) \\ &= \min(c - \sup_B \delta_F(B), 2n - 2d(S) - 2\dim_{\mathbb{C}} \pi^{-1}(p)) \end{aligned}$$

so

$$\begin{aligned} \Delta = c - h &\leq \max(\sup_B \delta_F(B), -c + 2\dim \pi^{-1}(p)) \\ &= \max(\sup_B \delta_F(B), \delta_F(S)) = \sup_B (\delta_F(B)) \end{aligned}$$

where the sup is taken over all strata $B \geq S$ such that $B \subset Z$. □

Proof of Case 2. Suppose $p \in \bar{Z} - Z$, i.e., that $\bar{\pi}^{-1}(p) \cap X = \phi$. Then it is not necessary to "puncture" the normal Morse data: by Sect. 3.4, Corollary 1, the pair

$$(\mathscr{L}_{Xz}^\pi, \partial\mathscr{L}_X^\pi) \times (I, \partial I)$$

is already homotopy normal Morse data. Thus, the same calculation as above gives

$$\Delta = c - h \leq \sup_{B>S} \delta_F(B)$$

where the sup is taken over all strata $B \subset Z$ such that $B \neq S$ and $\bar{B} \supset S$.

In summary, $\Delta = \sup_B (\delta_F(B))$, where the sup is taken over all strata $B \subset Z$ such that $p \in \bar{B}$, as claimed. □

Proof that $4.5.1_{m-1}$ implies $4.5.2_m$. Consider the following setup:

$$\begin{array}{ccc} X \cap \bar{\pi}^{-1} \mathscr{L}(A) = \mathscr{L}_X^\pi(S) \subset \bar{\pi}^{-1} \mathscr{L}(A) & & \\ \quad \searrow^{\pi} \qquad\qquad \downarrow{\scriptstyle \bar{\pi}} & & \\ \qquad\qquad \mathscr{L}(A) & \xrightarrow{\ f\ } & \mathbb{R} \end{array}$$

where $\mathscr{L}(A)$ denotes the complex link in $\bar{Z}$ of the stratum A. Let $f'(z)$ denote the function $\delta - r(z)$, where $r(z)$ denotes the square of the distance between the points p and z. Let f be a C^k close approximation to f' (for some $k \geq 2$) such that f is Morse function (see Part I, Sect. 2.2 or [P1]). By Corollary 4.4, for every critical point $q \in \mathscr{L}(A)$, the tangential defect is 0, i.e.,

$$\Gamma_f(q) = \Gamma^*_{-f}(q) = 0.$$

Now consider the effect on $\pi_*((\mathscr{L}_X^\pi)_{\leq a}, \partial\mathscr{L}_X^\pi)$ of a single critical point $q \in \mathscr{L}(A)$ of f. Suppose the interval $[a, b]$ contains no critical values except for the single critical value $f(q)$. By Proposition 4.3 we have

$$\pi_i((\mathscr{L}_X^\pi)_{\le b}, (\mathscr{L}_X^\pi)_{\le a})=0 \qquad \text{for all } i<\hat{n}$$

where

$$\hat{n}=\dim_{\mathbb{C}}(\mathscr{L}_X^\pi)-\Delta(q)$$

since $\Gamma(q)=0$. By Proposition $4.5.1_{m-1}$ we have

$$\Delta(q)\le\sup_B \delta_F(B)$$

where the sup is taken over all strata $B\subset Z$ such that $q\in\bar{B}$.

Applying this argument to each of the critical points of f on $\mathscr{L}(A)$ we obtain

$$\pi_i(\mathscr{L}_X^\pi, \partial\mathscr{L}_X^\pi)=0 \qquad \text{for all } i<\hat{n}$$

where

$$\hat{n}=\dim_{\mathbb{C}}(\mathscr{L}_X^\pi)-\sup_{B>A}\delta(B)=c(A)-1-\sup_{B>A}\delta_F(B)$$

where the sup is taken over all strata $B\neq A$ such that $B\subset Z$ and $A\subset\bar{B}$. □

4.5*. Estimates on the Dual Normal Defect for Proper π

In this section we show that the dual normal defect is bounded in terms of the same fibre defect $\delta_F(S)$, provided the map π is proper. We will assume $X=\bar{X}$, that $Z=\bar{Z}$, and that $\pi=\bar{\pi}$ throughout this section.

Definition. For each stratum $A\subset Z$ define the dual fibre defect

$$\delta_F^*(A)=2\dim_{\mathbb{C}}(\pi^{-1}(p))-(n-d(A))$$

where $p\in A$ and $d(A)$ denotes the complex dimension of A. This is the same as the fibre defect $\delta_F(A)$ of Sect. 4.5.

Proposition $4.5.1_m^*$. *Suppose* $\dim_{\mathbb{C}}(Z)=m$, *and* $\pi\colon X\to Z$ *is proper. Let* $p\in Z$ *be a critical point of the Morse function f. Then the dual normal defect of f at the point p satisfies*

$$\Delta^*(p)\le\sup_A \delta_F^*(A)$$

where the sup is taken over all strata A such that $p\in\bar{A}$. *In particular, if* π *is proper and finite, then the normal defect of each critical point is* 0.

Proposition $4.5.2_m^+$. *Suppose* $\dim_{\mathbb{C}}(Z)=m$ *and* π *is proper. Then for every stratum A, the space* $\mathscr{L}^\pi(A)$ *has the homotopy type of a CW complex of dimension less than or equal to* $\hat{n}^*$, *where*

$$\hat{n}^*=\dim_{\mathbb{C}}(\mathscr{L}^\pi(A))+\sup_{B>A}\delta_F^*(B)$$

where the sup is taken over all strata $B\neq A$ *such that* $A\subset\bar{B}$. *In particular, if* π *is finite then* $\mathscr{L}^\pi(A)$ *has the homotopy of a CW complex whose dimension is less than or equal to the complex dimension of* $\mathscr{L}^\pi(A)$.

Proof that 4.5.2$^*_m$ implies 4.5.1$^*_m$. We must estimate the "homotopy dimension" of the normal Morse data (J, K) at the point p. Let S denote the stratum of Z which contains the critical point p, and let $(\mathscr{L}^\pi, \partial\mathscr{L}^\pi)$ denote the relative complex link of the stratum S. By Sect. 3.4, Corollary 1*, the normal Morse data has the homotopy type of the pair

$$(\mathrm{cyl}(\mathscr{L}^\pi \to \pi^{-1}(p)), \mathscr{L}^\pi).$$

But, 4.5.2$^*_m$ implies that the mapping cylinder can be obtained from $\mathscr{L}^\pi$ by attaching cells of dimension less than or equal to h^* where

$$h^* = \max(2\dim_{\mathbb{C}}(\pi^{-1}(p)), \dim_{\mathbb{C}}(\mathscr{L}^\pi) + \sup_{B>S} \delta^*_F(B) + 1).$$

Thus,

$$\begin{aligned} \Delta^*(p) = h^* - c(S) &\leq \max(2\dim_{\mathbb{C}} \pi^{-1}(p) - c(S), \sup_{B>S} \delta^*_F(B)) \\ &= \sup_{B \geq S} \delta^*_F(B). \quad \square \end{aligned}$$

Proof that 4.5.1$^*_{m-1}$ implies 4.5.2$^*_m$. Consider the setup

$$\begin{array}{c} \mathscr{L}^\pi(A) \\ \downarrow \pi \\ \mathscr{L}(A) \xrightarrow{\ f\ } \mathbb{R}. \end{array}$$

Let f' denote the function $r(z)$ and let f be a close approximation to f' such that f is Morse function (see Part I, Sect. 2.2 or [P1]). By Corollary 4.4, for every critical point $q \in \mathscr{L}(A)$, the dual convexity defect is 0, i.e.,

$$\Gamma^*_f(q) = 0.$$

If the interval $[a, b]$ contains no critical values other than the single critical value $f(q)$, then by Proposition 4.3*, the space $\mathscr{L}^\pi(A)_{\leq b}$ is obtained from $\mathscr{L}^\pi(A)_{\leq a}$ by attaching cells of dimension $\leq \hat{n}^*$, where

$$\hat{n}^* = \dim_{\mathbb{C}} \mathscr{L}^\pi(A) + \Delta^*(q).$$

Using Proposition 4.5.1$^*_{m-1}$ we have

$$\Delta^*(q) \leq \sup_B \delta^*_F(B)$$

where the sup is taken over all strata B which contain the point p in their closures. Applying this argument to each of the critical points of f on $\mathscr{L}^\pi(A)$, we conclude that $\mathscr{L}^\pi(A)$ has the homotopy type of a CW complex of dimension $\leq \hat{n}^*$, where

$$\hat{n}^* = \dim_{\mathbb{C}} \mathscr{L}^\pi(A) + \sup_{B>A} \delta^*_F(B). \quad \square$$

Remark. Strictly speaking, the set $\partial\mathscr{L}^\pi(A)$ is a degenerate critical set of the function f. However, this boundary is collared in $\mathscr{L}(A)$ and (see Sect. 1) the Morse function f (which is called r in Sect. 1) has no critical points in this collar. Therefore, the homotopy type of $\mathscr{L}(A)$ is not affected by attaching its boundary.

Remark. The same method can be used to show the following: if X is an n-dimensional complex analytic subvariety of some complex analytic manifold M, if $p \in M$, and if $B_\delta(p)$ denotes a closed ball of radius δ centered at p (contained in some coordinate chart of M), then the intersection $X \cap B_\delta(p)$ has the homotopy type of a CW complex of dimension $\leq n$. (Use a Morse function which is a slight perturbation of the distance from the point p.)

4.6. Estimates on the Normal Defect if $\bar{\pi}$ is Finite

In this section we will show that the normal defect is bounded in terms of the number of equations which define X as a subvariety of some smooth variety M', when the map $\bar{\pi}$ is finite. We consider the setup of Sect. 4.1, i.e., $\bar{Z}$ is a complex analytic subvariety of M, $\bar{\pi}\colon \bar{X} \to \bar{Z}$ is a Whitney stratified finite proper map, $X \subset \bar{X}$ is an open union of strata, $\pi = \bar{\pi} | X$, and $Z = \pi(X)$. For any stratum A of $\bar{Z}$, let $c(A) = n - \dim_{\mathbb{C}}(A)$ be the complex dimension of the normal slice,

$$(\bar{N}(A), \partial \bar{N}(A)) = N' \cap \bar{Z} \cap (B_\delta(p), \partial B_\delta(p))$$

as in Sect. 4.5. Let $(\mathscr{L}(A), \partial \mathscr{L}(A))$ denote the complex link in $\bar{Z}$ of the stratum A. If $\bar{\pi}$ is finite, then the relative normal slice in X is a disjoint union

$$(N_X^\pi(A), \partial N_X^\pi(A)) = X \cap \bar{\pi}^{-1}(\bar{N}(A), \partial \bar{N}(A)) = \coprod_x (N_x, \partial N_x)$$

of the normal slices in X, where the union is taken over all points $x \in \bar{\pi}^{-1}(p)$. Similarly, the relative complex link is a disjoint union

$$(\mathscr{L}_X^\pi(A), \partial \mathscr{L}_X^\pi(A)) = X \cap \bar{\pi}^{-1}(\mathscr{L}(A), \partial \mathscr{L}(A)) = \coprod_x (\mathscr{L}_x, \partial \mathscr{L}_x)$$

of the complex links in X at each point $x \in \bar{\pi}^{-1}(p)$.

Definition. The *singularity defect* $\delta_S(A)$ at the stratum $A \subset \bar{Z}$ is the number

$$\delta_S(A) = \sup_x (c(A) - h(x))$$

where $h(x)$ is the degree of the first nonvanishing homotopy group of the pair $(N_x, \partial N_x)$, $c(A)$ is the (local) codimension of the stratum A in Z, $p \in A$, and the sup is taken over all points $x \in X \cap \bar{\pi}^{-1}(p)$ (which actually lie in the subspace X).

Remark. This means that the link ∂N_x is $c(A) - \delta_S(A) - 2$ connected. (Compare [Gro] Exp. 13, Sect. 6 or [HL2], Sect. 2.1.5. The number $\sup_A (n - \delta_S A)$ is the *rectified homotopical depth.*)

We recall the following theorem of Hamm [H1], a homology version of which will be provided in the remark in Sect. 4.6*.

Theorem. *Suppose X is a complex analytic subvariety of some complex m-dimensional analytic manifold M'. Let E denote the minimum number of equations needed to define a neighborhood of a point $x \in X$, and let $L(x) = X \cap \partial B_\delta(p)$ be the intersection of X with the boundary of a sufficiently small ball about the point p. Then $L(x)$ is $m - E - 2$ connected.*

Applying this theorem to a normal slice N through the stratum S of an analytic Whitney stratification of X, we find

$$h(x) \leq D + C - E$$

where C is the (complex) codimension of the stratum S in X and D is the (local) codimension of X in M' (near the point $p = N \cap S$). In particular, if for each $x \in \pi^{-1}(p)$ a neighborhood of x is defined by $\leq E$ equations and codimension$_x(X) \geq D$, then we have

$$\delta_S(A) \leq E - D.$$

Proposition 4.6.1$_\mathbf{m}$. *Suppose* $\dim(\bar{Z}) = m$, *and* $\bar{\pi}: \bar{X} \to \bar{Z}$ *is finite and* $p \in \bar{Z}$ *is a critical point of f. Then, the normal defect (Sect. 4.2) of f at the point p satisfies*

$$\Delta(p) \leq \sup_A \delta_S(p)$$

where the sup *is taken over all strata* $A \subset Z$ *such that* $p \in \bar{A}$. *In particular, if X is a local complete intersection, then the normal defect of every critical point is* 0.

Proposition 4.6.2$_\mathbf{m}$. *Suppose* $\dim(\bar{Z}) = m$, *and* $\bar{\pi}: \bar{X} \to \bar{Z}$ *is finite. Then, for every stratum* $A \subset \bar{Z}$ *we have*

$$\pi_i(\mathscr{L}_X^\pi(A), \partial \mathscr{L}_X^\pi(A)) = 0$$

whenever

$$i < c(A) - 1 - \sup_{B > A} \delta_S(B)$$

where the sup *is taken over all strata* $B \subset Z$ *such that* $\bar{B} \supset A$.

Proof that $4.6.2_m$ *implies* $4.6.1_m$. We must estimate the degree of the first nonvanishing homotopy group of the normal Morse data at the point $p \in \bar{Z}$,

$$(J, K) = X \cap \bar{\pi}^{-1}(f^{-1}[-\varepsilon, \varepsilon], f^{-1}(-\varepsilon)) \cap B_\delta(p) \cap N'$$

where N' is an analytic manifold which is transverse to the stratum S which contains the critical point p. We distinguish between two cases, (1) $p \in Z$ and (2) $p \in \bar{Z} - Z$.

Case 1. Suppose $p \in Z$, i.e., $\bar{\pi}^{-1}(p) \cap X \neq \phi$. We need the following lemma:

Lemma. *The homotopy groups of the Morse data* $\pi_i(J, K)$ *for* $f \circ \pi: X \to \mathbb{R}$ *coincide with the homotopy groups of the "punctured" normal Morse data*

$$\pi_i(J - \pi^{-1}(p), K)$$

for all $i < c(A) - \delta_S(A)$.

Proof of lemma. Since $\bar{\pi}$ is finite, the pair (J, K) breaks into a disjoint union of pieces,

$$(J, K) = \coprod_x (J_x, K_x)$$

of the normal Morse data for the map $f \circ \pi$ at the points $x \in \bar{\pi}^{-1}(p)$. For each such x which lies in X, we must compare the homotopy groups

$$\pi_i(J_x - x, K_x) \to \pi_i(J_x, K_x).$$

But this map is surjective if the third term in the exact sequence vanishes, i.e., if

$$\pi_i(J_x, J_x - x) = 0.$$

By Sect. 2.4, the pair $(J_x, J_x - x)$ is homeomorphic to the pair $(N_x, N_x - x)$ which is homotopy equivalent to the pair $(N_x, \partial N_x)$. The homotopy groups of this pair vanish in all degrees $i < c(A) - \delta_S(A)$. □

We now proceed as in the proof that 4.5.2 implies 4.5.1. By Sect. 3.4, Corollary 1, the punctured normal Morse data has the same connectivity as the pair

$$(\mathcal{L}_X^\pi, \partial \mathcal{L}_X^\pi) \times (I, \partial I).$$

But, Proposition $4.6.2_m$ implies that the first nonvanishing homotopy group of the pair $(\mathcal{L}_X^\pi, \partial \mathcal{L}_X^\pi)$ occurs in dimension $\geq \hat{n}$, where

$$\hat{n} = \dim_{\mathbb{C}}(\mathcal{L}_X^\pi) - \sup_{B > S} \delta_S(B).$$

Thus, the first nonvanishing homotopy group of the normal Morse data occurs in degree h, where

$$h \geq \min(\hat{n} + 1, \delta_S(S)).$$

Thus,

$$\Delta = c - h \leq \max(\sup_B (\delta_S(B), \delta_S(S))$$

which completes the proof of Case 1.

Case 2. Suppose $p \in \bar{Z} - Z$. By Sect. 3.4, Corollary 1, the pair

$$(\mathcal{L}_X^\pi, \partial \mathcal{L}_X^\pi) \times (I, \partial I)$$

is already homotopy normal Morse data. Applying $4.5.2_m$ as above, we have

$$\Delta \leq \sup_B (\delta_S(B))$$

where the sup is taken over all strata $B \subset Z$ such that $\bar{B} \supset S$, and $B \neq S$.

In summary we have

$$\Delta \leq \sup_B (\delta_S(B))$$

where the sup is taken over all strata $B \subset Z$ such that $p \in \bar{B}$. □

Proof that $4.6.1_{m-1}$ implies $4.6.2_m$. The proof is exactly the same as the proof that $4.5.1_{m-1}$ implies $4.5.2_m$, except the fibre defect δ_F must be replaced by the singularity defect δ_S. □

Remark. If $x \in X \cap \bar{\pi}^{-1}(p)$, then $\pi_k(\mathcal{L}_x) = 0$ for all k where

$$k \leq c(A) - 2 - \sup_{B \geq A} \delta_S(B).$$

Thus, if X is a local complete intersection then $\mathcal{L}_x$ has the homotopy type of a wedge of spheres of dimension $c(A) - 1$ (see [Lê1]).

Proof of remark. (We again make use of the trick of removing the critical point x and observing that the connectivity of the normal Morse data is unaffected by this operation.) If

$$k \leq c(A) - 2 - \sup_{B \geq A} \delta_S(B)$$

then $\pi_k(\partial N_x) = 0$, which gives rise to the surjection in the following sequence:

$$\begin{aligned} 0 &= \pi_k(\mathscr{L}_x, \partial \mathscr{L}_x) = \pi_{k+1}(\mathscr{L}_x, \partial \mathscr{L}_x) \times (I, \partial I) \\ &= \pi_{k+1}(J_x - x, K_x) \twoheadrightarrow \pi_{k+1}(J_x, K_x) \\ &= \pi_{k+1}(\text{cone}(\mathscr{L}_x), \mathscr{L}_x) \qquad \text{by Sect. 2.4} \\ &= \pi_k(\mathscr{L}_x). \quad \square \end{aligned}$$

4.6*. Local Geometry of the Complement of a Subvariety

In this section we will suppose that X is an n-dimensional complex subvariety of some complex analytic manifold M. We fix a subvariety $W \subset X$ which is locally determined in X by k equations. Fix a point $p \in W$ and let $\partial B_\delta(p)$ denote the boundary of a small ball of radius δ which is centered at the point p. Define

$$L_p(X) = \partial B_\delta(p) \cap X \qquad L_p(W) = \partial B_\delta(p) \cap W.$$

Proposition. *If δ is sufficiently small, then the space $L_p(X) - L_p(W)$ has the homotopy type of a CW complex of dimension $\leq n+k-1$.*

Remarks. This proposition can be obtained from [H5], since $L_p(X) - L_p(W)$ has the same homotopy type as $B_\delta(p) \cap (X - W)$, which, by [SoV], Sect. 2.6 is $(k-1)$-complete.

This proposition implies the homology version (see [Kp3]) of Hamm's theorem [H1] on the connectivity of the link of a local complete intersection as follows: if X is nonsingular, then $L_p(X) = S^{2n-1}$ is a sphere. Since $H_i(L_p(X) - L_p(W)) = 0$ for all $i > n+k-1$, we have (by Alexander duality) $H_i(L_p W) = 0$ for all $i \leq n-k-2$. Hamm's theorem can also be proven by Morse theory using the same method as in the following proof, but this is very close to his original proof. A slightly more refined version of Proposition 4.6* may be true, which would imply Hamm's theorem in homotopy as well as similar results in intersection homology:

Conjecture. The space $L_p(X) - L_p(W)$ deformation retracts (in a stratum preserving way) to a Whitney stratified subset of $L_p(X)$ which has (real) dimension $\leq n+k-1$, and which intersects each stratum of X in a subset S of dimension $\leq s+k-1$, where $s = \dim_{\mathbb{C}}(S)$. (See also [Lê2].)

Proof of proposition. First we consider the case $k=1$, i.e., W is defined as the zeroes of a single equation $f: X \to \mathbb{C}$. We can assume that $f(p) = 0$. We will make use of the results in the appendix Sect. 2.A, which are parallel to those of Proposition 2.4 even though the function f is not generic. Choose $0 < \varepsilon \ll \delta \ll 1$ as in Sect. 2.A.2, and consider the cylindrical neighborhood of p

and its boundary,

$$\tilde{C} = X \cap f^{-1}(D_\varepsilon) \cap B_\delta(p)$$
$$\partial\tilde{C} = X \cap [(f^{-1}(\partial D_\varepsilon) \cap B_\delta(p)) \cup (f^{-1}(D_\varepsilon) \cap \partial B_\delta(p))]$$

where $D_\varepsilon \subset \mathbb{C}$ denotes the closed disk of radius ε. The boundary $\partial\tilde{C}$ of the cylindrical neighborhood is divided into horizontal $\tilde{L}_h$ and vertical $\tilde{L}_v$ parts,

$$\tilde{L}_h = X \cap f^{-1}(D_\varepsilon) \cap \partial B_\delta(p) \qquad \tilde{L}_v = X \cap f^{-1}(\partial D_\varepsilon) \cap B_\delta(p).$$

By Proposition 2.A.3(b), the pair $(\tilde{C}, \partial\tilde{C})$ is homeomorphic (by a stratum preserving homeomorphism) to the pair

$$(X \cap B_\delta(p), X \cap \partial B_\delta(p)).$$

By Proposition 2.A.3(a), the space $\tilde{C} - f^{-1}(0)$ is a topological fibre bundle over $D_\varepsilon - \{0\}$, with fibre

$$\tilde{\mathscr{L}} = f^{-1}(\varepsilon + 0i) \cap X \cap B_\delta(p).$$

Thus, $\partial\tilde{C} - f^{-1}(0)$ (which deformation retracts to $\tilde{L}_v$) is a topological fibre bundle over $\partial D_\varepsilon = S^1$, with fibre $\tilde{\mathscr{L}}$. But $\tilde{\mathscr{L}}$ has the homotopy type of a CW complex of dimension $\leq n-1$ (by Proposition 4.5.2*, which applies to $\tilde{\mathscr{L}}$ as well as to $\mathscr{L}$, or by the remark following the proof of Proposition 4.5.1*). Thus, $\partial\tilde{C} - f^{-1}(0)$ has the homotopy type of a CW complex of dimension $\leq n$, which completes the proof of the case $k=1$.

Now suppose $k>1$. There is (locally near p) an analytic map $F: X \to \mathbb{C}^k$ such that $F(p)=0$ and $W=F^{-1}(0)$. Stratify this map, i.e., choose Whitney stratifications of X and of $Z=F(X)$ so that the restriction of F to each stratum is a submersion and so that W is a union of strata. It is possible to find a $k-1$ dimensional subspace (through the origin) $H \subset \mathbb{C}^k$ such that H is transverse to every stratum of $Z=F(X)$ except for the stratum $\{0\}$. (This follows from Part I, Sect. 1.8 because we can take H to be the kernel of any nondegenerate characteristic covector $\lambda \in T^*\mathbb{C}^k$.) Let $V=F^{-1}(H)$ and consider the triple

$$L_p(W) \subset L_p(V) \subset L_p(X).$$

The transversality condition on H guarantees that the inclusion $L_p(V) \subset L_p(X)$ is a normally nonsingular inclusion (see Part I, Sect. 1.11) of (real) codimension two. Thus, there is a tubular neighborhood $N(L_p(V))$ of $L_p(V)$ in $L_p(X)$ and a (topologically) locally trivial projection $\pi: N(L_p(V)) \to L_p(X)$ whose fibres are two-disks D^2. Thus,

$$\begin{aligned} L_p(X) - L_p(W) &= (L_p(X) - L_p(V)) \cup (N(L_p(V) - L_p(W)) \\ &= (L_p(X) - L_p(V)) \cup (N(L_p(V) - \pi^{-1}(L_p(W))) \\ &= (L_p(X) - L_p(V)) \cup N(L_p(V) - L_p(W)) \end{aligned}$$

where $N(L_p(V) - L_p(W)) = \pi^{-1}(L_p(V) - L_p(W))$ is a two-dimensional disk bundle over $L_p(V) - L_p(W)$. The intersection of these two spaces,

$$(L_p(X) - L_p(V)) \cap N(L_p(V) - L_p(W))$$

is homotopy equivalent to the bounding circle bundle,

$$\partial N(L_p(V) - L_p(W)) \to (L_p(V) - L_p(W)).$$

Therefore, $L_p(X)-L_p(W)$ has the homotopy type of a CW complex of dimension $\leq \hat{n}^*$, where

$$\begin{aligned}\hat{n}^* &= \sup(\dim(L_p(X)-L_p(V)), \dim(L_p(V)-L_p(W))+1+1)\\ &= \sup(n+1-1, (n-1)+k-1-1+2) \qquad \text{by induction and the case } k=1\\ &= \sup(n, n+k-1) = n+k-1. \quad \square\end{aligned}$$

4.A. Appendix. The Levi Form and the Morse Index

One of the miracles which occurs in complex Morse theory is the following: for a given complex manifold A and smooth function $f: A \to \mathbb{R}$ it is often possible to estimate the Morse index of *any* critical point of $f \mid B$, for *any* complex submanifold $B \subset A$ (see 4.A.4). This can be done whenever we have estimates on the signature of the Levi form of f, and such estimates can be found when f is the Euclidean (or Fubini-Study) distance from a linear subspace of Euclidean (or complex projective) space (4.A.5, 4.A.6). This miracle occurs because the Levi form for $f \mid B$ at a point $p \in B$ is the restriction to $T_p B$ of the Levi form for f at p. The analogous statement for the Hessian is false. In this section we review these basic and well-known facts about the Levi form.

4.A.1. Suppose A is a complex analytic manifold and $f: A \to \mathbb{R}$ is a smooth function. We shall use the notation $\partial f(p)$ to denote the complex linear map $T_p A \to \mathbb{C}$ which is given in local coordinates by

$$\partial f(p) = \left(\frac{\partial f}{\partial z_1}, \frac{\partial f}{\partial z_2}, \ldots, \frac{\partial f}{\partial z_n}\right).$$

Since f is real valued, the complex conjugate of $\partial f(p)$ is the antilinear map $\bar{\partial} f(p): T_p A \to \mathbb{C}$, which is given in local coordinates by

$$\bar{\partial} f(p) = \left(\frac{\partial f}{\partial \bar{z}_1}, \frac{\partial f}{\partial \bar{z}_2}, \ldots, \frac{\partial f}{\partial \bar{z}_n}\right).$$

The (real) differential $df(p): T_p A \to \mathbb{R}$ is the (real) linear map

$$df(p) = \tfrac{1}{2}(\bar{\partial} f(p) + \partial f(p)).$$

Definition. The *E.E. Levi form* of f at a point $p \in A$ is the (Hermitian) form,

$$L = \partial \bar{\partial} f(p): T_p A \times T_p A \to \mathbb{C}$$

which, in local coordinates about p, is given by the matrix of partial derivatives,

$$L_{ij} = \frac{\partial^2 f}{\partial z_i \partial \bar{z}_j}.$$

It is easy to see that if $B \subset A$ is a complex analytic submanifold which contains the point p, then the Levi form of $f \mid B: B \to \mathbb{R}$ at p coincides with the restriction of L to $T_p B \times T_p B$. (This is a remarkable fact, because the same statement does not hold for the Hessian H of f, unless p was a critical point of f.)

We associate with L the (real valued) quadratic form

$$\hat{L}(\xi)=L(\xi,\bar{\xi})$$

and define the signature $\sigma(L)$ to be the (complex) dimension of the largest subspace of T_pA on which $\hat{L}$ is negative definite. We also define the nullity $\nu(L)$ to be the complex dimension of the largest subspace on which L vanishes. Similarly, we associate to the Hessian H of f the quadratic form

$$\hat{H}(\xi)=H(\xi,\xi)$$

and define the signature $\sigma(H)$ to be the (real) dimension of the largest subspace of T_pA on which $\hat{H}$ is negative definite.

4.A.2. Lemma. *Suppose the Hessian H is nondegenerate. Then*

$$\sigma(H)\geq\sigma(L)+\nu(L)=n-\sigma(-L)$$
$$\sigma(-H)\geq\sigma(-L)+\nu(L)=n-\sigma(L)$$

where n denotes the complex dimension of the vectorspace T_pA.

Proof. First we verify the proposition in the case $A=\mathbb{C}$. We may write $z=x+iy$ and find that

$$L=\frac{\partial^2 f}{\partial z\,\partial\bar{z}}=\frac{\partial^2 f}{\partial x^2}+\frac{\partial^2 f}{\partial y^2}=\text{trace}(H).$$

Thus, L is the sum of the eigenvalues of H. If $\sigma(L)+\nu(L)=1$, i.e., if L is negative semidefinite, then H cannot be positive definite. Thus, signature $(H)\geq 1$. Similarly if $\sigma(-L)+\nu(-L)=1$, then L is positive semidefinite so H cannot be negative definite, so $\sigma(-H)\geq 1$.

We now consider the general case. Let L_n be a complex $\sigma(L)+\nu(L)$-dimensional subspace of T_pA on which L is negative semidefinite, and let H_p be a maximal real subspace of T_pA on which H is positive definite. Consider the subspace $V=H_p\cap L_n$ of L_n. This subspace V contains no complex line (for on such a complex line, H_p would be positive definite and L_n would be negative semidefinite). It follows that $\dim_{\mathbb{R}}(V)\leq\dim_{\mathbb{C}}(L_n)$ (otherwise $V\cap iV$ is a nontrivial complex subspace of V). Thus,

$$\dim_{\mathbb{C}}(L_n)\geq\dim(V)\geq\dim(H_p)+2\dim_{\mathbb{C}}(L_n)-2n$$

so

$$\sigma(H)=2n-\dim H_p\geq\dim_{\mathbb{C}}(L_n)=\sigma(L)+\nu(L)$$

as desired. □

Remark. If we drop the assumption that H is nondegenerate, then the above argument gives:

$$\sigma(H)+\nu(H)\geq\sigma(L)+\nu(L)=n-\sigma(-L)$$
$$\sigma(-H)+\nu(-H)\geq\sigma(-L)+\nu(-L)=n-\sigma(L).$$

4.A.3. Levi Form of a Composition. The signature of the Levi form of f can change by one if we compose f with a strictly increasing function $g\colon\mathbb{R}\to\mathbb{R}$. In this section we define the *restricted signature* of the Levi form, which does not change under such a composition. It is the restricted signature of the Levi form of $f\colon A\to\mathbb{R}$ which determines the Morse index of any nondegenerate critical point of $f|B$ (where B is a complex analytic submanifold of A). This number was considered by [Fr1] and later by [Sm1], in the context of Morse theory.

Let A be a complex analytic n-dimensional manifold and let $f\colon A\to\mathbb{R}$ be a smooth function. Let $E=\ker(\partial f(p))=\ker(\bar\partial f(p))$. If $df(p)\neq 0$, then E is the unique complex $n-1$ dimensional vector subspace of $\ker(df(p))$ (and this is the only case of nontrivial content).

Definition. The *restricted Levi form* $L'(f)$ is the restriction of the Levi form of f to the subspace $E=\ker(\partial f(p))$.

Definition. The *restricted signature* $\sigma'(f)$ (resp. *restricted nullity* $\nu'(f)$) at the point p is the signature (resp. nullity) of the restricted Levi form of f, i.e.,

$$\sigma'(f)=\sigma(L(f)|E)$$
$$\nu'(f)=\nu(L(f)|E).$$

Now suppose that $g\colon\mathbb{R}\to\mathbb{R}$ is a smooth strictly increasing function, and $h\colon A\to A$ is an invertible analytic isomorphism.

Proposition. *The restricted signature (and nullity) satisfy the following relations:*

$$\sigma'(f)(p)=\sigma'(g\circ f\circ h)(h^{-1}(p))$$
$$\nu'(f)(p)=\nu'(g\circ f\circ h)(h^{-1}(p))$$
$$\sigma'(f)(p)\leq\sigma(L(f)(p))\leq\sigma'(f)(p)+1.$$

It follows that $\sigma(L(f)(p))$ and $\sigma(L(g\circ f)(p))$ can differ at most by one.

Proof. A simple computation shows

$$L(g\circ f)(p)=\partial\bar\partial(g\circ f)(p)=g'(f(p))\,\partial\bar\partial f(p)+g''(f(p))\,\partial f(p)\otimes\bar\partial f(p).$$

Thus, the restriction to E of the Levi form of $g\circ f$ is a positive multiple of the restriction to E of the Levi form of f. Therefore, their signatures are the same. A similar calculation shows that

$$L(f\circ h)(q)=L(f)(H(q))\circ(\partial h(q)\otimes\bar\partial\bar h(q))$$

so it has the same signature, restricted signature, nullity, and restricted nullity as $L(f)(h(q))$. This proves the first two equations. The last part is obvious, since either E is a complex codimension one subspace of T_pA or else $E=T_pA$. □

Remark. This difference of one is a source of some confusion in Morse theory, because if f has a critical point p of index λ on some submanifold $B\subset A$, then $g\circ f$ also has a critical point of index λ on B. In order to use Proposition 4.A.2 to estimate the value of λ, most authors replace the function f by some composition $g\circ f$. A better technique is to estimate directly the restricted signature $\sigma'(f)$.

4.A.4. Theorem. *Suppose A^n is a complex manifold, $f\colon A\to\mathbb{R}$ is a smooth function, $B^k\subset A^n$ is a complex submanifold, and the restriction $f|B$ has a nondegen-*

erate critical point $p \in B$. Then the Morse index λ for $f \mid B$ at p is bounded as follows:

$$k+\sigma'(L_A) \geq \lambda \geq k-\sigma'(-L_A)$$

where $\sigma'(L_A)$ denotes the restricted signature of the Levi form of $f: A \rightarrow \mathbb{R}$ at the point p.

Remarks. (1) This means (in the language of Sect. 4.2) that the convexity defect Γ of $f \mid B$ is $\leq \sigma'(-L_A)$ and the dual convexity defect Γ^* of $f \mid B$ is $\leq \sigma'(L_A)$.

(2) It follows from 4.A.3 that the same result holds if we replace the restricted signature σ' by the signature σ in the above equation.

(3) If $f \mid B$ has a *degenerate* critical point $p \in B$, then there is function f', which is arbitrarily close to f (in the Whitney C^∞ topology) with nondegenerate critical points $q_1, q_2, \ldots, q_k$ near p. However, the restricted signature of the Levi form of f' at q_i may differ (by as much as $v'(L)$) from $\sigma'(L(f))$, because the eigenvalues of $L(f)$ which are zero may become nonzero. This gives the following corollary:

Corollary. *Suppose A is a complex manifold, $f: A \rightarrow \mathbb{R}$ is a smooth function and suppose there are numbers σ^+, σ^-, and v so that for each point $p \in A$ we have,*

$$\sigma'(L_A(p)) \leq \sigma^+, \quad \sigma'(-L_A(p)) \leq \sigma^-, \quad v'(L_A(p)) \leq v.$$

Let $B \subset A$ be any k-dimensional complex submanifold. Then there is a function $f': A \rightarrow \mathbb{R}$ which is arbitrarily close to f (in the Whitney C^∞-topology), so that $f' \mid B$ is a Morse function and the Morse index λ of any critical point of $f' \mid B$ is bounded as follows:

$$k+\sigma^+ + v \geq \lambda \geq k-\sigma^- - v.$$

4.A.4. Proof of theorem. Let $\sigma(L_B)$ denote the signature of the Levi form of the restriction $f \mid B: B \rightarrow \mathbb{R}$ at the point p. Since p is a critical point, $T_p B$ is a subspace of $\ker(df(p))$. Since the Hessian H_B of $f \mid B$ is nondegenerate, we have (by Lemma 4.A.3)

$$\begin{aligned} k+\sigma'(L_A) &\geq k+\sigma(L_B) \\ &= 2k-(k-\sigma(L_B)) \\ &\geq 2k-\sigma(-H_B)=\lambda=\sigma(H_B) \\ &\geq k-\sigma(-L_B) \\ &\geq k-\sigma'(-L_A). \quad \square \end{aligned}$$

Example. If $f: \mathbb{C}^n \rightarrow \mathbb{R}$ is a (real) linear function, or is the real part of a complex analytic function then its Levi form is zero. If B is a complex k-dimensional submanifold of $\mathbb{C}^n$, then the Morse index of any nondegenerate critical point of $f \mid B$ is exactly k.

4.A.5. Levi Form of the Distance Function on Affine Space. The (distance)^2 from a codimension r linear subspace in $\mathbb{C}^n$ is given in local coordinates by

$$f(z)=\sum_{i=1}^{r} z_i \bar{z}_i$$

so its Levi form L has $\sigma(L)=0$, $\sigma(-L)=r$, $\nu(L)=n-r$. Theorem 4.A.4 now states: If B is a k-dimensional complex analytic submanifold of $\mathbb{C}^n$, the index λ of any nondegenerate critical point of $f\,|\,B$ is bounded as follows:

$$k \geq \lambda \geq k-r.$$

4.A.6. Levi Form of the Distance Function in Projective Space. The (distance)^2 from a codimension r linear subspace H_1 of $\mathbb{CP}^n$ (with complementary $r-1$ dimensional linear subspace H_2) is given in homogeneous coordinates by (see Sect. 4.4)

$$f([z_0:z_1:\ldots:z_n])=\frac{\sum\limits_{i=n-r+1}^{n} z_i\bar{z}_i}{\sum\limits_{i=0}^{n} z_i\bar{z}_i}.$$

Proposition. *If $p\notin H_1\cup H_2$, then $\sigma'(f)(p)=n-r$, $\nu'(f)(p)=0$ and $\sigma'(-f)=r-1$. If $p\in H_1$, then $\sigma'(f)(p)=n-r$, $\nu'(f)(p)=r-1$ and $\sigma'(-f)(p)=0$. If $p\in H_2$, then $\sigma'(f)(p)=0$, $\nu'(f)(p)=n-r$ and $\sigma'(-f)(p)=r-1$.*

Proof. Let $p=[c_0:c_1:c_2:\ldots:c_n]\in\mathbb{CP}^n$. Consider the r-dimensional submanifold $K=K(p)$ which contains the point p and is given by the equations

$$z_0=c_0 \quad z_1=c_1 \quad \ldots \quad z_{n-r}=c_{n-r}$$

and consider the submanifold K^* which is given by the equations

$$z_{n-r+1}=c_{n-r+1} \quad z_{n-r+2}=c_{n-r+2} \quad \ldots \quad z_n=c_n.$$

Since $\partial f(p)$ does not vanish on T_pK or on T_pK^*, it suffices to show that $L(f)(p)\,|\,(T_pK\cap\ker\partial f(p))$ is positive definite and $L(f)(p)\,|\,(T_pK^*\cap\ker\partial f(p))$ is negative definite. These statements are verified using the following trick: since the function $x/(1-x)$ is monotonically increasing on $[0,1)$ we can (by 4.A.3) replace the function f with the function $g=f/(1-f)$ when computing the Levi form. But, the restriction of g to the submanifold K is given by:

$$g\,|\,K_p=\sum_{i=n-r+1}^{n} z_i\bar{z}_i/\text{const}.$$

Thus, $L(g\,|\,K)(p)$ is positive definite, so $L(g)\,|\,(T_pK\cap\ker\partial g(p))$ is also positive definite. But, this form is a positive multiple of $L(f)(p)\,|\,(T_pK\cap\ker\partial f(p))$. The other verifications are similar. $\square$

Corollary. *Suppose $B\subset\mathbb{CP}^n-(H_1\cup H_2)$ is a k-dimensional submanifold. Let $\tilde{f}\colon\mathbb{CP}^n\to\mathbb{R}$ be a function whose restriction to B has only nondegenerate critical points. If $\tilde{f}$ is sufficiently close (in the C^∞ topology) to f, then the index λ of each critical point of $f\,|\,B$ is bounded as followed:*

$$k+n-r\geq\lambda\geq k-r+1.$$

Proof. Apply Corollary 4.A.3. $\square$

Chapter 5. Proof of the Main Theorems

5.1. Proof of Theorem 1.1: Relative Lefschetz Theorem with Large Fibres

Theorem. *Let X be a purely n-dimensional nonsingular connected algebraic variety. Let $\pi: X \to \mathbb{CP}^N$ be an algebraic map and let $H \subset \mathbb{CP}^N$ be a linear subspace of codimension c. Let H_δ be the δ-neighborhood of H with respect to some real analytic Riemannian metric. Define $\phi(k)$ to be the dimension of the set of points $z \in \mathbb{CP}^N - H$ such that the fibre $\pi^{-1}(z)$ has dimension k. (If this set is empty define $\phi(k) = -\infty$.) If δ is sufficiently small, then then the homomorphism induced by inclusion, $\pi_i(\pi^{-1}(H_\delta)) \to \pi_i(X)$ is an isomorphism for all $i < \hat{n}$ and is a surjection for $i = \hat{n}$, where*

$$\hat{n} = n - \sup_k (2k - (n - \phi(k)) + \inf(\phi(k), c-1)) - 1.$$

Furthermore: *In this theorem, π is not necessarily poper, and $\pi^{-1}(H_\delta)$ may be replaced by $\pi^{-1}(H)$ if H is generic or if π is proper. The assumption that X is algebraic may be replaced by the assumption that X is the complement of a closed subvariety of a complex analytic variety $\bar{X}$ and the π extends to a proper analytic map $\bar{\pi}: \bar{X} \to \mathbb{CP}^N$.*

Preliminaries to the proof. Since π is algebraic, it extends to a proper algebraic map

$$\bar{\pi}: \bar{X} \to \mathbb{CP}^N$$

on some algebraic variety $\bar{X}$ which contains X as an open subset. It is possible (Part I, Sect. 1.7) to stratify $\bar{X}$ and

$$\bar{Z} = \bar{\pi}(\bar{X}) \subset \mathbb{CP}^N$$

so that $\bar{\pi}$ is a stratified map, so that X is a union of strata of $\bar{X}$, and so that $Z = \pi(X)$ is a union of strata in $\bar{Z}$. (It follows that the function $z \mapsto \dim \bar{\pi}^{-1}(z)$ is constant on each stratum of $\bar{Z}$.) Let $G \subset \mathbb{CP}^N$ be a linear subspace which is complementary to H (i.e., $H \cap G = \phi$ and $\dim(G) = c-1$) and which is transverse to each stratum of $\bar{Z}$.

By Sect. 4.4, the distance $\bar{f}: \mathbb{CP}^N \to \mathbb{C}$ from H (with respect to G) is a real analytic function which satisfies: $0 \le \bar{f}(x) \le 1$, $\bar{f}^{-1}(0) = H$, and $\bar{f}^{-1}(1) = G$. Since $\bar{f} \circ \bar{\pi}: \bar{X} \to \mathbb{R}$ has finitely many critical values, there is a number $\varepsilon > 0$ so that the interval $(0, \varepsilon]$ contains no critical values of $\bar{f} \circ \bar{\pi}$. Choose $\delta > 0$ so that $\bar{f}(H_\delta) \subset [0, \varepsilon]$. It follows (see Proposition 5.A.2) that the inclusion $\pi^{-1}(H_\delta) \cap X \to \pi^{-1}(\bar{f}^{-1}[0, \varepsilon]) \cap X$ induces isomorphisms on homotopy groups of all dimensions. Thus, we can replace the neighborhood H_δ by the set $H_\varepsilon = \bar{f}^{-1}[0, \varepsilon]$. Now approximate $\bar{f}$ by a (sufficiently C^∞ close) function f which coincides with $\bar{f}$ in the region $\bar{f}^{-1}[0, \varepsilon]$. Thus,

(a) $0 \le f(x) \le 1$, $f^{-1}(0) = H$, and $f^{-1}(1) = G$.
(b) $f^{-1}[0, \varepsilon) = H_\varepsilon$.
(c) f is a Morse function for the above stratification of $\bar{Z} \cap f^{-1}(\varepsilon/2, 1)$ with distinct critical values (Part I, Sect. 2.2.1).
(d) For every critical point $p \in \bar{Z}$ of f we have $\Gamma(p) \le \min(d(S), c-1)$ where $d(S)$ is the dimension of the stratum which contains the critical point p (4.A.4 and 4.A.6).

Such an approximation exists by [P1]. (In fact, (Part I, Sect. 2.2) by choosing the complementary subspace G generically, we can even guarantee that the original function $\bar{f}$ restricts to a Morse function on $\bar{Z}$.)
Consider the following setup:

$$\begin{array}{ccccc} X & \subset & \bar{X} & & \\ \downarrow \pi & & \downarrow \bar{\pi} & & \\ Z & \subset & \bar{Z} & \xrightarrow[f]{} & \mathbb{R}. \end{array}$$

We will apply Morse theory to this function, building X from $\pi^{-1}(H_\varepsilon)$ by crossing critical values of f which correspond to critical points $p \in \bar{Z} - H_\varepsilon$.

Proof of theorem. First we rewrite the formula for $\hat{n}$ as $\hat{n} = \inf_A (n(A))$, where

$$n(A) = n - (2\dim_{\mathbb{C}} \pi^{-1}(a) - (n - \dim_{\mathbb{C}}(A)) + \min(\dim_{\mathbb{C}}(A), c-1)) - 1$$

where the inf is taken over all strata $A \subset Z$, and where $a \in A$. We estimate the connectivity of the pair $(X_{\le 1-\theta}, X \cap \pi^{-1}(H_\varepsilon))$ (for sufficiently small θ) using the relative Morse theory of the function f. If the interval $[a, b]$ contains a single critical value of f corresponding to a critical point $p \in \bar{Z}$, and if $f(p) \in (a, b)$, then by Proposition 4.3 we have $\pi_i(X_{\le b}, X_{\le a}) = 0$ for all $i \le \hat{m}$, where

$$\hat{m} = n - (\Delta(p) + \Gamma(p)) - 1.$$

However Sect. 4.4, Corollary 1 gives

$$\Gamma(p) \le \min(d(S), c-1)$$

where $d(S)$ is the dimension of the stratum containing p. Furthermore, 4.5.1 gives

$$\Delta(p) \le \sup_{A > S} (0, 2\dim_{\mathbb{C}} \bar{\pi}^{-1}(p) - (n - d(A))).$$

Where the sup is taken over all strata $A \subset Z$ such that $\bar{A} \supset S$. Thus,

$$\hat{m} \ge n - (\sup_{A > S} (2\dim \bar{\pi}^{-1}(p) - (n - d(A))) + \min(d(S), c-1)) - 1.$$

But, $\hat{n}$ is just the infemum of this expression when S is allowed to vary over all the strata in Z. Thus, $\hat{m} \geq \hat{n}$. We conclude that $\pi_i(X_{\leq 1-\theta}, X \cap \pi^{-1}(H_\varepsilon)) = 0$ for all $i \leq \hat{n}$.

There are two ways to handle the addition of the degenerate maximum $\pi^{-1} f^{-1}(1)$:

Method 1. Find a perturbation (near $f^{-1}(1)$) of the function f, so that $f|\bar{Z}$ is a Morse function with nondegenerate critical points (see Part I, Sect. 2.2, or [P1]). Estimate the Morse index of each such critical point using Corollary 4.A.4, noting that the Levi form of f at any point $p \in f^{-1}(1)$ has $\dim_{\mathbb{C}}(G) = c-1$ eigenvalues which are zero, and all the rest are negative. Now proceed as above, showing that each of these new critical points does not affect the relative homotopy groups $\pi_i(X_{\leq 1-\eta}, X \cap H_\varepsilon)$ in the range of dimensions $i \leq \hat{n}$.

Method 2. Show directly that the addition of the degenerate maximum $\pi^{-1} f^{-1}(1)$ does not affect the connectivity of $(X_{\leq 1-\theta}, X \cap \pi^{-1}(H_\varepsilon))$ in the range of dimensions which we are considering. Here are the details:

Since $f^{-1}(1)$ is transverse to all the strata of $\bar{Z}$, the inclusion $\bar{Z} \cap f^{-1}(1) \subset \bar{Z}$ is normally nonsingular (Part I, Sect. 1.11). Thus, the neighborhood and its boundary

$$\bar{Z} \cap (f^{-1}[1-\theta, 1], f^{-1}(1-\theta))$$

is homeomorphic (by a stratum preserving homeomorphism) to the disk bundle and boundary sphere bundle of a vectorbundle over $\bar{Z} \cap f^{-1}(1)$, whose fibre is $\mathbb{C}^{N-c+1}$ (provided θ is chosen sufficiently small). Since this homeomorphism preserves strata, the same is true for the pair

$$(X \cap \bar{\pi}^{-1} f^{-1}[1-\theta, 1], X \cap \bar{\pi}^{-1} f^{-1}(1-\theta)).$$

The long exact sequence in homotopy for this pair coincides with the long exact sequence for the fibration of the boundary sphere bundle,

$$X \cap \bar{\pi}^{-1} f^{-1}(1-\theta) \to f^{-1}(1)$$

so we obtain

$$\pi_i(X \cap \bar{\pi}^{-1} f^{-1}[1-\theta, 1], X \cap \bar{\pi}^{-1} f^{-1}(1-\theta)) \cong \pi_{i-1}(S^{2N-2c+1})$$

which vanishes for all $i < 2N-2c+1$. So, it suffices to show that $\hat{n} \leq 2N-2c+1$.

Since X is n-dimensional, there is a stratum A in Z such that $\dim(\pi^{-1}(A) \cap X) = n$. If $\phi_0 = \dim(A)$ and $k_0 = \dim(\pi^{-1}(a) \cap X)$ for any $a \in A$, then $n = \phi_0 + k_0$. The formula for $\hat{n}$ is an infemum over all strata in Z, so

$$\hat{n} \leq n - (2k_0 - (n-\phi_0) + \inf(\phi_0, c-1)) - 1 = \sup(-1, \phi_0 - c).$$

Clearly, $\phi_0 \leq N$ and $c \leq N$, so

$$\phi_0 - c \leq N - c \leq 2N - 2c \leq 2N - 2c + 1$$

as desired.

Proof of furthermore. The furthermore is a direct consequence of Proposition 5.A.1 or 5.A.3: under either hypothesis, the inclusion

$$X \cap \pi^{-1}(H) \to X \cap \pi^{-1}(H_\delta) \to X \cap \pi^{-1}(H_\varepsilon)$$

induces isomorphisms on homotopy groups of all dimension. (In fact, a combination of these two arguments can be used to show that H_δ may be replaced by H provided H is transverse to the strata of $\bar{Z}$ over which π is not proper.) □

5.1*. Proof of Theorem 1.1*: Homotopy Dimension with Large Fibres

Theorem. *Let X be an n-dimensional (possibly singular) complex analytic variety. Let $\pi: X \to \mathbb{CP}^N - H$ be a proper analytic map, where H is a linear subspace of codimension c. Let $\phi(k)$ denote the dimension of the set of points $y \in \pi(X)$ such that the fibre $\pi^{-1}(y)$ has dimension k. (If this set is empty define $\phi(k) = -\infty$.) Then, X has the homotopy type of a CW complex of (real) dimension less than or equal to*

$$\hat{n}^* = n + \sup_k (2k - (n - \phi(k)) + \inf(\phi(k), c-1)).$$

Proof. Let $Z = \bar{Z} = \pi(X) \subset \mathbb{CP}^N - H$. Stratify the map $\pi: X \to Z$. Let $G \subset \mathbb{CP}^N$ be a linear subspace of dimension $c-1$ which is complementary to H (i.e., $G \cap H = \phi$) and which is transverse to all the strata of Z. Let $(1-f): \mathbb{CP}^N \to \mathbb{R}$ denote the Morse function which is created in Sect. 5.1, i.e., $f^{-1}(0) = G$, $f^{-1}(1) = H$ and f has finitely many nondegenerate critical points on Z with distinct critical values.

As in Sect. 5.1 we rewrite the expression for $\hat{n}^*$ as follows: $\hat{n}^* = \sup_A (n^*(A))$, where

$$n^*(A) = n + (2\dim_{\mathbb{C}} \pi^{-1}(a) - (n - \dim(A)) + \inf(\dim(A), c-1))$$

where the sup is taken over all strata $A \subset Z$, and where $a \in A$.

We wish to build X from $\pi^{-1}(G \cap Z)$ by attaching CW complexes of dimension $\leq \hat{n}^*$ at each critical point of f. By Proposition 5.A.3, the inclusion $\pi^{-1}(G \cap Z) \to \pi^{-1}(G_\varepsilon \cap Z)$ is a homotopy equivalence for sufficiently small ε, where $G_\varepsilon = f^{-1}[0, \varepsilon]$. Furthermore, we note that $\dim_{\mathbb{R}}(\pi^{-1}(G \cap Z)) \leq \hat{n}^*$. This is because for any stratum $A \subset Z$, if $\phi_0 = \dim_{\mathbb{C}}(A)$ and if $k_0 = \dim_{\mathbb{C}}(\pi^{-1}(a))$ for any $a \in A$, then by transversality,

$$\dim_{\mathbb{R}}(\pi^{-1}(G \cap A)) \leq 2\phi_0 + 2k_0 + 2(c-1) - 2N.$$

But $\phi_0 \leq N$ and $c-1 \leq N$, so

$$\dim_{\mathbb{R}}(\pi^{-1}(G \cap A)) \leq \inf(2\phi_0 + 2k_0, \phi_0 + 2k_0 + c - 1).$$

However,

$$\hat{n}^* \geq n + (2k_0 - (n - \phi_0) + \inf(\phi_0, c-1)) \geq \inf(2k_0 + 2\phi_0, 2k_0 + \phi_0 + c - 1)$$

which establishes the claim.

Now consider the effect of passing an isolated critical value $v \in \mathbb{R}$ of f, which corresponds to a nondegenerate critical point $p \in Z$. If the interval $[a, b]$ contains no critical value other than v, then by Proposition 4.3* the space $X_{\leq b}$ is obtained from the space $X_{\leq a}$ by attaching cells of dimension less than or equal to

$$\hat{m}^* = n + \Delta^*(p) + \Gamma^*(p).$$

By Proposition 4.4, we have

$$\Gamma^*(p) \leq \min(d(S), c-1)$$

where S is the stratum of Z which contains the point p (since codimension$(G) = N-c+1$). By Proposition 4.5.1* we have

$$\Delta^*(p) \leq \sup_A (2\dim_{\mathbb{C}} \pi^{-1}(q) - (n-d(A)))$$

where $q \in A$ and the sup is taken over all strata A such that $p \in \bar{A}$. Thus,

$$\hat{m}^* \leq n + \sup_A (2\dim_{\mathbb{C}}(\pi^{-1}(q)) - (n-d(A)) + \inf(d(S), c-1))$$

where the sup is taken over all strata A of Z such that $S \subset \bar{A}$. Taking the sup of this expression over all strata in Z we conclude that $\hat{m}^* \leq \hat{n}^*$. □

5.2. Proof of Theorem 1.2: Lefschetz Theorem with Singularities

Theorem. *Let X be an algebraic subvariety of some algebraic manifold M. Let $\pi: X \to \mathbb{CP}^N$ be a (not necessarily proper) algebraic map withy finite fibres. Let H be a linear subspace of codimension c in $\mathbb{CP}^N$, and let H_δ be a δ-neighborhood of H (with respect to some real analytic Riemannian metric, as in Theorem 5.1). Let $\phi(k)$ denote the dimension of the set of points $p \in X - \pi^{-1}(H)$ such that a neighborhood of p (in X) can be defined (in M) by k equations, and no fewer. (If this set is empty define $\phi(k) = -\infty$.) If $\delta > 0$ is sufficiently small, then the homomorphism*

$$\pi_i(\pi^{-1}(H_\delta)) \to \pi_i(X)$$

is an isomorphism for all $i < \hat{n}$ and is a surjection for $i = \hat{n}$, where

$$\hat{n} = \inf_k (\dim_{\mathbb{C}}(M) - k - \inf(\phi(k), c-1)) - 1.$$

Furthermore: *It is possible to replace $\pi^{-1}(H_\delta)$ by $\pi^{-1}(H)$ if H is generic or if π is proper.*

Preliminaries to the proof. Since π is algebraic and finite, it extends to a *proper finite* algebraic map $\bar{\pi}: \bar{X} \to \mathbb{CP}^N$ on some algebraic variety $\bar{X}$ which contains X as an open dense subvariety. Choose Whitney stratifications of $\bar{X}$ and of $\bar{Z} = \bar{\pi}(\bar{X})$ so that $\bar{\pi}$ is a stratified map and so that X is a union of strata.

Proof. The proof (in the case that X is purely n-dimensional) is exactly the same as that in Sect. 5.1, however, we first rewrite the formula for $\hat{n}$ as follows: $\hat{n} = \inf_A(\hat{n}(A))$ where the inf is taken over all strata $A \subset Z$, and

$$\hat{n}(A) = n - (E(A) - D + \inf(d(A), c-1)) - 1$$

where $E(A)$ denotes the (minimum) number of equations needed to define a neighborhood of any point $x \in X \cap \bar{\pi}^{-1}(A)$, $d(A)$ is the complex dimension of the stratum A, and D is the codimension of X in M. Now follow the proof in Sect. 5.1, but instead of 4.5.1 (which applies only to nonsingular X) we use the following estimate which applies to finite π:

$$\Delta(p) \leq \sup_{A} (E(A) - D)$$

from 4.6.1 and Hamm's theorem (Sect. 4.6). Here, the sup is taken over all strata $A \subset Z$ which contain the point p in their closures. (If X is not purely n-dimensional, then these estimates must be applied separately to each of the points $x \in \pi^{-1}(p)$ whenever p is a critical point of the Morse function.) □

Proof of furthermore. This is the same as the proof of the furthermore in Sect. 5.1.

5.2*. Proof of Theorem 1.2*: Homotopy Dimension of Nonproper Varieties

In this section we fix an n-dimensional complex analytic variety X and let $\pi: X \to \mathbb{CP}^N - H$ be a proper finite analytic map, where H is a linear subspace of codimension c. Let W be a subvariety of X. Choose a Whitney stratification of the map π, and for each stratum A of X define $d(A) = \dim_{\mathbb{C}}(A)$. Let $k(A)$ be the number of equations needed to define $W \cap T(A)$, where $T(A)$ is a neighborhood of the stratum A, and set $k(A) = 0$ if $W \cap A = \phi$.

Theorem. *The space $X - W$ has the homotopy type of a CW complex of dimension $\leq \hat{n}^*$, where*

$$\hat{n}^* = \sup_{A} (n + \sup(0, k(A) - 1) + \inf(d(A), c - 1))$$

where the sup is taken over all strata $A \subset X$.

Proof. Let $Z = \pi(X) \subset \mathbb{CP}^N - H$. Stratify X, Z, and W so that W is a union of strata in X and so that the map $\pi: X \to Z$ is a stratified map. Let $G \subset \mathbb{CP}^N$ be a linear subspace of dimension $c - 1$, which is complementary to H and is transverse to all the strata of Z. Choose a Morse function $f: \mathbb{CP}^N \to R$ which is an approximation to the distance function from G, so that f has nondegenerate critical points on Z and so that $f^{-1}(0) = G$ and $f^{-1}(1) = H$. (See Proposition Sect. 4.4. In fact, (Part I, Sect. 2.2) by choosing the complementary subspace G generically, we can guarantee that the distance function from G restricts to a Morse function on $\bar{Z}$.) Consider the following setup:

$$\begin{array}{ccc} X - W \subset X & & \\ \Big\downarrow \pi & & \\ Z & \xrightarrow{\ f\ } & \mathbb{R}. \end{array}$$

We wish to build $X - W$ from $\pi^{-1}(G \cap Z) \cap (X - W)$ by attaching CW complexes of dimension $\leq \hat{n}^*$ at each critical point of f.

It is a triviality that $\dim_{\mathbb{R}}(\pi^{-1}(G \cap Z) \cap (X - W)) \leq \hat{n}^*$: Since π is finite and G is transverse to each stratum of Z we have

$$\begin{aligned}\dim(\pi^{-1}(G\cap Z)\cap(X-W)) &\leq \sup_A \dim(G\cap A)\\ &\leq \sup_A 2(\dim_{\mathbb{C}} A-(N-c+1))\\ &\leq \sup_A (\dim_{\mathbb{C}}(X)+\dim_{\mathbb{C}}(A)-2(N-c+1))\\ &\leq \sup_A (\dim_{\mathbb{C}}(X)+\inf(\dim_{\mathbb{C}}(A), c-1))\\ &\leq \hat{n}^*.\end{aligned}$$

Furthermore, (by 5.A.3) for sufficiently small $\varepsilon>0$, the set $(X-W)\cap(\pi\circ f)^{-1}[0,\varepsilon]$ is homotopy equivalent to the set $(X-W)\cap(\pi\circ f)^{-1}(0)=(X-W)\cap\pi^{-1}(G\cap Z)$. Now consider the effect of passing a critical value $v\in\mathbb{R}$ of f, which corresponds to a nondegenerate critical point $p\in Z$. As in 4.6.2, the relative normal Morse data (J, K) for f at the point p breaks into a disjoint union of pieces,

$$(J, K)=\coprod_x (J_x, K_x)$$

of the normal Morse data for the map $f\circ\pi$ at the points $x\in\pi^{-1}(p)$. Similarly the relative complex link $\mathscr{L}^\pi$ for f at p breaks into a disjoint union of pieces

$$\mathscr{L}^\pi=\coprod_x \mathscr{L}_x$$

of the complex links at each point $x\in\pi^{-1}(p)$. We consider each of these pieces separately.

If $x\in\pi^{-1}(p)$ is an element of $X-W$, then (by Sects. 2.4 and 3.3) the normal Morse data (J_x, K_x) is homotopy equivalent to the pair $(\mathrm{cone}(\mathscr{L}_x), \mathscr{L}_x)$ which (by 4.5.2*) has the homotopy type of a CW complex of dimension $\leq\dim_{\mathbb{C}}\mathscr{L}+1$. If A denotes the stratum which contains the critical point p, then by Sect. 4.4, the Morse index of $f\,|\,A$ at p satisfies

$$\lambda\leq\inf(2\dim_{\mathbb{C}}(A), \dim_{\mathbb{C}}(A)+c-1).$$

Thus, the Morse data (which is the normal Morse data $\times$ the tangential Morse data) at p has the homotopy type of a CW complex of dimension no more than

$$n-\dim_{\mathbb{C}}(A)-1+1+\inf(2\dim(A), \dim(A)+c-1)$$

which is less than or equal to $\hat{n}^*$.

The problem arises when we pass a critical value which corresponds to a critical point $x\in W$. In this case, it suffices to show that the quotient J_x/K_x has the homotopy type of a CW complex of dimension $\leq n-\dim_{\mathbb{C}}(A)+k-1$, where A is the stratum which contains the critical point, and where k is (locally) the number of equations needed to define W as a subset of X (near the point x).

In the dual case (Sect. 4.6) we used Hamm's theorem to show that it was possible to remove the critical fibre $\pi^{-1}(p)$ without affecting the connectivity of the normal Morse data. This time we will insert the critical fibre and check that the homotopy dimension is unaffected by this procedure, using our generalization (Sect. 4.6*) of Hamm's theorem.

Lemma. *Suppose the quotient* $(J_x \cup \{x\}/K_x)$ *has the homotopy type of a CW complex of dimension* $\leq m_1$, *and suppose that the link in* $X-W$ *of the stratum* A,

$$L_x(X-W) = N \cap B_\delta(x) \cap (X-W)$$

has the homotopy type of a CW complex of dimension $\leq m_2$. *Then, the quotient* J_x/K_x *has the homotopy type of a CW complex of dimension* $\leq \sup(m_1, m_2)$.

Proof. A neighborhood of $\{x\}$ in J_x is homeomorphic to the cone over L_x. Puncturing this neighborhood by removing the point x can increase the homotopy dimension to (at most) m_2.

We now estimate the above numbers m_1 and m_2. By Proposition 4.6* the number m_2 = "homotopy dimension" of $L_x(X-W) = L_x(X) - L_x(W)$ is less than or equal to $n - \dim_{\mathbb{C}}(A) + k - 1$. By Proposition 3.2 the pair $(J_x \cup \{x\}, K_x)$ is homotopy equivalent to the pair $(\text{cone}(\mathscr{L}_x), \mathscr{L}_x)$, where $\mathscr{L}_x$ denotes the complex link in $X-W$ of the stratum A. We use induction (and Morse theory on $\mathscr{L}_x$) to conclude that $\mathscr{L}_x$ has the homotopy type of a CW complex of dimension $\leq n - \dim_{\mathbb{C}}(A) - 1 + k - 1$. Therefore, the quotient $(J_x \cup \{x\}/K_x)$ has the homotopy type of a CW complex of dimension $\leq n - \dim_{\mathbb{C}}(A) + k - 1$.

In summary, the normal Morse data J_x/K_x has the homotopy type of a CW complex of dimension $\leq n - \dim_{\mathbb{C}}(A) + k - 1$. As above, we note that at each critical point, the tangential Morse data has the homotopy type of a pair $(D^\lambda, \partial D^\lambda)$, where the number λ is the Morse index of $f \mid A$ and is bounded above by $\inf(2\dim_{\mathbb{C}}(A), \dim_{\mathbb{C}}(A) + c - 1)$ (see Sect. 4.4). Therefore, at each critical point a space with the homotopy type of a CW complex of dimension $\leq \hat{n}^*$ has been added. □

5.3. Proof of Theorem 1.3: Local Lefschetz Theorems

5.3.1. Statement of theorem. In this section, X will denote a complex algebraic subvariety of some nonsingular variety M, and $\pi: X \to P$ will be a complex algebraic map, where P is a nonsingular algebraic variety. Fix $p \in \overline{\pi(X)}$ and let $\partial B_\delta(p)$ denote the boundary of a ball of radius δ about the point p (with respect to some Riemannian metric on P). Let H be an affine linear subspace of codimension c in P (with respect to some local coordinate system about p) which passes through the point p, and let H_ε denote an ε-neighborhood of H (with respect to some real analytic Riemannian metric on P).

Theorem 1. *Suppose X is nonsingular, connected, and purely n-dimensional. Let $\phi(k)$ denote the dimension of the set of points $z \in P - H$ such that the fibre $\pi^{-1}(z)$ has dimension k. (If this set is empty define $\phi(k) = -\infty$.) If δ is sufficiently small, then for any $\varepsilon > 0$ sufficiently small, the homomorphism induced by inclusion,*

$$\pi_i(X \cap \pi^{-1}(\partial B_\delta(p) \cap H_\varepsilon)) \to \pi_i(X \cap \pi^{-1}(\partial B_\delta(p)))$$

is an isomorphism for all $i < \hat{n}$ and is a surjection for $i = \hat{n}$, where

$$\hat{n} = n - \sup_k (2k - (n - \phi(k)) + \inf(\phi(k), c-1)) - 2.$$

Furthermore: *If H is a generic affine subspace or if π is proper, then the neighborhood H_ε may be replaced by H in the above formula. The assumption that X is algebraic may be replaced by the assumption that P is a nonsingular complex analytic variety and X is the complement of a closed subvariety of a complex analytic variety $\bar{X}$ and that π extends to a proper analytic map $\bar{\pi}: \bar{X} \to P$.*

Theorem 2. *Suppose π has finite fibres (but is not necessarily proper). Let $\phi(k)$ denote the dimension of the set of points $x \in X - \pi^{-1}(H)$ such that a neighborhood of x (in X) can be defined by k equations, and no fewer. (If this set is empty define $\phi(k) = -\infty$.) If δ is sufficiently small, then for all $\varepsilon > 0$ sufficiently small, the homomorphism induced by inclusion*

$$\pi_i(X \cap \pi^{-1}(\partial B_\delta(p) \cap H_\varepsilon)) \to \pi_i(X \cap \pi^{-1}(\partial B_\delta(p)))$$

is an isomorphism for all $i < \hat{n}$ and is a surjection for $i = \hat{n}$, where

$$\hat{n} = \inf_k (\dim_{\mathbb{C}}(M) - k - \inf(\phi(k), c-1)) - 2.$$

Furthermore: *If H is a generic affine subspace or if π is proper, then the neighborhood H_ε may be replaced by H in the above formula. The assumption that X is algebraic may be replaced by the assumption that P is a complex analytic variety, and X is the complement of a closed subvariety of a complex analytic variety $\bar{X}$, and that π extends to a proper analytic map $\bar{\pi}: \bar{X} \to P$.*

5.3.2. Lemma. *The pair $(\mathcal{L}_X^\pi, \partial\mathcal{L}_X^\pi)$ is i connected $\Leftrightarrow$ the pair $(L_X^\pi, \mathcal{L}_X^\pi)$ is $i+1$ connected.*

Proof. Suppose the pair $(\mathcal{L}_X^\pi, \partial\mathcal{L}_X^\pi)$ is i connected. Then, for any $j \leq i$ we have

$$\begin{aligned} 0 &= \pi_j(\mathcal{L}_X^\pi, \partial\mathcal{L}_X^\pi) \\ &= \pi_{j+1}(\ell_X^{\pi+}, \partial\ell_X^\pi) \quad \text{by part (d) of Proposition 2.6} \\ &\quad \text{(see also Proposition 2.4)} \\ &= \pi_{j+1}(\ell_X^{\pi+} \cup_{\partial\ell_X^\pi} \ell_X^{\pi-}) \quad \text{by excision} \\ &= \pi_{j+1}(L, \ell_X^{\pi-}) \quad \text{by Part I, Sect. 3.11} \\ &= \pi_{j+1}(L, \mathcal{L}_X^\pi) \quad \text{by part (d) of Proposition 2.6} \\ &\quad \text{(see also Proposition 2.4).} \end{aligned}$$

The reverse implication is similar. □

5.3.3. Proof of Theorem 1. We shall prove this result for generic H and then at the end (5.3.5) indicate how to modify the proof for arbitrary H. Since the problem is local, we may replace the algebraic manifold P by affine space $\mathbb{C}^N$.

As in Sect. 5.1, since the map π is algebraic, it extends to a proper algebraic map

$$\bar{\pi}: \bar{X} \to \mathbb{C}^N$$

on some algebraic variety $\bar{X}$ which contains X as an open subset. Let $\bar{Z} = \bar{\pi}(\bar{X}) \subset \mathbb{C}^N$ and let $Z = \pi(X)$. It is possible (Part I, Sect. 1.7) to stratify $\bar{X}$

and $\bar{Z}$ so that $\bar{\pi}$ is a stratified map, X is an open union of strata in $\bar{X}$, Z is a union of strata in $\bar{Z}$, and so that the point $\{p\}$ is a separate stratum of $\bar{Z}$. As in Sect. 5.1, we rewrite the expression for $\hat{n}$ as follows: $\hat{n}=\inf_A(n(A))$ where

$$n(A)=n-(2\dim\pi^{-1}(a)-(n-\dim(A))+\inf(\dim(A), c-1))-2$$

where the inf is taken over all strata A of Z, and where $a\in A$.

We will prove Theorem 1 by induction on c, the codimension of H. First we consider the case $c=1$. For almost every linear projection $f\colon \mathbb{C}^N\to\mathbb{C}$ the point $p\in\bar{Z}$ is a nondegenerate critical point of f (in the sense of Sect. 2.1). Choose such a projection f so that $f(p)=0$, and let $H=f^{-1}(0)$. Since the point $\{p\}$ is a single stratum, the variety $\bar{Z}$ is also a normal slice through p. By Sect. 2.4(e), the stratified set $\bar{Z}\cap H\cap\partial B_\delta(p)$ is homeomorphic (by a stratum preserving homeomorphism) to the boundary of the complex link (in $\bar{Z}$) at the point p,

$$\partial\mathscr{L}=\bar{Z}\cap f^{-1}(\varepsilon+0i)\cap\partial B_\delta(p)$$

provided $\varepsilon>0$ is sufficiently small. Since $\bar{\pi}$ is a stratified map and X is a union of strata, we have a homeomorphism

$$X\cap\bar{\pi}^{-1}(H\cap\partial B_\delta(p))\cong\partial\mathscr{L}_X^\pi\cong X\cap\bar{\pi}^{-1}(f^{-1}(\varepsilon+0i)\cap\partial B_\delta(p)).$$

It therefore suffices to show that

$$\pi_i(L_X^\pi, \partial\mathscr{L}_X^\pi)=0 \qquad \text{for all } i\le\hat{n}$$

where $L_X^\pi=X\cap\bar{\pi}^{-1}(\bar{Z}\cap\partial B_\delta(p))$ is the relative link for X at the point p. To complete the proof of the case $c=1$, we verify that $\pi_i(L_X^\pi, \partial\mathscr{L}_X^\pi)=0$ in two steps:

(a) $\pi_i(L_X^\pi, \mathscr{L}_X^\pi)=0$ for all $i\le\hat{n}$.

(b) $\pi_i(\mathscr{L}_X^\pi, \partial\mathscr{L}_X^\pi)=0$ for all $i\le\hat{n}$.

By Lemma 5.3.2, we have part (b) implies part (a). However, part (b) is simply a restatement of Proposition 4.5.2, where the stratum A consists of the single point $\{p\}$ and $c(A)=n$.

Inductive step. We now consider the case $c>1$, i.e., we suppose that H is a codimension c affine subspace of $\mathbb{C}^N$ which passes through the point p and which is transverse to all the strata of $\bar{Z}$ except for the stratum $\{p\}$. It follows that there is a codimension $c-1$ affine subspace $J\subset\mathbb{C}^N$ which contains H and is also transverse to all the strata of $\bar{Z}$. Applying the inductive hypothesis to the affine subspace J, we find that

$$\pi_i(X\cap\pi^{-1}(\partial B_\delta(p)\cap J))\to\pi_i(X\cap\pi^{-1}(\partial B_\delta(p)))$$

is an isormorphism for all $i<\hat{m}$ and is a surjection for $i=\hat{m}$, where

$$\hat{m}=n-\sup_A(2\dim\pi^{-1}(a)-(n-\dim(A))+\inf(\dim(A), c-2))-2$$

where the sup is taken over all strata $A\subset Z$. Let $X'=X\cap\pi^{-1}(J)$ and $P'=P\cap J$. Consider the local Lefschetz theorem for the problem $\pi\colon X'\to P'$. We obtain a homomorphism

$$\pi_i(X'\cap\pi^{-1}(\partial B_\delta(p)\cap H))\to\pi_i(X'\cap\pi^{-1}(\partial B_\delta(p)))$$

which is an isomorphism for all $i<\hat{\ell}$ and is a surjection for $i=\hat{\ell}$, where

$$\hat{\ell}=n-(c-1)-\sup_A(2\dim\pi^{-1}(a)-(n-c+1-\dim(A)))-2$$

where the sup is taken over all strata $A' \subset Z \cap J$. Since J is transverse to Z, this means that such strata A' are in one to one correspondence with strata A of Z such that $\dim(A) \geq c-1$, i.e.,

$$\hat{\ell} = n - \sup_{A} (2 \dim \pi^{-1}(a) - (n - \dim(A)) + c - 1) - 2$$

where the sup is taken over all strata $A \subset Z$ such that $\dim(A) \geq c-1$. It is easy to see that $\hat{n} = \min(\hat{m}, \hat{\ell})$ by considering the strata one at a time: For each stratum $A \subset Z$, let

$$m(A) = n - (2 \dim \pi^{-1}(a) - (n - \dim(A)) + \inf(\dim(A), c-2)) - 2$$

$$\ell(A) = n - (2 \dim \pi^{-1}(a) - (n - \dim(A) + c - 1) - 2.$$

Thus,

$$\hat{m} = \inf_{A \subset Z} (m(A)) \quad \text{and} \quad \hat{\ell} = \inf_{d(A) \geq c-1} (\ell(A)).$$

There are two cases to consider: if $\dim(A) < c-1$, then A does not occur in the formula for $\hat{\ell}$, but it does occur in the formula for $\hat{m}$, and $\inf(\dim(A), c-1) = \dim(A)$. Thus, $n(A) = m(A)$. The second case is if A is a stratum of Z and $\dim(A) \geq c-1$. Then, A occurs in both the formulas for $\hat{\ell}$ and $\hat{m}$. However, $\ell(A) \leq m(A)$, and $\inf(\dim(A), c-1) = c-1$. Thus, $n(A) = \ell(A)$. In summary we have,

$$\hat{n} = \inf_{A} (n(A)) = \inf(\hat{\ell}, \hat{m}). \quad \square$$

5.3.4. Proof of Theorem 2. The proof is exactly the same as the proof of Theorem 1 except that the extension $\bar{\pi}$ must be chosen so as to be a finite map and Proposition 4.6.2 must be used instead of Proposition 4.5.2. □

5.3.5. For *nongeneric subspaces* H (or for a local complete intersection H), it is necessary to modify the preceding proof as follows:

Case $c=1$: Realize $H = f^{-1}(0)$ for a specific complex analytic function $f: \mathbb{C}^N \to \mathbb{C}$. Then, (as in 2.A.2) replace the spaces $\mathscr{L}_X^\pi$ and $\partial \mathscr{L}_X^\pi$ with the spaces

$$(\tilde{\mathscr{L}}_X^\pi, \partial \tilde{\mathscr{L}}_X^\pi) = X \cap \bar{\pi}^{-1}(\tilde{\mathscr{L}}, \partial \tilde{\mathscr{L}})$$

where

$$(\tilde{\mathscr{L}}, \partial \tilde{\mathscr{L}}) = \bar{Z} \cap f^{-1}(\varepsilon + 0i) \cap (B_\delta(p), \partial B_\delta(p)).$$

By 2.A.3, Lemma 5.3.2 continues to hold when we replace each $\mathscr{L}_X^\pi$ and $\partial \mathscr{L}_X^\pi$ with $\tilde{\mathscr{L}}_X^\pi$ and $\partial \tilde{\mathscr{L}}_X^\pi$.

We are thus reduced to 5.3.3 statements (a) and (b), with (b)⇒(a). However, statement (a) may be proven the same way that Proposition 4.5.2 was proven: by applying Morse theory to a Morse perturbation of the distance from the critical point $\{p\}$.

Inductive step. Even if H is not transverse to the strata of $\bar{Z}$, it is possible to choose the subspace J to be transverse to the strata of $\bar{Z}$ (by Part I, Sect. 1.3). Therefore, no changes are necessary in the inductive step.

5.3*. Proof of Theorem 1.3*: Local Homotopy Dimension

5.3.1*. Statement of theorem. In this section we suppose that X is an n-dimensional connected analytic variety and that $\pi: X \to P \subset \mathbb{C}^N$ is an analytic map to some open subset P of $\mathbb{C}^N$. Fix $p \in Z = \pi(X)$ and let ∂B_δ denote the boundary of a ball of radius δ about the point p. Let H be an affine linear subspace of codimension c in $\mathbb{C}^N$ which passes through the point p.

Theorem. *Suppose π is proper. Let $\phi(k)$ denote the dimension of the set of all points $z \in Z$ such that the fibre $\pi^{-1}(z)$ has dimension k. (If this set is empty, define $\phi(k) = -\infty$.) If δ is chosen sufficiently small, then the space $\pi^{-1}(Z \cap \partial B_\delta(p) - H)$ has the homotopy type of a CW complex of dimension less than or equal to*

$$\hat{n} = n + \sup_k (2k - (n - \phi(k)) + \inf(\phi(k), c-1)).$$

5.3.2*. Lemma. *The space $\mathcal{L}_X^\pi$ has the homotopy type of a CW complex of dimension $i \Leftrightarrow$ the space $L_X^\pi - \mathcal{L}_X^\pi$ has the homotopy type of a CW complex of dimension $i+1$.*

Proof of lemma. By Proposition 2.6 parts (a) and (e) i, the difference $L_X^\pi - \mathcal{L}_X^\pi$ is homotopy equivalent to L_v (the vertical part of the link), which is a fibre bundle over the circle, with fibre $\mathcal{L}_X^\pi$.* □

5.3.3*. Proof of Theorem 5.3*. We prove this result for generic H of codimension one, leaving the general case as an exercise. As in Sect. 5.3, the set $\pi^{-1}(Z \cap H \cap \partial B_\delta(p))$ can be identified with the relative complex link $\mathcal{L}_X^\pi$ for X at the point p, and the set $\pi^{-1}(Z \cap \partial B_\delta(p))$ can be identified with the relative link L_X^π for X at the point p. We have already computed the homotopy dimension of the relative complex link to be less than or equal to

$$\dim_{\mathbb{C}} \mathcal{L}_X^\pi + \sup_{B > A} (2 \dim_{\mathbb{C}}(\pi^{-1}(q)) - (n - d(B)))$$

(see Proposition 4.5.2*), where A is the stratum which contains the point p and where the sup is taken over all strata B which contain A in their closures (and $q \in B$). By Lemma 5.3.2*, we must add 1 to this number in order to obtain the homotopy dimension of L_v. The result is $\hat{n}^*$. □

5.A. Appendix: Analytic Neighborhoods of an Analytic Set

The point of this section is to show that under certain conditions, any sufficiently small δ-neighborhood H_δ of a hyperplane section H of a complex analytic variety X is independent of δ and of the Riemannian metric which was used in defining H_δ.

5.A.1. Suppose $\bar{X}$ is a Whitney stratified real analytic subset of some real analytic manifold M. Let $f: M \to \mathbb{R}$ be a proper real analytic function. Since the interval $[-1, 1]$ contains finitely many critical values of the restriction $f|\bar{X}: \bar{X} \to \mathbb{R}$, there is a number $\delta(f) > 0$ so that the interval $[-\delta(f), \delta(f)]$ contains no critical values of $f|\bar{X}$ other than (possibly) 0.

Proposition. *If $\delta \le \delta(f)$, then the inclusion*

$$\bar{X} \cap f^{-1}(0) \to \bar{X} \cap f^{-1}[-\delta, \delta]$$

induces an isomorphism on homotopy groups π_i for all degrees i.

Proof. (This proof follows the method of [Sm4].) It follows from Thom's first isotopy lemma (Part I, Sect. 1.5) that for any $\delta' \leq \delta$, the inclusion

$$i\colon \bar{X} \cap f^{-1}[-\delta', \delta'] \to \bar{X} \cap f^{-1}[-\delta, \delta]$$

is homotopic (by a stratum preserving homotopy) to a homeomorphism (which also preserves strata). Furthermore, $\bar{X} \cap f^{-1}(0)$ has a neighborhood basis of "good" open sets $U \subset \bar{X}$ which are homotopy equivalent to $\bar{X} \cap f^{-1}(0)$. Choose such a good neighborhood $U \subset \bar{X} \cap f^{-1}[-\delta, \delta]$. Choose $\delta' < \delta$ so that $\bar{X} \cap f^{-1}[-\delta', \delta'] \subset U$. Consider the three inclusion

$$\bar{X} \cap f^{-1}(0) \xrightarrow{\alpha} \bar{X} \cap f^{-1}[-\delta', \delta'] \xrightarrow{\beta} U \xrightarrow{\gamma} \bar{X} \cap f^{-1}[-\delta, \delta].$$

Since $\beta \circ \alpha$ induces isomorphisms on homotopy groups of all dimensions and since $i = \gamma \circ \beta$ is a homotopy equivalence, it follows that β induces isomorphisms on homotopy groups in all dimensions, and so the same is true for α and γ.

5.A.2. Now let $X \subset \bar{X}$ be a union of strata, where $\bar{X} \subset M$ and $f\colon M \to \mathbb{R}$ are defined as above, and suppose that $g\colon M \to \mathbb{R}$ is another proper real analytic function such that $g^{-1}(0) = f^{-1}(0)$. Let $\delta(g)$ be so small that $[-\delta(g), \delta(g)]$ contains no critical values of $g \mid \bar{X}$ except (possibly) 0.

Proposition. *If* $0 < a \leq \delta(f)$ *and if* $0 < b \leq \delta(g)$ *are chosen so that* $f^{-1}[-a, a] \subset g^{-1}[-b, b]$, *then the inclusion*

$$X \cap f^{-1}[-a, a] \to X \cap g^{-1}[-b, b]$$

induces isomorphisms on homotopy groups in all dimensions.

Proof. The proof is essentially the same as above: choose $b' < b$ so that $g^{-1}[-b', b'] \subset f^{-1}[-a, a]$. Choose $a' < a$ so that $f^{-1}[-a', a'] \subset g^{-1}[-b', b']$. Consider the three inclusions,

$$X \cap f^{-1}[-a', a'] \xrightarrow{\alpha} X \cap g^{-1}[-b', b'] \xrightarrow{\beta} X \cap f^{-1}[-a, a] \xrightarrow{\gamma} X \cap g^{-1}[-b, b].$$

The composition $\beta \circ \alpha$ is a homotopy equivalence and the composition $\gamma \circ \beta$ is a homotopy equivalence. Thus, each of the maps α, β, and γ must induce isomorphisms on homotopy groups in all dimension. □

5.A.3. Suppose $\bar{X}$ is a Whitney stratified subset of a smooth manifold M and suppose $X \subset \bar{X}$ is a union of strata of $\bar{X}$. Let $f\colon M \to \mathbb{R}$ be a smooth function so that $f^{-1}(0) = N$ is a submanifold which is transverse to each stratum of $\bar{X}$.

Proposition. *Suppose that f is given by the distance from N, with respect to some Riemannian metric on the fibres of the normal bundle TM/TN. Then (for any sufficiently small $\delta > 0$), the inclusion*

$$X \cap f^{-1}(0) \to X \cap f^{-1}[-\delta, \delta]$$

is a homotopy equivalence.

Proof. This is precisely Part I, Sect. 1.11. □

Chapter 6. Morse Theory and Intersection Homology

Many of the results in this chapter appeared first in [GM3].

6.0. Introduction

The ("middle") intersection homology of a singular complex algebraic variety exhibits many of the same properties as the ordinary homology of a nonsingular variety. For example, it satisfies Poincaré duality, the hard Lefschetz formula, and probably has a pure Hodge decomposition. In this chapter we add a further property to the list: a critical point of a Morse function has a Morse *index* for intersection homology. This means that if X is a Whitney stratified complex analytic variety and if $f: X \to \mathbb{R}$ is a proper Morse function with a critical point $p \in X$ and critical value $v = f(p)$, then there is a single integer i such that, for sufficiently small $\varepsilon > 0$,

$$IH_k(X_{\leq v+\varepsilon}, X_{\leq v-\varepsilon}) = \begin{cases} 0 & \text{for } k \neq i \\ A_p & \text{for } k = i. \end{cases}$$

Furthermore, the group A_p is torsion-free and depends only on the connected component of the stratum S containing p, but does not depend on the function f. (However, the index i depends on the function f, and in fact $i = \lambda + c$, where c is the complex codimension of the stratum S and λ is the Morse index at p of the restriction $f \mid S$.) The existence of a Morse index is false for ordinary homology in the singular case: the group $H_k(X_{\leq v+\varepsilon}, X_{\leq v-\varepsilon})$ may be nonzero for several different values of k.

We use this basic fact to obtain four results:

(a) The Lefschetz theorem and local Lefschetz theorems hold for the intersection homology of an arbitrary quasi-projective algebraic variety.

(b) The intersection homology of a complex n-dimensional Stein space vanishes in dimensions greater than n (and is torsion-free in dimension n). (This result does not follow from the homotopy statement in Sect. 1.1* because intersection homology is not a homotopy invariant.)

(c) Intersection homology satisfies the Morse inequalities.

(d) The sheaf of intersection chains on a general fibre specializes (over a curve) to a perverse object ([BBD]) on the special fibre.

Many of the results in this chapter can be applied to arbitrary complexes of sheaves, or to perverse sheaves, as will be remarked in the appendix.

6.1. Intersection Homology

In this section we will denote by Y a (not necessarily compact) purely n-dimensional subanalytic stratified pseudomanifold with even codimension strata and collared boundary ∂Y ([GM3], [GM4], [GM5]). We fix a point $q \in Y$ in some stratum S of (real) codimension $2c$. We will use the symbol $IH_k(Y)$ to denote the intersection homology group (with compact support, as in [GM4]) of "middle perversity" and integer coefficients, which is based on the chain complex $IC_k(Y)$ of $(\bar{m}, k)$-allowable chains. We shall use the following facts about intersection homology:

Intersection homology of a pair. The group $IH_k(Y, \partial Y)$ may be unambiguously defined as the homology of the chain complex $IC_*(Y)/IC_*(\partial Y)$. The usual long exact sequence on homology results. Relative intersection homology can also be defined for the pair (Y, U) where U is any open subset of Y, and in this case the following *excision formula* holds: if V is closed in U then the inclusion $(Y-V, U-V) \to (Y, U)$ induces an isomorphism on intersection homology. We also have the usual long exact sequence for a triple (Y, U_1, U_2) provided U_2 is open in U_1 and U_1 is open in Y.

Kunneth formula. Let $(D^a, \partial D^a)$ denote the closed a-dimensional disk and its boundary sphere. Then,

$$IH_i(Y, \partial Y) \cong IH_i(Y \times D^a, \partial Y \times D^a)$$
$$IH_i(Y, \partial Y) \cong IH_{i-a}(Y \times D^a, \partial Y \times D^a \cup Y \times \partial D^a).$$

Local calculation. Any point $q \in Y$ has a fundamental system of neighborhoods homeomorphic to

$$U = D^a \times \operatorname{cone}(L)$$

where $a = n - 2c$ is the dimension of the stratum S which contains the point q and L is the link of the stratum S (see Part I, Sect. 1.4). Then,

$$IH_i(U) = \begin{cases} 0 & \text{if } i \geq c \\ IH_i(L) & \text{if } i < c \end{cases}$$

$$IH_i(U, \partial U) = IH_i(U, U - q) = \begin{cases} I\tilde{H}_{i-n+2c-1}(L) & \text{if } i \geq n - c + 1 \\ 0 & \text{if } i < n - c + 1 \end{cases}$$

where $I\tilde{H}$ denotes reduced homology.

Twisted coefficients. If Σ denotes the singular set of Y, and T is a local coefficient system on $Y - \Sigma$ (in the sense of Steenrod [St]), then $IH_k(Y; T)$ may be defined as the homology group of the complex of $(\bar{m}, k)$-allowable chains with coefficients in T. These groups exhibit the formal properties listed above: long exact sequence for a pair and a triple, excision formula, Kunneth formula and local calculation.

6.2. The Set-up and the Bundle of Complex Links

For the remainder of this chapter we will assume that Z is a Whitney stratified complex analytic variety of pure dimension, which is embedded as a subvariety of some smooth complex analytic manifold M. We will study the Morse theory of a subvariety X which is an open dense union of strata of Z, i.e., $X=Z-Y$, where Y is a closed subvariety which is a union of strata. Fix a connected components S of some stratum of Z. For any nondegenerate conormal vector $\omega\in T_S^*M$ there is associated (in a canonical way, up to isotopy) a (transverse) cylindrical neighborhood, cut off neighborhood, and complex link with boundary:

$$(C, C_{<0})=X\cap\tilde{N}\cap B_\delta(p)\cap(f^{-1}(D), f^{-1}(D_{<0}))$$
$$(\mathscr{L}_X, \partial\mathscr{L}_X)=X\cap\tilde{N}\cap(B_\delta(p), \partial B_\delta(p))\cap f^{-1}(\varepsilon+0\cdot i)$$

where $\tilde{N}$ is a complex submanifold which is transverse to S and intersects S in the single point $\{p\}$; $f\colon M\to\mathbb{C}$ is a locally defined analytic function such that $df(p)|\tilde{N}=\omega$, and where $D\subset\mathbb{C}$ is the disk of radius $\varepsilon>0$, with

$$D_{<0}=\{\zeta\in D\,|\,\mathrm{Re}(\zeta)>0\}$$

and where $\varepsilon\ll\delta$ (see Sects. 2.2, 2.3, and 2.6). Choices of the complex link $\mathscr{L}_X$ (resp. the cylindrical neighborhood C, resp. the cut off neighborhood $C_{<0}$) may be made so that the collection of all complex links (resp. cylindrical neighborhoods, resp. cut off neighborhoods) form a fibre bundle over the connected space C_S of nondegenerate conormal vectors to S. This is proven in Sect. 2.3.2. (Thus, $\mathscr{L}_X$ is independent of ω up to noncanonical homeomorphism.)

6.3. The Variation

Suppose the stratum S is contained in X, and is a singular stratum of X. Fix a nondegenerate conormal vector $\omega\in T_S^*M$.

Definition. The variation map is the boundary homomorphism

$$\mathrm{Var}_k\colon IH_k(C-\{p\}, C_{<0})\xrightarrow{\partial_*} IH_{k-1}(C_{<0}).$$

Interpretation. Let $\mathscr{L}=\mathscr{L}_X=\mathscr{L}_Z$ denote the corresponding complex link. By Sect. 2.6, a choice of square root of -1 determines a monodromy homeomorphism

$$\mu\colon \mathscr{L}\to\mathscr{L}$$

which is well defined up to isotopy (modulo some neighborhood of the boundary) and may be chosen so as to be the identity on some neighborhood of $\partial\mathscr{L}$. It follows that for any $\xi\in IC_k(\mathscr{L}, \partial\mathscr{L})$ the chain $\xi-\mu(\xi)$ is an element of $IC_k(\mathscr{L})$. In other words, $I-\mu$ induces a homomorphism

$$(I-\mu)_*\colon IH_k(\mathscr{L}, \partial\mathscr{L}; \mathbb{Z})\to IH_k(\mathscr{L}; \mathbb{Z}).$$

By [GM3] Sect. 3.5, Corollary 1, the group $IH_k(C-\{p\}, C_{<0})$ is canonically identified with $IH_{k-1}(\mathscr{L}, \partial\mathscr{L})$. By Sect. 2(a) or [GM3] Sect. 3.5, Corollary 3, the group $IH_{k-1}(C_{<0})$ is canonically identified with $IH_{k-1}(\mathscr{L}^-, \partial\mathscr{L}^-)$, where $(\mathscr{L}^-, \partial\mathscr{L}^-)$ is the complex link which corresponds canonically to the conormal vector $-\omega$. A choice of path from $-\varepsilon+0i$ to $+\varepsilon+0i$ in $\mathbb{C}-\{0\}$ gives a homeomorphism

$$(\mathscr{L}^-, \partial\mathscr{L}^-) \to (\mathscr{L}, \partial\mathscr{L}).$$

With these identifications, we have

$$\mathrm{Var}_k\colon IH_{k-1}(\mathscr{L}, \partial\mathscr{L}; \mathbb{Z}) \to IH_{k-1}(\mathscr{L}; \mathbb{Z}).$$

Fact. For an appropriate choice of square root of minus one, the homomorphism Var_k coincides with the homomorphism $(I-\mu)_*$.

Definition. The *Morse group* A_ω is the image of VAR_c where c is the complex codimension of the stratum S. If ω is a conormal vector to the nonsingular part of X, then we define A_ω to be the integers, $\mathbb{Z}$.

Remarks. It follows from Lemma 2.3.2 that the groups A_ω form a local system over the space C_S of nondegenerate covectors. Thus, if $\eta \in C_S$ is another nondegenerate covector, a choice of path between ω and η determines an isomorphism between A_ω and A_η. In particular, there are two distinguished isomorphisms $A_\omega \cong A_{-\omega}$ whose composition $A_\omega \to A_{-\omega} \to A_\omega$ gives an action of the monodromy μ on A_ω. Furthermore, the intersection pairing from intersection homology,

$$IH_i(\text{Morse data for } \omega) \times IH_{2n-i}(\text{Morse data for } -\omega) \to \mathbb{Z}$$

induces a bilinear pairing

$$A_\omega \times A_{-\omega} \to \mathbb{Z}$$

which is nondegenerate over $\mathbb{Q}$.

6.4. Theorem $\mathbf{A_n}$. *Let X be an open dense union of strata in a complex n-dimensional Whitney stratified analytic variety Z, and let $f\colon Z \to \mathbb{R}$ be a proper Morse function with a nondegenerate critical point $p \in Z$ in some stratum $S \subset Z$ of complex codimension c. Let λ denote the Morse index of $f|S$ at the point p and let $\omega \in T_S^* M$ denote the conormal vector determined by $df(p)$. Suppose that the interval $[a, b]$ contains no critical values of $f|Z$ other than $v = f(p)$, and that $v \in (a, b)$. Then,*

$$IH_i(X_{\le b}, X_{\le a}; \mathbb{Z}) = \begin{cases} IH_{i-\lambda-1}(\mathscr{L}_X, \partial\mathscr{L}_X; \mathbb{Z}) & \text{if } p \notin X \\ A_\omega & \text{if } i = \lambda + c \text{ and } p \in X \\ 0 & \text{if } i \neq \lambda + c \text{ and } p \in X \end{cases}$$

and this identification is canonical. Furthermore, the group A_ω is torsion-free, and the group $IH_{i-\lambda-1}(\mathscr{L}_X, \partial\mathscr{L}_X; \mathbb{Z})$ vanishes if $i < \lambda + c$.

Remarks. (1) The same result holds for intersection homology with twisted coefficients, but it is necessary to define the variation using intersection homology with twisted coefficients, and the group A_ω at a nonsingular point must be replaced by the stalk of the local system at that point.

(2) If we replace intersection homology by ordinary homology, then by Sect. 3.2 we have

$$H_i(X_{\le b}, X_{\le a}) \cong \begin{cases} H_{i-\lambda-1}(\mathscr{L}_X, \partial\mathscr{L}_X; \mathbb{Z}) & \text{if } p \notin X \\ \tilde{H}_{i-\lambda-1}(\mathscr{L}; \mathbb{Z}) & \text{if } p \in X \end{cases}$$

(where $\tilde{H}$ denotes reduced homology) and so it is still independent of the Morse function, although it may be nontrivial for many different values of i.

(3) If X is an algebraic curve with a singular point p, then the Morse group A_p has rank $m-b$, where m is the multiplicity of X at p and b is the number of analytic branches of X at p.

(4) The proof which follows in Sect. 6.7 may be used to study the Morse group for intersection homology with different perversities. If the perversity lies between the "sub-logarithmic" ($p(2k)=k-2$) and the "logarithmic" ($p(2k)=k$) and if $p\in X$, then the corresponding Morse group $IH_i^{\bar{p}}(X_{\leq b}, X_{\leq a})$ is nonzero in one degree only ($i=c+\lambda$), but the value of the group depends on the perversity. For the logarithmic perversity, if $p\in X$ then

$$IH_{\lambda+c}^{\bar{p}}(X_{\leq b}, X_{\leq a}; \mathbb{Z}) = IH_{c-1}^{\bar{p}}(\mathscr{L}; \mathbb{Z})$$

and is torsion-free. For the sublogarithmic perversity, if $p\in X$, then

$$IH_{\lambda+c}^{\bar{p}}(X_{\leq b}, X_{\leq a}) = IH_{c-1}^{\bar{p}}(\mathscr{L}, \partial\mathscr{L}; \mathbb{Z}).$$

The detailed calculations for general perversities, together with applications to Lefschetz theorems and vanishing theorems with general perversities may be found in the sequence of papers [FK1] to [FK6]. The perversities between the sublogarithmic and the logarithmic are those for which the intersection homology is a *perverse sheaf* [BBD], i.e., it is the homology of the solution complex of a holonomic $\mathscr{D}$-module with regular singularities. See [KS] for a $\mathscr{D}$-module proof that the Morse group exists in a single dimension.

6.5. Vanishing of the Morse Group

Recall (Part I, Sect. 1.8) that if Z is a Whitney stratified complex analytic subvariety of a smooth analytic manifold M, then a covector $\theta\in T_p^*(M)$ is *characteristic* if it annihilates the tangent space T_pS to the stratum S of Z which contains the point p. In this case, θ is called *degenerate* if it annihilates some generalized tangent space at p (i.e., a limit of tangent spaces from some stratum $R>S$). The set of degenerate characteristic covectors is a homogeneous cone of complex codimension ≥ 1 in the space of characteristic covectors.

Let $X\subset Z$ be an open dense union of strata in Z.

Proposition. *Suppose $p\in X$ and the set D of degenerate characteristic covectors has codimension ≥ 2 in the space C of characteristic covectors at the point p. Then for any $\omega\notin D$, the Morse group A_ω is* 0.

Proof. It is possible to find two complex analytic functions $(f, g)\colon X\to\mathbb{C}^2$ such that the pair (f, g) is nondegenerate, i.e., $\omega=df(p)$ and

$$(df(p), dg(p))(Q)=\mathbb{C}^2$$

for every generalized tangent space Q. (This is because the projectivization $\mathbb{P}(D)$ of the set of degenerate covectors is a set of complex codimension two in the projective space $\mathbb{P}(C)$ of all characteristic covectors. Therefore, almost every linear $\mathbb{CP}^1\subset\mathbb{P}(C)$ misses this set $\mathbb{P}(D)$.)

By Sect. 2.7, this implies that the monodromy is isotopic to the identity. Therefore, the variation map is 0. □

Example. This happens for the Schubert variety $X=\Omega[2,4]$ of all two-planes P in $\mathbb{C}^4$ such that $\dim(P\cap\mathbb{C}^2)\geq 1$. This variety has a single singular point p and has a small resolution $\pi\colon \tilde{X}\to X$ such that the composition $f\circ\pi$ has no critical points on $f^{-1}(p)$ ([CGM], [GM4]). It follows that the Morse group A_p is 0. ($\tilde{X}$ consists of all pairs (L, P) such that $P\in\Omega[2,4]$ and L is a line in $P\cap\mathbb{C}^2$).)

6.6. Intuition Behind Theorem A_n

If $p\notin X$, then the result is the same as in Sect. 3.3. The interesting case is when $p\in X$. By restricting to a normal slice through p (using Part I, Sect. 3.7), we can reduce to the case that $\{p\}$ is a zero-dimensional stratum of X and $\lambda=0$. It is possible (locally near p) to construct a controlled vectorfield (Part I, Sect. 1.5) V on $X-\{p\}$ such that $df(x)(V(x))<0$ for all x, i.e., V flows in the direction of $-\nabla f$. We shall denote the flow of this vectorfield by φ, i.e.,

$$\varphi\colon (X-\{p\})\times\mathbb{R}\to X-\{p\}$$

and we define the stable and unstable sets,

$$p^+=\{x\in X\mid \lim_{t\to\infty}\varphi(x,t)=p\}$$

$$p^-=\{x\in X\mid \lim_{t\to-\infty}\varphi(x,t)=p\}.$$

Conjecture. A vectorfield V exists so that p^+ and p^- are Whitney stratified subsets of X, and so that for each stratum B of X we have

$$\dim_{\mathbb{R}}(p^\pm\cap B)\leq\dim_{\mathbb{C}}(B).$$

We have been informed that Lê D.T. ([Lê2]) has succeeded in finding a vectorfield V of this type such that the sets $p^\pm$ are homeomorphic to simplicial complexes and satisfy the above dimensionality estimates. The idea is to approximate f (near p) by the real part of a complex analytic function $g\colon X\to\mathbb{C}$, which is a submersion except at the point $p\in X$. Then, find a controlled lift of the vectorfield

$$W=-\frac{\partial}{\partial x}+0\cdot\frac{\partial}{\partial y}.$$

Now consider a cycle $\xi\in IC_i(X_{\leq v+\varepsilon}, X_{\leq v-\varepsilon})$ where $i<n$. Since ξ is $(\bar{m}, i)$-allowable ([GM4]), we have

$$\dim(|\xi|\cap B)<n-c-1=\dim_{\mathbb{C}}(B)-1$$

for any stratum B of X (and also $|\xi|\cap\{p\}=\phi$). Therefore, by transversality, ξ is homologous to a cycle ξ' such that $|\xi'|\cap p^+=\phi$. This means that if we flow ξ by the vectorfield V it can be pushed into the subspace $X_{\leq v-\varepsilon}$ without ever intersecting the singular point $\{p\}$, i.e., the "flowout" of this cycle is an allowable homology between ξ' and a cycle ξ'' in $IC_i(X_{\leq v-\varepsilon})$. This is why $IH_i(X_{\leq v+\varepsilon}, X_{\leq v-\varepsilon})=0$.

Now consider a cycle $\xi \in IC_i(X_{\leq v+\varepsilon}, X_{\leq v-\varepsilon})$ where $i>n$. Although it may happen that $|\xi| \cap \{p\} \neq \phi$ we can still push ξ by the vectorfield V into some subset of p^-. But, the set p^- has dimension equal to n, so ξ is homologous to 0. It is a pleasant exercise to verify that this homology to 0 satisfies the "allowability conditions" for the middle perversity.

6.7. Proof of Theorem A_n

If $p \notin X$, then the result follows from Sect. 3.3. Thus, we may assume that $p \in X$ (i.e., that the Morse function f is proper near p). The proof is by induction on n, the complex dimension of the variety X and proceeds concurrently with the following:

Proposition B_n. *Let Y be a complex analytic Whitney stratified space of any dimension and let S be a stratum of Y whose complex codimension (in Y) is $c=n+1>0$. Choose a point $y_0 \in S$ and let $\mathscr{L}$ be the complex link of S at the point y_0. Then,*

$IH_i(\mathscr{L}; \mathbb{Z})=0$ for all $i>n$ and $IH_n(\mathscr{L}; \mathbb{Z})$ is torsion-free

$IH_i(\mathscr{L}, \partial\mathscr{L}; \mathbb{Z})=0$ for all $i<n$.

Proof that A_n (for proper Morse functions) implies B_n. Consider a Morse function $f: \mathscr{L} \to \mathbb{R}$, which is a ($C^\infty$ close) approximation to the function

$$\tilde{f}(y)=(\text{distance}(y, y_0))^2.$$

It follows (4.A.4) that for any stratum A of $\mathscr{L}$, index $(f|A)=\dim_{\mathbb{C}}(A)$ at any critical point in A. Applying proposition A_n to this Morse function gives, for each critical value v,

$$IH_i(\mathscr{L}_{\leq v+\varepsilon}, \mathscr{L}_{\leq v-\varepsilon})=0 \qquad \text{for all } i>\dim_{\mathbb{C}}(A)+\operatorname{cod}_{\mathbb{C}}(A)=n$$

where A is the stratum which contains the critical point. (Note that f has a degenerate maximum on $\partial\mathscr{L}$, but this is a collared boundary so adding it to $\mathscr{L}-\partial\mathscr{L}$ does not change the intersection homology). Furthermore, if $i \leq n$ then $IH_i(\mathscr{L}_{\leq v+\varepsilon}, \mathscr{L}_{\leq v-\varepsilon}; \mathbb{Z})$ is torsion-free, by proposition A_n. Using the long exact sequence for the pair $(\mathscr{L}_{\leq v+\varepsilon}, \mathscr{L}_{\leq v-\varepsilon})$ and induction, it follows that $IH_n(\mathscr{L}_{\leq v+\varepsilon}; \mathbb{Z})$ is torsion-free. Since this holds for each critical value v, we have $IH_n(\mathscr{L}; \mathbb{Z})$ is torsion-free.

The second statement may be proven by considering instead the Morse function $\delta-f$, where $\delta=f(\partial\mathscr{L})$. This has a minimum on $\partial\mathscr{L}$ and critical points which, in any stratum A have Morse index $\geq \dim_{\mathbb{C}}(A)$. Therefore, by proposition A_n,

$$IH_i(\mathscr{L}_{\leq v+\varepsilon}, \mathscr{L}_{\leq v-\varepsilon})=0 \qquad \text{for all } i<n.$$

Proof that B_k for all $k<n$ implies A_n for proper Morse functions. By Part I, Sect. 3.2 and Part I, Sect. 10.2, we may assume that $a=v-\varepsilon$ and $b=v+\varepsilon$ where ε is arbitrarily small. If the critical point p lies in the nonsingular part of X, then the conclusion of A_n is clear since $X_{\leq v+\varepsilon}$ is obtained from $X_{\leq v-\varepsilon}$ by attach-

ing $(D_\lambda \times D^{n-\lambda})$ along $(\partial D^\lambda \times D^{n-\lambda})$. Thus, we may assume the critical point lies in a singular stratum S of X. Let c be the complex codimension of S and let λ be the Morse index of $f\,|\,S$ at p. By Part I, Sect. 3.5.4, the space $X_{\leq v+\varepsilon}$ is homeomorphic to the space $X_{\leq v-\varepsilon} \cup (J, K)$ where the pair (J, K) is the local Morse data,

$$(J, K) = (D^{s-\lambda} \times D^\lambda, D^{s-\lambda} \times \partial D^\lambda) \times (\text{Normal Morse data}).$$

By excision and the Kunneth formula for intersection homology, we have

$$M_i = IH_i(X_{\leq v+\varepsilon}, X_{\leq v-\varepsilon}) = IH_{i-\lambda}(\text{Normal Morse data}).$$

We can (locally near p) embed X as an analytic subset of $\mathbb{C}^N$ and extend the function f to some smooth function on a neighborhood of X in $\mathbb{C}^N$. Define $F = F_1 + iF_2 \colon \mathbb{C}^N \to \mathbb{C}$ as follows:

$$F(z) = df(p)(z) - i\,df(p)(iz).$$

Thus, F is a complex analytic map such that $\operatorname{Re}(dF(p)) = df(p)$. By Part I, Sect. 7.5.1 the normal Morse data for f is homeomorphic to the normal Morse data for $F_1 = \operatorname{Re}(F)$ which (by Sect. 2.4(c)) is homeomorphic to the pair

$$(C \cap F_1^{-1}[-\eta, \eta], C \cap F_1^{-1}(-\eta))$$

(for η sufficiently small) where C is the cylindrical (transverse) neighborhood (see Sect. 2.4),

$$C = N \cap B_\delta(p) \cap F^{-1}(D_\varepsilon).$$

In [GM3] Sect. 3.5, Corollary 2, a further deformation and the Kunneth theorem is used in order to identify the intersection homology of the normal Morse data as follows:

$$M_i = IH_{i-\lambda}(C \cap F_1^{-1}[-\eta, \eta], C \cap F_1^{-1}(-\eta)) \cong IH_{i-\lambda}(C, C_{<0})$$

where $C_{<0} = C \cap F_1^{-1}(-\infty, 0)$. Therefore, to calculate the Morse group we should examine the long exact sequence associated to the triple of spaces

$$C_{<0} \subset C - \{p\} \subset C.$$

This long exact sequence, together with the exact sequences for the pairs $(C - \{p\}, C_{<0})$, $(C, C_{<0})$, and $(C, C - \{p\})$ fit together in the following "braid diagram" with exact sinusoidal rows (compare [Cl]):

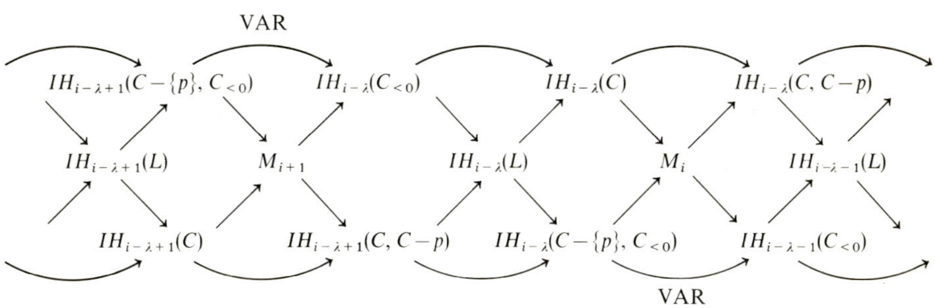

where we have made the following identifications:

$$IH_{i-\lambda}(C, C_{<0}) = M_i = IH_i(X_{\leq v+\varepsilon}, X_{\leq v-\varepsilon})$$

as above

$$IH_{i-\lambda}(C-\{p\}) = IH_{i-\lambda}(L) \quad \text{by Sect. 2.4(b).}$$

The proposition now follows from a simple diagram chase together with the following facts:

$$\left.\begin{array}{lll} IH_{i-\lambda}(C, C-\{p\}) = 0 & \text{for } i-\lambda \leq c \\ IH_{i-\lambda}(C) \qquad = 0 & \text{for } i-\lambda \geq c \end{array}\right\} \text{by the local calculation and Sect. 2.4(b)}$$

$$\left.\begin{array}{l} IH_{i-\lambda}(C-\{p\}, C_{<0}) = IH_{i-\lambda-1}(\mathscr{L}, \partial\mathscr{L}) \\ IH_{i-\lambda}(C_{<0}) \qquad = IH_{i-\lambda}(\mathscr{L}) \end{array}\right\} \text{by Sect. 2(a) or [GM3] Sect. 3.5}$$

$$\left.\begin{array}{ll} IH_{i-\lambda}(\mathscr{L}) \qquad = 0 & \text{for } i-\lambda > c-1 \\ IH_{i-\lambda}(\mathscr{L}, \partial\mathscr{L}) = 0 & \text{for } i-\lambda < c-1 \end{array}\right\} \text{from Proposition } B_c.$$

Thus, $M_i = 0$ unless $i = \lambda + c$ and $M_{\lambda+c} = A_\omega = \text{Image(VAR)}$. Since this is a subgroup of $IH_{c-1}(\mathscr{L}; \mathbb{Z})$, it is torsion-free (by Proposition B_c). □

6.8. Intersection Homology of the Link

Suppose X is a Whitney stratified complex analytic variety and $p \in X$ is a point in a stratum $S \subset X$ of complex codimension $c > 0$. Then,

$$IH_i(L) = \begin{cases} IH_i(\mathscr{L}, \partial\mathscr{L}) & \text{for } i > c \\ \ker(I-\mu) & \text{for } i = c \\ \operatorname{coker}(I-\mu) & \text{for } i = c-1 \\ IH_i(\mathscr{L}) & \text{for } i < c-1 \end{cases}$$

where L is the link of S at p, $\mathscr{L}$ is the complex link of S at p, and

$$(I-\mu)\colon IH_{c-1}(\mathscr{L}, \partial\mathscr{L}) \to IH_{c-1}(\mathscr{L})$$

is the variation.

Proof. The proof is immediate from the braid diagram. □

6.9. Intersection Homology of a Stein Space

Theorem. Let X be an n-dimensional Stein space. Then, $IH_i(X; \mathbb{Z}) = 0$ for all $i > n$, and $IH_n(X; \mathbb{Z})$ is torsion-free. (See also [GM3] and [FK5].)

Proof. Choose a homeomorphism between X and a closed analytic subspace (which we will also denote by X) of $\mathbb{C}^N$. Choose a generic point $q \in \mathbb{C}^N$ and let $f\colon X \to \mathbb{R}$ be the Morse function

$$f(x) = (\text{distance}(q, x))^2.$$

For any stratum A of X and for any critical point $p \in A$ we have (4.A.4)

$$\operatorname{index}(f \mid A) \text{ at } p \leq \dim_{\mathbb{C}}(A)$$

so $IH_i(X_{\leq v+\varepsilon}, X_{\leq v-\varepsilon})=0$ for all $i>n$, by proposition A_n (where $v=f(p)$), and $IH_i(X_{\leq v+\varepsilon}, X_{\leq v-\varepsilon})$ is torsion-free for all $i \leq n$. Using induction and the long exact sequence for the pair $(X_{\leq v+\varepsilon}, X_{\leq v-\varepsilon})$ we conclude that $IH_i(X;\mathbb{Z})=0$ for all $i>n$, and $IH_n(X;\mathbb{Z})$ is torsion-free. □

6.10. Lefschetz Hyperplane Theorem

The following Lefschetz hyperplane theorem was our original motivation for developing Morse theory on singular spaces. It was discovered independently by P. Deligne, who used sheaf theory and the method of Artin [Art] in his proof (see [GM5]). Later refinements appear in [FK1] through [FK6].

Let X be a purely n-dimensional algebraic variety and suppose that π: $X \to \mathbb{CP}^N$ is a (not necessarily proper) algebraic map with finite fibres. Let $H \subset \mathbb{CP}^N$ be a linear subspace of codimension c. Let H_ε denote an ε-neighborhood of H, with respect to some Riemannian metric on $\mathbb{CP}^N$.

Theorem. *If $\varepsilon>0$ is sufficiently small, then the inclusion $\pi^{-1}(H_\varepsilon) \to X$ induces an isomorphism $IH_i(\pi^{-1}(H_\varepsilon);\mathbb{Z}) \cong IH_i(X;\mathbb{Z})$ for all $i<n-c$ and a surjection $IH_{n-c}(\pi^{-1}(H_\varepsilon);\mathbb{Z}) \to IH_{n-c}(X;\mathbb{Z}) \to 0$.*

Furthermore. *If H is generic, then H_ε may be replaced by H in the above theorem.*

Remarks. If $\mathbf{IC}_X^{\bullet}$ denotes the intersection homology complex of sheaves ([GM5]) on X, then there is a canonical isomorphism

$$IH_i(\pi^{-1}(H_\varepsilon)) \cong H_c^{2n-i}(\mathbf{IC}_X^{\bullet} \mid (X \cap \pi^{-1}(H)))$$

i.e., it is the (hyper) cohomology with compact supports of the restriction of the sheaf $\mathbf{IC}_X^{\bullet}$ to $X \cap \pi^{-1}(H)$. In this sense, the neighborhood H_ε may be replaced with H provided we also replace the intersection homology of $H \cap X$ with the cohomology of the restriction of the intersection homology sheaf to $H \cap X$. For *generic* H, these are the same, i.e., there is a canonical isomorphism between $IH_i(\pi^{-1}(H))$ and $IH_i(\pi^{-1}(H_\varepsilon))$ (see below).

Preliminaries to the proof. The map π extends to a finite and proper algebraic morphism $\bar{\pi}$: $\bar{X} \to \mathbb{CP}^N$, where $\bar{X}$ contains X as a dense open subset. Choose a Whitney stratification of $\bar{X}$ and of $\bar{Z}=\bar{\pi}(\bar{X})$ so that X is a union of strata and so that $\bar{\pi}$ is a stratified map (Part I, Sect. 1.7). Choose H to be transverse to each of the strata of $\bar{Z}$. (This is the genericity assumption on H.) Let f: $\bar{Z} \to \mathbb{R}$ be a Morse perturbation of the square of the distance from H (see Sect. 4.A).

$$\begin{array}{ccccc} X & \subset & \bar{X} & & \\ \downarrow{\scriptstyle \pi} & & \downarrow{\scriptstyle \bar{\pi}} & & \\ Z & \subset & \bar{Z} & \xrightarrow[f]{} & \mathbb{R}. \end{array}$$

Proof of theorem. It suffices to prove the theorem for H a linear hyperplane, since the higher codimension cases follow from this by induction. By the same trick as appears in Proposition 5.A.1, we can assume that $U=\pi^{-1}f^{-1}[0,\varepsilon]$ for $\varepsilon>0$ sufficiently small. Thus, (by the long exact sequence for the pair (X, U)) it suffices to show that $IH_i(X, U; \mathbb{Z})=0$ for all $i\leq n-1$. For any stratum A of $\bar{Z}$ and for any critical point $p\in A$ we have

$$\operatorname{index}(f\mid A, p)\geq \dim_{\mathbb{C}}(A).$$

Now consider the relative Morse theory of the function $f\pi\colon X\to\mathbb{R}$. For each critical point $p\in\bar{Z}$ we have (as in Sect. 4.6) a decomposition of the relative normal Morse data for $f\pi$ at p into a disjoint union

$$\coprod_{x\in\bar{\pi}^{-1}(p)} (J_x, K_x)$$

where (J_x, K_x) is the normal Morse data for $f\pi$. If $[a,b]\subset\mathbb{R}$ contains no critical values except $v=f(p)\in(a,b)$, then

$$IH_i(X_{\leq b}, X_{\leq a}) = \bigoplus_{z\in\bar{\pi}^{-1}(p)} IH_{i-\lambda}(J_x, K_x).$$

By Proposition 6.4, whether or not $x\in X$, this group vanishes for all $i<\operatorname{cod}(A)$. Thus, $IH_i(X_{\leq b}, X_{\leq a}; \mathbb{Z})=0$ for all $i<n$. Apply this formula to each critical point in $X-U$ to obtain the result.

Proof of Furthermore. The transversality assumption implies that $\pi^{-1}(H)$ has a vectorbundle neighborhood U in X (Part I, Sect. 1.11). By the Kunneth formula and Mayer Vietoris we have

$$IH_i(\pi^{-1}(H))\cong IH_i(U)$$

for all i. By the same trick as appears in 5.A.1, we can replace U by $\pi^{-1}(H_\varepsilon)$ or by $\pi^{-1}f^{-1}[0,\varepsilon]$ for $\varepsilon>0$ sufficiently small. □

6.11. Local Lefschetz Theorem for Intersection Homology

Theorem. *Suppose X is a purely n-dimensional complex algebraic subvariety of some nonsingular complex algebraic variety M. Suppose P is a nonsingular complex algebraic variety, and that $\pi\colon X\to P$ is a (not necessarily proper) algebraic morphism with finite fibres. Fix a point $p\in\overline{\pi(X)}$ and let $\partial B_\delta(p)$ denote the boundary of a ball of radius δ about the point p, with respect to some smooth Riemannian metric on P. Let H be an affine linear subspace of codimension c in P (with respect to some local coordinate system about p) which passes through the point p, and is generic among all affine linear subspaces through p. If $\delta>0$ is sufficiently small, then the homomorphism*

$$IH_i(X\cap\pi^{-1}(\partial B_\delta(p)\cap H); \mathbb{Z})\to IH_i(X\cap\pi^{-1}(\partial B_\delta(p)); \mathbb{Z})$$

is an isomorphism for all $i<n-c-1$ and is a surjection for $i=n-c-1$.

Proof. The proof is exactly the same as the one in Sect. 5.3, so we will just sketch it here: we reduce to the case $c=1$ by induction. Since the problem is local, we may replace the manifold P by $\mathbb{C}^N$. Extend π to a proper finite map $\bar{\pi}: \bar{X} \to \mathbb{C}^N$ where $\bar{X}$ contains X as a dense open subset. Choose Whitney stratifications of $\bar{X}$ and of $\bar{Z} = \overline{\pi(X)}$ so that the point $\{p\}$ is a separate stratum of $\bar{Z}$, the map $\bar{\pi}$ is a stratified map, and so that X is a union of strata (Part I, Sect. 1.7). Take H to be transverse to the strata of $\bar{Z} - \{p\}$. (This is the genericity assumption on H.) If $f: \mathbb{C}^N \to \mathbb{C}$ is a linear projection such that $H = f^{-1}(0)$, then $df(p)$ is nondegenerate (Part I, Sect. 2.1). If

$$L_X^\pi = X \cap \bar{\pi}^{-1}(\bar{Z} \cap \partial B_\delta(p))$$

denotes the "relative link" of the point p, then we must show that

$$IH_i(L_X^\pi, \partial \mathscr{L}_X^\pi; \mathbb{Z}) = 0 \qquad \text{for } i \leq n-2.$$

Since the map $\bar{\pi}$ is finite, this group is a sum of groups $IH_i(L_x, \partial \mathscr{L}_x; \mathbb{Z})$ over points $x \in \bar{\pi}^{-1}(p)$. As in 5.3.2, we verify that these groups vanish in two steps:

(a) $IH_i(L_x, \mathscr{L}_x) = 0$ for all $i \leq n-1$

(b) $IH_i(\mathscr{L}_x, \partial \mathscr{L}_x) = 0$ for all $i \leq n-2$

and apply the long exact sequence for the triple $(L_x, \mathscr{L}_x, \partial \mathscr{L}_x)$. However, part (b) implies part (a) by [GM3] Sect. 3.5, Corollary 1, or by the same proof as appears in 5.3.2 (with π_i replaced by IH_i). In fact, both groups are calculated in the braid diagram of Sect. 6.7 and are found to be zero. □

6.12. Morse Inequalities

An argument identical to the standard one (e.g., [Mi1]) can be used to derive Morse inequalities for the intersection homology groups. However, each critical point must be counted with a multiplicity which is the rank of the Morse group at that point.

Suppose X is a complex analytic variety with a Whitney stratification. For each $x \in X$ define the rank of the variation at x,

$$\ell(x) = \operatorname{rank}(A_\omega)$$

where ω is any nondegenerate covector at x. If x is a nonsingular point of X we set $\ell(x) = 1$.

Theorem. *Suppose X is compact and $f: X \to \mathbb{R}$ is a Morse function. Define $Ib_j = \operatorname{rank}(IH_j(X))$ and for each nonnegative integer m define*

$$R_m = \sum \{\ell(x) \mid x \text{ is a critical point and } \operatorname{codim}(S) + \operatorname{index}(f \mid S, x) = m\}$$

(where S is the stratum containing x). Then the following Morse inequalities hold:

$$R_0 \geq Ib_0$$
$$R_1 - R_0 \geq Ib_1 - Ib_0$$
$$R_2 - R_1 + R_0 \geq Ib_2 - Ib_1 + Ib_0 \quad \text{etc.}$$

and

$$\sum_{i=0}^{2n} (-1)^i R_i = \sum_{i=0}^{2n} (-1)^i Ib_i$$

where n is the complex dimension of X. □

6.13. Specialization Over a Curve

This section is largely a summary of [GM3] § 6.

6.13.1. Introduction. If $f: X \to D^0$ is a Whitney stratified proper analytic map to the open unit disk, with fibre $f^{-1}(t)=X_t$, then there is a "canonical retraction" $\psi: U \to X_0$ of some neighborhood U of the central fibre to the central fibre. Restricting ψ to a nearby fibre X_t defines the specialization map,

$$\psi_t: X_t \to X_o.$$

We study the fibres of this map. Each fibre $\psi_t^{-1}(x)$ may be given (in a noncanonical way) the structure of a Stein space with boundary, whose dimension is equal to the codimension of the stratum which contains x. (We will show in 6.13.6 using Morse theory that the specialization

$$R\psi_{t*}(\mathbf{IC}^{\cdot})$$

of the intersection homology sheaf is a *perverse sheaf* in the sense of [BBD].)

6.13.2. The setup. We suppose $f: X \to D^0$ is a proper analytic map of an irreducible variety X to the open disk in the complex plane. We will assume that f is the restriction of a smooth proper nonsingular map

$$\tilde{f}: M \to D^0.$$

We denote by X_t the fibre $f^{-1}(t)$ over a point $t \in D^0$ and we will suppose that X and D^0 have been stratified so that:

(a) The origin $0 \in D^0$ is the only zero-dimensional stratum in the target
(b) f takes each stratum of X submersively to a stratum of D^0
(c) $X_0 = f^{-1}(0)$ is a union of strata.

We also choose a system of control data ([Ma1], [T5]), $\{T_A, \pi_A, \rho_A\}$ on X, i.e., a tubular neighborhood (in M), $\pi_A: T_A \to A$ for each stratum A of X, and a tubular distance function $\rho_A: T_A \to [0, 2\varepsilon)$ such that whenever $B > A$ is another stratum we have:

(1) $(\pi_A, \rho_A) | B: B \to A \times (0, 2\varepsilon)$ is a submersion
(2) $\pi_A \pi_B = \pi_B \pi_A$
(3) $\rho_A \pi_B = \rho_A$.

Consider the following neighborhood of X_0 consisting of the unions of the tubular neighborhoods of all the strata in X_0:

$$U(\varepsilon) = \bigcup_{A \subset X_0} \{y \in T_A \mid \rho_A(y) \leq \varepsilon\}.$$

There is a continuous "retraction" [G1], [G2]

$$\psi: U(\varepsilon) \to X_0$$

which collapses the fibres of each π_A to points. (ψ is not actually a retraction because its restriction to X_0 is not the identity, but is rather homotopic to the identity.)

6.13.3. Definition. Let $\psi_t\colon X_t \to X_0$ be the restriction of ψ to the fibre X_t, where t is chosen so small that $X_t \subset U(\varepsilon)$.

It is a fact (that we do not prove or use here) that the topological type of the map ψ_t is independent of the choice of control data, family of lines, or parameter value t.

6.13.4. Structure of the fibres. It is easy to see from the construction of ψ that for any stratum A of X_0 and for any $x \in A$, there is a unique point $x' \in A$ such that

$$\psi_t^{-1}(x) = \pi_A^{-1}(x') \cap U(\varepsilon) \cap X_t.$$

Furthermore if $|t|$ is sufficiently small, then X_t is transvese to the set

$$\pi_A^{-1}(x') \cap \rho_A^{-1}(\varepsilon).$$

(Note that $\pi_A^{-1}(x')$ is a smooth submanifold of M which meets A transversally in the single point $\{x'\}$.)

6.13.5. Proposition. *Suppose $x \in X_0$ lies in some stratum A whose codimension in X_0 is c. Then, the fibre over x of the specialization map $\psi_t\colon X_t \to X_0$ is homeomorphic to the closure (in X_t) of an affine complex analytic space of dimension c. (The closure is obtained by adding a collared boundary to this analytic space.) Fibres over nearby points in the same stratum can be given compatible complex analytic structures.*

Proof. Careful details of this proposition were published in [GM3] Sect. 6.4, so we give here an outline only. Find x' as above. The fibre $N = \pi_A^{-1}(x')$ of the tubular neigborhood $T_A \to A$ is a smooth submanifold of M whose dimension is complementary to that of A. If N were complex analytic, then the proposition would follow directly from 6.13.4, because $\psi_t^{-1}(x)$ is the transverse intersection of the sets

$$N,\ X_t, \quad \text{and} \quad B_\varepsilon(x')$$

where $B_\varepsilon(x')$ denotes a (closed) ball of radius ε (with respect to some Riemannian metric on M) centered at the point x'. At least we can say (by transversality) that $\psi_t^{-1}(x)$ is a compact Whitney stratified set of (real) dimension $2c$, with a collared boundary $N \cap X_t \cap \partial B_\delta(x')$.

Although N may not be complex analytic, it suffices to find a complex analytic manifold N' with the same dimension as N which is transverse to the stratum A, such that the spaces

$$B_\delta(x') \cap N' \cap X_t \quad \text{and} \quad B_\delta(x') \cap N \cap X_t$$

are homeomorphic. This homeomorphism is found using Thom's first isotopy lemma by embedding the manifolds N and N' into a one-parameter family of slices N_θ transversal to A at the point x'. (As the value of θ varies, it may be necessary to reduce δ and $|t|$ to guarantee that the sets

$$N_\theta,\ B_\delta(x), \quad \text{and} \quad X_t$$

remain transverse to each other. This kind of argument is much better made with the language of "fringed sets" and "moving the wall"; we will not repeat the details here.) □

6.13.6. Morse theory of the fibre. Since $Y=\psi_t^{-1}(x)$ is homeomorphic to some affine analytic space,

$$N'\cap B_\delta(x')\cap X_t$$

there is a Morse perturbation g of the function

$$\tilde{g}(y)=(\text{distance}\,(y,x'))^2$$

with the following properties: g has a maximum on

$$\partial Y=N'\cap\partial B_\delta(x')\cap X_t$$

and for any stratum $S\cap N'$ of $X_t\cap N'$, and for any critical point $q\in S\cap N'$ of the function g, we have

$$\operatorname{index}(g\,|\,S\cap N',q)\leq\dim_{\mathbb{C}}(S\cap N')=\dim_{\mathbb{C}}(S)-\dim_{\mathbb{C}}(A)$$

(see 4.A.4). For each critical value $v\in\mathbb{R}$ of the Morse function $g\colon Y\to\mathbb{R}$, we have:

(a) $Y_{\leq v+\eta}$ is obtained from $Y_{\leq v-\eta}$ by attaching cells of dimension less than or equal to $\dim_{\mathbb{C}}(Y)=\operatorname{cod}_X(A)$ (see Sect. 4.3).

(b) $IH_i(Y_{\leq v+\eta}, Y_{\leq v-\eta})=0$ for all $i>\dim_{\mathbb{C}}(Y)=\operatorname{cod}_X(A)$ (see Sect. 6.4).

This proves the following:

Proposition. *Let $Y=\psi_t^{-1}(x)$ be a fibre of the specialization map over a point x which lies in some stratum $A\subset X_0$. Let c denote the complex codimension of the stratum A in the set X_0. Then,*

(a) *Y has the homotopy type of a CW complex of dimension $\leq c$*

(b) *$IH_i(Y)=0$ for all $i>c$*

(c) *$IH_i(Y,\partial Y)=0$ for all $i<c-1$.*

Corollary. *The sheaf $R\psi_{t*}(\mathbf{IC}^{\bullet}_{X_t})$ is perverse on X_0.*

Proof of corollary. The stalk cohomology at a point $x\in X_0$ is precisely $IH_*(\psi_t^{-1}(x))$, which satisfies the above vanishing condition. This verifies the "support condition" of a perverse sheaf. The "cosupport condition" is similarly a consequence of the vanishing of the relative intersection homology, $IH_*(Y,\partial Y)$. □

6.A. Appendix: Remarks on Morse Theory, Perverse Sheaves, and $\mathscr{D}$-Modules

The following remarks stated without proof, are included for the convenience of the reader who is familiar with sheaf theoretic and $\mathscr{D}$-module theoretic techniques and wishes to know their relation with Morse theory.

6.A.1. Morse theory and sheaf theory. The main results of stratified Morse theory apply to arbitrary sheaf coefficients without change: suppose X is a Whitney stratified complex analytic subvariety of some smooth complex analytic manifold M and that $\mathscr{F}^{\bullet}$ is a complex of sheaves (of abelian groups or of complex vectorspaces) on X which is constructible with respect to the given stratification [GM5]. (This means that each cohomology sheaf $\mathscr{H}^i(\mathscr{F}^{\bullet})$ is locally constant on each stratum.) Suppose S is a stratum in X, $p\in S$, and $\xi\in T^*_{S,p}M$

is a conormal vector to S which is nondegenerate. It is possible to define Morse groups $A_\xi^i(\mathscr{F}^\bullet)$ associated to this data,

$$A_\xi^i(\mathscr{F}^\bullet)=H^i(J, K; \mathscr{F}^\bullet)$$

where the pair (J, K) is normal Morse data corresponding to any smooth function $f: M \to \mathbb{R}$ such that $df(p)=\xi$. In other words, for any choice of complex analytic submanifold $N \subset M$ which intersects S transversally in the single point p, and for any $\varepsilon \ll \delta$ sufficiently small (and chosen in accordance with Part I, Sect. 3.6) we have

$$(J, K)=(X \cap N \cap B_\delta(p) \cap f^{-1}[v-\varepsilon, v+\varepsilon], X \cap N \cap B_\delta(p) \cap f^{-1}(v-\varepsilon))$$

where $v=f(p)$. The main results of stratified Morse theory now states:

Proposition. *The cohomology groups $A_\xi^i(\mathscr{F}^\bullet)=H^i(J, K; \mathscr{F}^\bullet)$ are independent of the choices of f, N, ε, δ, or the Riemannian metric involved in the definition of the ball $B_\delta(p)$. Furthermore, any path between nondegenerate conormal vectors ξ and $\eta \in T_{S,q}^* M$ determines a canonical isomorphism*

$$A_\xi^i(\mathscr{F}^\bullet) \cong A_\eta^i(\mathscr{F}^\bullet)$$

and in particular this Morse group for the sheaf $\mathscr{F}^\bullet$ is independent of the Morse function. If $f: M \to \mathbb{R}$ is a Morse function such that $df(p)=\xi$ and if the interval $[a, b]$ contains no critical values other than $v=f(p)$, then there is an isomorphism which is canonical up to a choice of orientation for the tangential Morse data,

$$H^i(X_{\leq b}, X_{\leq a}; \mathscr{F}^\bullet) \cong A_\xi^{i-\lambda}(\mathscr{F}^\bullet)$$

where λ is the Morse index of the restriction $f \mid S$ at the point p.

This proposition may be proven using the same technique as Sect. 6.7 ($B_k \Rightarrow A_n$).

6.A.2. Morse theory and vanishing cycles. Suppose $\mathscr{F}^\bullet$ is a constructible complex of sheaves of complex vectorspaces on the variety X (as in 6.A.1), that X is Whitney stratified so that the cohomology sheaves of $\mathscr{F}^\bullet$ are locally constant on the strata, that S is a stratum, $p \in S$, and $\xi \in T_{S,p}^* M$ is a nondegenerate conormal vector. Let $F: X \to \mathbb{C}$ be a locally defined complex analytic map such that $F(S)=0$ and $dF(p)=\xi$, where we use the canonical identification

$$T_p^* M \cong \operatorname{Hom}_{\mathbb{C}}(T_p M, \mathbb{C}) \cong \operatorname{Hom}_{\mathbb{R}}(T_p M, \mathbb{R}).$$

Then, the Morse group $A_\xi^i(\mathscr{F}^\bullet)$ is canonically isomorphic to the "$R\Phi$ vanishing cycles" $R^i \Phi(\mathscr{F}^\bullet)_p$ of [D3].

6.A.3. Pure sheaves. In [KS], Kashiwara and Schapira define a *pure complex* $\mathscr{F}^\bullet$ to be a constructible complex of sheaves (of complex vectorspaces) on a variety X, such that the Morse group $A_\xi^i(\mathscr{F}^\bullet)$ vanishes in all degrees except (possibly) for one. Thus, our result Sect. 6.4 states that the (middle) intersection homology complex $\mathscr{F}^\bullet=\mathbf{IC}^\bullet(X)$ is pure (with shift 0) in the sense of [KS].

6.A.4. Morse theory and holonomic $\mathscr{D}$-modules. Suppose $\mathscr{M}$ is a holonomic $\mathscr{D}$-module with regular singularities whose singular support is contained in the analytic subvariety $X \subset M$. Then, the sheaf $\mathscr{F}^\bullet=DR(\mathscr{M})$ is pure in the sense of [KS] (where DR denotes the de Rham functor). (This may be proven as in Sect. 6.7, or using 6.A.5 below together with the main result Sect. 6.4, and is also shown in [KS] Theorem 9.5.2.) Fix a stratification of X with respect to which $\mathscr{F}^{\bullet}$ is constructible, and suppose that $\xi \in T^*_{S,p}M$ is a nondegenerate covector. Then,

(a) The dimension of the single (possibly) nonzero Morse group $A^i_\xi(\mathscr{F}^{\bullet})$ is equal to the multiplicity (at ξ) of the characteristic variety $ch(\mathscr{M})$ of the $\mathscr{D}$-module $\mathscr{M}$ [Sab1], [Gin].

(b) The Morse group $A^i_\xi(\mathscr{F}^{\bullet})$ can be identified with the micro-solutions of the $\mathscr{D}$-module $\mathscr{M}$, at the point ξ.

6.A.5. Holonomic $\mathscr{D}$-modules and perverse sheaves. The (analytically constructible) perverse sheaves on X are those complex $\mathscr{F}^{\bullet}$ of sheaves of complex vectorspaces for which there exists a complex analytic Whitney stratification of X with respect to which $\mathscr{F}^{\bullet}$ is constructible, and such that for each stratum S of X we have

(a) $H^k(s^* \mathscr{F}^{\bullet})=0$ for all $k > -\dim_{\mathbb{C}}(S)$ and
(b) $H^k(s^! \mathscr{F}^{\bullet})=0$ for all $k < \dim_{\mathbb{C}}(S)$

where $s: S \to X$ denotes the inclusion. The (analytically constructible) perverse sheaves on X form an abelian category whose simple objects are (up to a shift in degree) the intersection homology complex with irreducible local coefficient systems on subvarieties of X [BBD]. Thus, every perverse sheaf has a filtration whose graded pieces are the intersection homology sheaves of subvarieties. In particular, the three main results in this chapter on intersection homology also apply to perverse sheaves.

The de Rham functor gives an equivalence of categories between the category of holonomic $\mathscr{D}_M$-modules with regular singularities whose singular support is contained in X, and the category of (analytically constructible) perverse sheaves on X. If we fix a particular complex analytic Whitney stratification of X, then this functor restricts to an equivalence of categories between the category of holonomic regular $\mathscr{D}$-modules whose characteristic variety is contained in the union of the conormal bundles to the strata of X, and the category of perverse sheaves on X which are constructible with respect to the given stratification.

6.A.6. Specialization of $\mathscr{D}$-modules. Our theorem that perverse sheaves specialize to perverse sheaves is equivalent to the statement that there exists a specialization functor on the category of $\mathscr{D}$-modules, which commutes with the functor DR. This specialization functor was explicitly constructed by Kashiwara and Malgrange.

Chapter 7. Connectivity Theorems for q-Defective Pairs

7.0. Introduction

The four results in this chapter are generalizations of Theorems 1.1, 1.1*, 1.2, and 1.2*. We will replace the pair $(\mathbb{CP}^N, L)$ = (projective space, linear subspace) by the pair (Y, A) = (complex manifold, compact complex submanifold) in the statement of these four theorems. We define the notion of a "q-defective pair", which implies that the Lefschetz theorem holds for (Y, A) in a range of dimensions which is smaller (by q) than for the pair $(\mathbb{CP}^N, \mathbb{CP}^{N-1})$.

It was first observed by Bott [Bo1] (in the case (Y, A) = (a positive line bundle, zero section)) that the usual Morse-theoretic proof of the Lefschetz theorem also worked in this setting. Later Griffiths [Gr1] extended these results to positive vectorbundles, and Sommese [Sm1] further extended the results to "q-convex" vectorbundles.

In this chapter we will show that our Morse theory techniques can be used to further generalize the results of Sommese to include singular and noncompact varieties, and also to give "relative" Lefschetz theorems for maps with large fibres. These four results are proven in exactly the same manner as Theorems 1.1, 1.1*, 1.2, and 1.2*, so we will only outline the proofs here.

7.1. q-Defective Pairs

Let $A \subset Y$ be a compact complex submanifold of a complex manifold. If $f\colon Y \to \mathbb{R}$ is a smooth function and $y \in Y$, we define (see [Fr1], p. 258, and also 4.A.3)

$$H_y(f) = \{\xi \in T_y Y \mid \partial f(y)(\xi) = 0\}$$

to be the unique maximal complex subspace of $\ker df(y)$.

Definition. The pair (Y, A) is *q-defective* (resp. q-codefective) if there exists a proper continuous function $f\colon Y \to \mathbb{R}$ such that

(a) $f \mid (Y - A)$ is smooth.

(b) For all $y \in Y$, $f(y) \geq 0$, and $f^{-1}(0) = A$.

(c) For all $y \in Y - A$, the restriction of the Levi form $L(f)(y)$ to the subspace $H_y(f)$ has at most q nonnegative (resp. nonpositive) eigenvalues.

Remark. The pair (Y, A) is q-defective if and only if $Y - A$ admits an "asymptotically $q+1$ convex exhaustion function" in the sense of [Sm4]. It follows ([Sm3], Sect. 0.3.3) that $Y - A$ is "$q+1$ convex" in the sense of [AG]. See also the related concepts of topologically q-complete [FK5]; q-complete, cohomologically q-complete, q-convex, and cohomologically q-convex ([Fe1], [Fr2], [AG]).

Proposition. *If (Y, A) is q-defective (resp. q-codefective) and if $f\colon Y\to\mathbb{R}$ is a function which realizes this defect, and if $B\subset A-Y$ is any complex submanifold, then f can be approximated by a function f' whose restriction to B is a Morse function with convexity defect $\leq q$ (resp. dual convexity defect $\leq q$). Thus, the Morse index λ of any critical point of $f|B$ satisfies the bound $\lambda\geq\dim_{\mathbb{C}}(B)-q$ (resp. $\lambda\leq\dim_{\mathbb{C}}(B)+q$).*

Proof. This is a restatement of Corollary 4.A.4. □

7.2. Defective Vectorbundles

Let $\pi\colon E\to M$ be a holomorphic vectorbundle over a compact complex analytic manifold M. We shall identify M with the zero section.

7.2.1. Definition. A *Finsler metric* on E is a continuous nonnegative function $r\colon E\to\mathbb{R}$ such that

(a) $r^{-1}(0)=M$,
(b) r is smooth (C^∞) on $E-M$,
(c) the restriction of r to each fibre $E_x=\pi^{-1}(x)$ is strictly convex.
(d) for any $\lambda\in\mathbb{C}$ and for any $e\in E$, we have

$$r(\lambda e)=|\lambda|^2\, r(e).$$

(Hermitian metrics give rise to linear Finsler metrics, in an obvious way.)

7.2.2. Definition. The vectorbundle $\pi\colon E\to M$ is *q defective* (resp. *q-codefective*) if there exists a Finsler metric $r\colon E\to\mathbb{R}$ on E such that at each point $v\in E-M$, the restriction of the Levi form of r to the subspace

$$H_v(r)=\{\xi\in T_v E\mid \partial r(v)(\xi)=0\}$$

has at most q nonnegative (resp. nonpositive) eigenvalues.

7.2.3. Examples. The vectorbundle $qL=L\oplus L\oplus\ldots\oplus L$ is $q-1$ defective, where $L\to\mathbb{CP}^N$ is the canonical line bundle. Bott [Bo1] shows that a positive line bundle is zero-defective. Griffiths [Gr1] shows that a "positive" vectorbundle E of rank r is $r-1$ defective. (These papers study the Levi form of $\log(r)$ rather than the Levi form of r, but by 4.A.3 their restrictions to $H_v(r)$ will have the same signatures.) Sommese ([Sm4], Sect. 3.6) proves the following: if $\pi\colon E\to M$ is an ample vectorbundle (in the sense of Hartshorne [Hr]) which is generated by its global holomorphic sections, then E is $r-1$ defective, where $r=\operatorname{rank}(E)$. See also the related concepts of q-ample vectorbundles ([Sm6], [Oko1]), strictly q-positive, and Finsler q-positive vectorbundles ([Sm1], [Oko3]), q-convex and strongly q-convex vectorbundles ([Sm1]).

Problem. What is the defect of an ample vectorbundle ([Hr]) of rank r?

7.2.4. Proposition. *If $\pi: E \to M$ is a q-defective vectorbundle and if $s: M \to E$ is a global holomorphic section of E with image $S = s(M)$, then the pair (E, S) is q-defective.*

Proof. If r is a Finsler metric on E which achieves the q-defect, then the function $f: E \to \mathbb{R}$ given by

$$f(v) = r(v - s\pi(v))$$

achieves the q-defect on $E - S$, by Corollary 4.A.4. □

7.3. Lefschetz Theorems for Defective Pairs

In this section we will suppose that Y is a complex algebraic manifold, and $A \subset Y$ is a compact complex algebraic submanifold. We will assume the pair (Y, A) is $(c-1)$-defective. Let A_δ be the δ-neighborhood of A with respect to some real analytic Riemannian metric on Y.

Theorem. *The results in Theorems 1.1 and 1.2 (and the remark following the statements of these theorems) continue to hold when we replace the pair* $(\mathbb{CP}^N, H)$ *by the pair* (Y, A).

Special cases. Theorems 1.1 and 1.2 follow from Theorem 7.3: let $E \to \mathbb{CP}^N$ be the bundle $E = cL = L \oplus L \oplus \ldots \oplus L$ (where L is the canonical line bundle on projective space). This bundle admits a Hermitian metric (,) such that the associated distance function $r(\xi) = (\xi, \bar{\xi})$ is $c-1$ defective. Choose a section s of E which vanishes on the linear subspace H (of codimension c). Let S denote the image of the section s, and let $z: \mathbb{CP}^N \to E$ denote the zero section. Consider the following diagram of spaces:

$$\begin{array}{ccc} X & & \\ \downarrow{\scriptstyle \pi} & & \\ \mathbb{CP}^N & \xrightarrow{z} & E \\ & & \cup \\ & & S \end{array}$$

By Proposition 7.2.4, the pair (E, S) is $(c-1)$-defective. Apply Theorem 7.3 to the map $z \circ \pi: X \to E$, noting that $z^{-1}(S) = H$. This gives Theorems 1.1 and 1.2.

Remark. Similar results can be obtained for intersection homology instead of homotopy, using the techniques of Sect. 6. See also [FK5] and [FK6].

Proof of Theorem 7.3. We shall give the proof of the analogue to Theorem 1.1 (the analogue of Theorem 1.2 being entirely parallel). This proof parallels Sect. 5.1.

Extend the map π to a proper algebraic map $\bar{\pi}: \bar{X} \to Y$. Stratify $\bar{X}$ and $\bar{Z} = \bar{\pi}(\bar{X})$ so that $\bar{\pi}$ is a stratified map and so that X is a union of strata of $\bar{X}$. Let $\bar{f}: Y \to \mathbb{R}$ be a function (as in Sect. 7.1) which realizes the $c-1$ defective nature of the pair (Y, A), and let f be a small (close in the Whitney $C^\infty(Y - A, \mathbb{R})$

topology) perturbation of $\bar{f}|(Y-A)$ so that $\bar{f}|(\bar{Z}-A)$ has nondegenerate critical points, and distinct critical values (see Part I, Sect. 2.2 or [P1]). Even though f may have infinitely many critical points near A, for each interval $[a, b]$ which contains a single critical value (corresponding to a critical point $p \in \bar{Z}$) we have, by Proposition 4.3,

$$\pi_i(X_{\leq b}, X_{\leq a}) = 0 \quad \text{if} \quad i \leq \hat{m} = n - (\Delta(p) + \Gamma(p)) - 1$$

where $\Delta(p)$ and $\Gamma(p)$ are the "normal and tangential defects" for f at the point p (Sect. 4.3). By Corollary 4.A.4 we have $\Gamma(p) \leq c-1$, while 4.5.1 gives an estimate for $\Delta(p)$ from which we conclude that $\hat{m} \geq \hat{n}$. Applying this estimate to each critical value, we have $\pi_i(X, X_{\leq \varepsilon}) = 0$ for arbitrarily small $\varepsilon > 0$. It remains to show that there exists $\varepsilon > 0$ and $\delta > 0$ so that $A_\delta \subset f^{-1}[0, \varepsilon]$ and so that

$$\pi_i(X_{\leq \varepsilon}, X \cap \pi^{-1}(A_\delta)) = 0 \qquad \text{for } i \leq \hat{n}.$$

This follows from the same argument as 5.A.2. Choose $\delta' < \delta \ll 1$ and $\varepsilon' < \varepsilon \ll 1$ so that

$$A_{\delta'} \subset f^{-1}[0, \varepsilon'] \subset A_\delta \subset f^{-1}[0, \varepsilon]$$

and so that $X \cap \pi^{-1}(A_\delta)$ deformation retracts to $X \cap \pi^{-1}(A_{\delta'})$ and so that $\pi_i(X_{\leq \varepsilon'}, X_{\leq \varepsilon}) = 0$ for all $i \leq \hat{n}$. The composition of the first two inclusions induces isomorphisms on homotopy groups of all dimensions, while the composition of the second two inclusions induces isomorphisms on homotopy groups of all dimensions $\leq \hat{n}$ and a surjection in dimension $\hat{n}$. The result follows immediately. □

7.3*. Homotopy Dimension of Codefective Pairs

Let Y be a complex manifold and $A \subset Y$ be a compact complex submanifold. Suppose the pair (Y, A) is $(q-1)$-codefective and $f\colon Y \to \mathbb{R}$ is a function (as in Sect. 7.1) which achieves this $(q-1)$-codefect. Let $A_\varepsilon = f^{-1}[0, \varepsilon]$ be an ε-neighborhood of A. Let X be an n-dimensional (possibly singular) complex analytic variety. Let $\pi\colon X \to Y$ be a proper analytic map.

7.3*.1. Theorem. *Let $\phi(k)$ denote the dimension of the set of points $y \in \pi(X)$ such that the fibre $\pi^{-1}(y)$ has dimension k. (If this set is empty, we set $\phi(k) = -\infty$.) Then X has the homotopy type of a CW complex, which is obtained from the space $\pi^{-1}(A_\varepsilon)$ by attaching cells of (real) dimension less than or equal to*

$$\hat{n}^* = n + \sup_k (2k - (n - \phi(k)) + \inf(\phi(k), q-1)).$$

Special cases. Theorem 1.1* follows from Theorem 7.3*.1 by the trick of Sect. 7.3. Choose a section $s\colon \mathbb{CP}^N \to E$ of the vectorbundle $E = (n-c)L \to \mathbb{CP}^n$, so that s vanishes on a generic $c-1$ dimensional linear space $G \subset \mathbb{CP}^n$ which is complementary to H and which is transverse to each stratum of a Whitney stratification of $\pi(X)$. If S denotes the image of s, then the pair $(E|(\mathbb{CP}^n - H), S)$ is $c-1$ codefective. Let $z\colon \mathbb{CP}^n - H \to E|(\mathbb{CP}^n - H)$ denote the zero section and apply Proposition 7.3*.1 to the composition $z \circ \pi\colon X \to E|(\mathbb{CP}^n - H)$. By 5.A.3,

$(z\circ\pi)^{-1}(S_\varepsilon)$ is homotopy equivalent to $\pi^{-1}(G)=(z\circ\pi)^{-1}(S)$, and this has dimension $\leq\hat{n}^*$ by direct computation. Theorem 1.1* follows immediately. □

Remark. Related results on the vanishing of the intersection homology of the pair (X, A) may be obtained, using the techniques of Sect. 6. See also [FK5] and [FK6].

7.3*.2. Theorem. *Let W be a subvariety of X. We consider the extent to which the inclusion $W\subset X$ fails to be a local complete intersection morphism by defining for each k the number $\phi(k)$ to be the dimension of the set of all points $p\in W$ such that a neighborhood of p (in W) can be defined (as a subset of X) by $n-\dim_p(W)+k$ equations, and no fewer. (If this set is empty, we set $\phi(k)=-\infty$.) Then the space $X-W$ has the homotopy type of a CW complex, which can be obtained from the space $(X-W)\cap\pi^{-1}(A_\varepsilon)$ by attaching cells of dimension $\leq\hat{n}^*$, where*

$$\hat{n}^*=\sup_{k\geq 1}(n+k-1+\inf(\phi(k), q-1)).$$

Special cases. Theorem 1.2* follows from Theorem 7.3*.2 by the same trick as above.

Proof of 7.3.* Stratify X and $Z=\pi(X)$ so that π is a stratified map. Approximate f by a Morse function g so that g has distinct critical values, $g^{-1}[0,\varepsilon]=f^{-1}[0,\varepsilon]$, and so that the Morse index λ of each critical point p on any stratum S of Z is bounded as follows: $\lambda\leq\dim_{\mathbb{C}}(S)+q-1$. (See 4.A.3.) By Proposition 4.3*, for each critical value $v=f(p)$, the space $X_{\leq v+\varepsilon}$ has the homotopy type of a space obtained from $X_{\leq v-\varepsilon}$ by attaching a CW complex of dimension $\leq\hat{m}^*=n+(\Delta^*(p)+\Gamma^*(p))$ where $\Gamma^*(p)$ is the "dual convexity defect" and is $\leq q-1$; and where $\Delta^*(p)$ is the "normal defect" which is estimated in 4.5.1*. From these estimates, we conclude that $\hat{m}^*\leq\hat{n}^*$. □

Chapter 8. Counterexamples

8.1. The estimates of fibre dimension in Theorem 1.1 are sharp: let $Y = \mathbb{C}^3 \subset \mathbb{CP}^3$ and let $\pi: X \to Y$ be the result of blowing up the origin in Y. Let $H \subset \mathbb{CP}^3$ be a generic hyperplane which does not contain the origin. Then $\pi_2(\pi^{-1}(H)) = 0$ since $\pi^{-1}(H)$ is contractible, but X has the homotopy type of $\mathbb{CP}^2$.

8.1*. The estimates on fibre dimension are sharp in Theorem 1.1*. Let X be the result of blowing up a point in $\mathbb{C}^3$ and consider the projection $\pi: X \to \mathbb{C}^3 = \mathbb{CP}^3 - H$. Then, X has the homotopy type of a CW complex of dimension four instead of three.

8.2. The estimates in Theorem 1.2 are sharp: let Z be the union of two transversally intersecting copies of $\mathbb{CP}^2$ in $\mathbb{CP}^4$. (This is not a local complete intersection.) Let H be a hyperplane in $\mathbb{CP}^4$ which is transverse to each of the components of Z. Then,

$$\pi_0(Z \cap H) \neq \pi_0(Z).$$

The assumption that H is generic cannot be removed from the Lefschetz theorem for affine varieties (remark in Sect. 1.2): let X be an affine subspace of $\mathbb{C}^n$ and take H to be a parallel hyperplane. Then $H \cap X = \phi$. (However, as noted in Sect. 1.2, H need only be transverse to the strata at infinity which occur in a Whitney stratification of $\bar{X} \subset \mathbb{CP}^n$.)

8.2*. The estimates in Theorem 1.2* are sharp: let $Z = \mathbb{C}^2 - \{\text{point}\}$. Then, Z has the homotopy type of a CW complex of dimension three.

8.3. The Lefschetz theorem does not hold for constructible sets: let $X = \mathbb{CP}^n - \mathbb{CP}^{n-1} + \{x\}$, where x is a point in $\mathbb{CP}^{n-1}$. Let Y be a generic hyperplane section of X. Then, $Y \cong \mathbb{C}^{n-1}$ so it has no homology except in dimension zero. However, $H_2(X) \cong \mathbb{Z}$ since X is obtained (up to homotopy) from $\mathbb{C}^n$ by attaching a two-cell.

8.4. One might hope that the "fibre defect" (Sect. 4.5) can be added to the "singularity defect" (Sect. 4.6) to obtain a Lefschetz theorem for a map $\pi: X \to \mathbb{CP}^n$, where X has singularities. However, this is false: even if X is a local complete intersection (so the "singularity defect" is zero), and the map π is "small" (so the "fibre defect" is zero), the Lefschetz theorem fails:

Fix a two-dimensional plane T (through the origin) in $\mathbb{C}^8$. First we define a family F (parametrized by $\mathbb{CP}^2$) of four-dimensional planes (through the origin) in $\mathbb{C}^8$ with the property that the intersection of any two of them is precisely the subspace T: consider the projection

$$r\colon \mathbb{C}^8 \to \mathbb{C}^8/T \cong \mathbb{C}^6 \cong \mathbb{C}^3 \oplus \mathbb{C}^3.$$

Choose an isomorphism $I\colon \mathbb{C}^3 \cong \mathbb{C}^3$ between the first and second copy of $\mathbb{C}^3$. The family F will be thought of as an embedding $\phi\colon \mathbb{CP}^2 \to G_4(\mathbb{C}^8)$. This map associates to any line L through the origin in $\mathbb{C}^3$ the four-plane

$$\phi(L) = r^{-1}(L \oplus I(L)).$$

Now define the space X to be the set of pairs

$$X = \{(f, g) \in G_4\,\mathbb{C}^8 \times G_2\,\mathbb{C}^8 \mid f \in F \text{ and } g \subset f \text{ and } \dim(g \cap T) \geq 1\}.$$

Let π denote the projection to the second factor $\pi\colon X \to G_2(\mathbb{C}^8)$ and let Y denote the image $\pi(X)$. We will show that $\pi\colon X \to Y$ is a small map, X is a local complete intersection, but the relative Lefschetz theorem is false.

X can be stratified with two strata: the open (five-dimensional) part, and the two-dimensional stratum $S = \pi^{-1}(T)$, which is isomorphic to $\mathbb{CP}^2$ by construction. The fibre $\pi^{-1}(g)$ (for any other point $g \neq T$) consists of a single point. Thus, π is a small map (Sect. 1.1). Furthermore, X is a local complete intersection since the projection to the first factor $\pi_1\colon X \to G_4(\mathbb{C}^8)$ is a fibration over the image of ϕ (which is a smooth $\mathbb{CP}^2$), and the fibre is the Schubert variety $\Omega[2, 4]$ which is a local complete intersection.

To see that the relative Lefschetz theorem fails, let H be a generic hyperplane section of Y (which misses the point $T \in Y$). Then, $\pi^{-1}(H)$ is simultaneously a (relative) hyperplane section of X and of $X - S$. By Proposition 1.1, the inclusion $\pi^{-1}(H) \to \pi^{-1}(X - S)$ induces an isomorphism on homotopy groups of dimensions 0, 1, 2, 3, and a surjection in dimension four (since $X - S$ is nonsingular). However, $\pi_i(X - S) \to \pi_i(X)$ fails to be an isomorphism in these dimensions, as can be seen from the long exact sequence in homotopy for the fibration $L \to S$, where L is the boundary of a tubular neighborhood of S in X.

Remark. The variety X has a small resolution $\theta\colon \tilde{X} \to X$ which is obtained by resolving each of the copies $\pi_1^{-1}(p) \cong \Omega[2, 4]$ separately, as in Sect. 6.5. However, the composition $\pi \circ \theta\colon \tilde{X} \to Y$ is not small or even semismall.

8.5. The topological results of Sect. 2.4 and 2.A.3 do not hold for arbitrary stratified maps to the disk. The following example is due to Thom [T5] and Hironaka [Hi2]. Let $Y = \mathbb{R}^2$ and $f\colon X \to Y$ be the (real) Hopf blow up of the origin. This is a real algebraic map, but is most easily pictured locally in polar coordinates: $y_1 = r\cos\theta$, $y_2 = r\sin\theta$. We stratify Y with two strata: $S_1 = \{0\}$, and $S_2 = Y - \{0\}$; and we stratify X with the strata $T_1 = f^{-1}(S_1)$ and $T_2 = X - T_1$. The map f is then a stratified map (and it is a submersion over each stratum of X), but it does not satisfy Thom's condition A_f. Furthermore, the local structure statements of Sect. 2.4 and 2.A.3 fail. For example, fix $p \in T_1$, and choose $\varepsilon \ll \delta \ll 1$ sufficiently small and let $y \in D_\varepsilon(0) \subset Y$. Then the set

$$\mathscr{L} = f^{-1}(y) \cap B_\delta(p)$$

is not independent of the point y, and the map

$$f \mid \tilde{L}_v\colon \tilde{L}_v \to \partial D_\varepsilon(0)$$

is not a fibre bundle. (Here, $\tilde{L}_v = f^{-1}(\partial D_\varepsilon(0)) \cap B_\delta(p)$.)

Part III. Complements of Affine Subspaces

Chapter 0. Introduction

In this section we consider the topology of the complement

$$M = \mathbb{R}^n - \bigcup_{i=1}^{m} A_i$$

of a finite collection $\mathscr{A} = \{A_1, A_2, \ldots, A_m\}$ of linear affine subspaces of Euclidean space $\mathbb{R}^n$. (The subspaces can be of arbitrary dimension, and do not necessarily contain the origin.) We give a combinatorial formula for the homology of this complement, which depends only on the partially ordered set $\mathscr{P}$ whose elements v are the intersections of the subspaces (ordered by inclusion), and on the dimensions of these "flats" (see Sect. 1.1, "statement of results"). Complements of *hyperplanes* have received considerable attention during the last ten years ([Ar], [Br], [Ca], [D2], [OS1], [OS2], [Zas]) and a nice survey article on the subject is [Ca]. In the case that the collection $\mathscr{A}$ consists of real hyperplanes, our formula reduces to Zaslavsky's formula [Zas] for the number of connected components of M. If $\mathscr{A}$ is the underlying real arrangement of an arrangement of complex hyperplanes in $\mathbb{C}^{n/2}$ then our formula reduces to the Orlik-Solomon formula ([OS1]). It is interesting that in this case, the complex structure is irrelevant except that it guarantees that each flat is even-dimensional. In particular, our approach clarifies the connection (as observed in [OS1] and [OS4]) between real and complex arrangements in the following corollary:

Corollary. *If $\mathscr{A} \otimes \mathbb{C}$ denotes the corresponding (complexified) arrangement of complex affine subspaces of $\mathbb{C}^n$, then*

$$\sum_{i=0}^{n} b_i\left(\mathbb{R}^n - \bigcup_{i=1}^{m} A_i\right) = \sum_{i=0}^{2n} b_i\left(\mathbb{C}^n - \bigcup_{i=1}^{m} A_i \otimes \mathbb{C}\right)$$

where b_i denotes the ith Betti number.

These Morse theoretic techniques do *not* easily allow one to analyze the product structure on the cohomology of M, although this has been computed for complex arrangements of hyperplanes in [Br], [OS1].

We also solve the following related problem (as suggested to us by R. Thomason): to give a combinatorial formula for the homology of the complement (in $\mathbb{RP}^n$ or $\mathbb{CP}^n$) of a finite collection of linear *projective* subspaces. This is accomplished by considering the corresponding arrangement of linear subspaces of $\mathbb{R}^{n+1}$ (resp. $\mathbb{C}^{n+1}$).

The main idea behind our computation of the homology of M is the following: We consider $\mathbb{R}^n$ to be a "singular space", which is Whitney stratified by the flats $|v|$ for each $v \in \mathscr{P}$. We choose a generic point $p \in \mathbb{R}^n$ and define a Morse function $f(x) = \text{distance}^2(x, p)$. Using Morse theory for nonproper maps (Part I, Sect. 10), we find that f is a *perfect* Morse function (when restricted to the complement M of the flats) which has a unique critical point (a minimum) on each flat. (This means that for each critical value v, the long exact homology sequence for the pair $(M_{\leq v+\varepsilon}, M_{\leq v-\varepsilon})$ splits into short exact sequences.) It is a remarkable fact that this Morse function is perfect for *local* reasons near each critical point, i.e., each relative homology class in $H_i(M_{\leq v+\varepsilon}, M_{\leq v-\varepsilon})$ can be locally completed to an absolute homology class. This local phenomenon has no analogue in classical Morse theory (except for the case of a minimim).

We briefly describe the main steps in the identification of the local homological contribution of each critical point. The local contribution at the critical point x in a flat $|v|$ is $H_*(\ell^+, \partial\ell)$, where ℓ denotes the *halflink* (Part I, Sect. 3.9) in M of the stratum containing the critical point x. Moreover, the space ℓ itself has the structure of the complement of a collection of linear subspace $\mathscr{B} = \{B_1, B_2, \ldots, B_k\}$ of $\mathbb{R}^p$, whose flats are indexed by the partially ordered subset $\mathscr{P}_{>v}$ of all elements bigger than v. (Here $p = n - \dim(|v|) - 1$.) By Poincaré duality we identify $H_i(\ell^+, \partial\ell)$ with $H^{p-i}(\mathbb{R}^p, \bigcup\mathscr{B})$. A geometric argument shows that this pair $(\mathbb{R}^p, \bigcup\mathscr{B})$ is homotopy equivalent to a pair of simplicial complexes $(K(\mathscr{P}_{>v}), K(\mathscr{P}_{(v,T)}))$ which can be constructed combinatorially from the partially ordered set $\mathscr{P}$, and whose homology groups can be defined combinatorially from $\mathscr{P}$.

Chapter 1. Statement of Results

1.1. Notation

(Unless otherwise noted, homology with integer coefficients will be used throughout.) Let $A_1, A_2, \ldots, A_m \subset \mathbb{R}^n$ be a finite set of affine subspaces (of possibly various dimensions) of Euclidean space and let

$$M = \mathbb{R}^n - \bigcup_{i=1}^{m} A_i$$

be the complement of the union of these subspaces. Associated to this collection of subspaces $\mathscr{A}$ there is a partially ordered set $\mathscr{P}$ whose elements v correspond to the "flats",

$$|v| = A_{i_1} \cap A_{i_2} \cap \ldots \cap A_{i_r}$$

partially ordered by *inclusion*, with one maximal element T corresponding to the ambient space $\mathbb{R}^n$. We shall use the notation $v<w$ if v and w are distinct elements of $\mathscr{P}$ such that the flat $|v|$ is contained in the flat $|w|$. We define a "ranking function" d, whose value on the flat v is the dimension

$$d(v) = \dim_{\mathbb{R}}(|v|).$$

Homology and cohomology with *integer* coefficients will be used throughout this chapter.

1.2. The Order Complex

For any partially ordered set $\mathscr{S}$ we may consider ([Fo], [Al]) the *order complex* $K(\mathscr{S})$. This is a simplicial complex with one vertex for every element $v \in \mathscr{S}$ and one k-simplex for every chain $v_0 < v_1 < v_2 < \ldots < v_k$ of elements of $\mathscr{S}$. (The boundary of this k-simplex consists of all corresponding subchains.) (Thus, $K(\mathscr{S})$ is the classifying space of the category whose objects are the elements $v \in \mathscr{S}$ and whose arrows $v \to w$ correspond to the order relations, $v<w$.) Clearly if $\mathscr{S}_1 \subset \mathscr{S}_2$ then $K(\mathscr{S}_1) \subset K(\mathscr{S}_2)$. We remark that $K(\mathscr{S}) = K(\mathscr{S}^*)$, where $\mathscr{S}^*$ denotes the dual partially ordered set, which is obtained from $\mathscr{S}$ by reversing the partial order. If $\mathscr{S}$ has a minimal or a maximal element, then $K(\mathscr{S})$ is contractible. (In fact, it is a cone.) We assume $\mathscr{S}$ has a unique maximal element T.

For any two comparable elements $v<w$ of a partially ordered set $\mathscr{S}$, we define the following subsets:

$$\mathcal{S}_{[v,w]} = \{x \in \mathcal{S} \mid v \le x \le w\}$$
$$\mathcal{S}_{(v,w)} = \{x \in \mathcal{S} \mid v < x < w\}$$
$$\mathcal{S}_{>v} = \{x \in \mathcal{S} \mid x > v\}.$$

We also define a "Möbius function" $\mu\colon \mathcal{S} \to \mathbb{Z}$ recursively, by $\mu(T)=1$, and

$$\mu(x) = -\sum_{v>x} \mu(v).$$

(Our Möbius function is derived from the usual one [Bi], [Ro1], [Ro2] by reversing the partial ordering on $\mathcal{S}$ and evaluating the second variable on T. This reversal will be necessary because we have partially ordered the flats in an arrangement of affine spaces by *inclusion*, rather than by reverse inclusion as in [OS1], [OS2].)

1.3. Theorem A. *The homology of the complement* $M = \mathbb{R}^n - \bigcup_{i=1}^{m} A_i$ *is given by*

$$H_i(M; \mathbb{Z}) \cong \bigoplus_{v \in \mathcal{P}} H^{n-d(v)-i-1}(K(\mathcal{P}_{>v}), K(\mathcal{P}_{(v,T)}); \mathbb{Z})$$

where we make the convention that $H^{-1}(\phi, \phi) = \mathbb{Z}$, *i.e., the top vertex* $v = T$ *contributes a copy of* $\mathbb{Z}$ *to the homology group* $H_0(M)$.

1.4. Corollary. *If* $\mathcal{A} \otimes \mathbb{C}$ *denotes the corresponding (complexified) arrangement of complex affine subspaces of* $\mathbb{C}^n$, *then*

$$\sum_{i=0}^{n} b_i\left(\mathbb{R}^n - \bigcup_{i=1}^{m} A_i\right) = \sum_{i=0}^{2n} b_i\left(\mathbb{C}^n - \bigcup_{i=1}^{m} A_i \otimes \mathbb{C}\right)$$

where b_i *denotes the* i^{th} *Betti number, i.e., the rank of the homology group in degree i.*

Proof of Corollary. The complexified arrangement $\mathcal{A} \otimes \mathbb{C}$ has the same partially ordered set $\mathcal{P}$ of flats as does the original arrangement $\mathcal{A}$, but has a different ranking function. □

1.5. Remarks

1. Cohomology and relative cohomology of a simplicial complex can be computed using simplicial cochains. This gives an *algorithm* for finding $H_*(M)$ in terms of the partially ordered set $\mathcal{P}$ and the ranking function.

2. Since $K(\mathcal{P}_{\ge v})$ is always contractible, using the long exact cohomology sequence for the pair $(K(\mathcal{P}_{\ge v}), K(\mathcal{P}_{(v,T)}))$, we may rewrite the formula for $H_*(M)$ as follows: let us say that an element $v \in \mathcal{P}$ is semimaximal if there are no elements $w > v$ except for T. Then $H_i(M) \cong \bigoplus_{v \in \mathcal{P}} G_i(v)$ where

$$G_i(v) = \begin{cases} H^{-i}(\text{point}) & \text{if } v = T \\ H^{n-d(v)-i-1}(\text{point}) & \text{if } v \text{ is semimaximal} \\ \tilde{H}^{n-d(v)-i-2}(K(\mathcal{P}_{(v,T)})) & \text{otherwise,} \end{cases}$$

where $\tilde{H}$ denotes reduced cohomology.

3. Using Poincaré duality we can rewrite the formula as follows:

$$H^i(M) \cong \bigoplus_{v \in \mathscr{P}} H_{n-d(v)-i-1}(K(\mathscr{P}_{>v}), K(\mathscr{P}_{(v,T)})).$$

1.6. Theorem B. *Let $\mathscr{A} = \{A_1, A_2, \ldots, A_m\}$ be an arrangement of affine subspaces of $\mathbb{R}^n$, as above. Suppose that there is a positive integer c such that each $A_i \subset \mathbb{R}^n$ has codimension c, and such that each flat $|v| = A_{i_1} \cap A_{i_2} \cap \ldots \cap A_{i_r}$ has a codimension which is a multiple of c. Then the homology of the complement M is torsion-free, and its Poincaré polynomial is given by*

$$1 + \sum_{\substack{v \in \mathscr{P} \\ v \neq T}} t^{f(v)} |\mu(v)|$$

where

$$f(v) = \left(\frac{c-1}{c}\right)(n - d(v))$$

and where $\mu(v)$ is the Moebius function defined above (Sect. 1.2).

Examples. If the A_i are real hyperplanes then $c = 1$, and all the homology of M appears in degree 0. The number of connected components of M is, thus,

$$\sum_{v \in \mathscr{P}} |\mu(v)|$$

a result of Zaslavsky [Zas]. If the A_i are complex hyperplanes in $\mathbb{C}^n = \mathbb{R}^{2n}$, then $c = 2$, each $n - d(v)$ is even, and we regain the formula of Orlik and Solomon [OS1]. If the A_i are quaternionic hyperplanes in $\mathbb{H}^n = \mathbb{R}^{4n}$, then $c = 4$ and all the homology of M lies in degrees $3k$.

1.7. Complements of Real Projective Spaces

Let $\mathscr{A} = \{A_1, A_2, \ldots, A_r\}$ denote a collection of real linear subspaces of real projective space $\mathbb{RP}^n$ and let $\mathscr{P}$ denote the corresponding partially ordered set of flats. For each element $v \in \mathscr{P}$ we denote by $|v|$ the corresponding projective subspace of $\mathbb{RP}^n$, and we define the same dimension function,

$$d(v) = \dim_{\mathbb{R}}(|v|).$$

For each integer u, we denote by $\mathscr{P}^{(u)}$ the partially ordered set obtained from $\mathscr{P}$,

$$\mathscr{P}^{(u)} = \{v \in \mathscr{P} \mid d(v) \geq u\}.$$

Let m be the dimension of the largest flat, $m = \sup_{1 \leq i \leq r} d(A_i)$.

Theorem C. *The homology (with coefficients in $\mathbb{Z}/(2)$) of the complement is given by*

$$H_j(\mathbb{RP}^n - \bigcup \mathscr{A}; \mathbb{Z}/(2)) \cong \bigoplus_{u=0}^{m} H^{n-u-j-1}(K(\mathscr{P}^{(u)}), K(\mathscr{P}^{(u)}_{<T}); \mathbb{Z}/(2))$$
$$\oplus H_j(\mathbb{RP}^{n-m-1}; \mathbb{Z}/(2)).$$

1.8. Complements of Complex Projective Spaces

If $\{A_i\}$ denotes a collection of complex linear subspaces of complex projective space, we shall denote by $\mathscr{P}$ the corresponding partially ordered set of flats. Let $d(v) = \dim_{\mathbb{C}}(|v|)$ and, for each u define

$$\mathscr{P}^{(u)} = \{v \in \mathbb{P} \mid d(v) \geq u\}.$$

Let m be the complex dimension of the largest flat.

Theorem D. *The homology (with $\mathbb{Z}$ coefficients) of the complement is given by*

$$H_j(\mathbb{CP}^n - \bigcup \mathscr{A}) \cong \bigoplus_{u=0}^{m} H^{2n-2u-j}(K(\mathscr{P}^{(i)}), K(\mathscr{P}^{(i)}_{<T})) \oplus H_j(\mathbb{CP}^{n-m-1}).$$

(It is an interesting combinatorial exercise to show that the formula in Theorem C or D reduces to the formula in Theoren A, when the arrangement $\mathscr{A}$ of projective subspaces contains at least one hyperplane.)

Chapter 2. Geometry of the Order Complex

2.1. $\mathscr{S}$-filtered Stratified Spaces

Suppose $\mathscr{S}$ is a partially ordered set with a unique maximal element T.

Definition. An $\mathscr{S}$-filtration of a topological space X is a collection of closed subsets $\{X_{\leq v}\}_{v\in\mathscr{S}}$ such that $X_{\leq T}=X$ and $v<w \Rightarrow X_{\leq v}\subset X_{\leq w}$. An $\mathscr{S}$-filtered map $f: X\to Y$ between two $\mathscr{S}$-filtered spaces X and Y is a continuous map such that $f(X_{\leq v})\subset Y_{\leq v}$ for each $v\in\mathscr{S}$. An $\mathscr{S}$-filtered homotopy between two $\mathscr{S}$-filtered maps f_0 and $f_1: X\to Y$ is an $\mathscr{S}$-filtered map $F: X\times[0,1]\to Y$ such that $F(x,0)=f_0(x)$ and $F(x,1)=f_1(x)$. For an $\mathscr{S}$-filtered space X, we define

$$X_{<v}=\bigcup_{w<v} X_{\leq w}.$$

2.2. The Complex $C(\mathscr{A})$

Fix an arrangement $\mathscr{A}$ of affine subspaces of $\mathbb{R}^n$, and let $\mathscr{P}$ denote the partially ordered set of flats of intersections. Let T denote the unique maximal element, i.e., $|T|=\mathbb{R}^n$. The order complex $K(\mathscr{P})$ has a natural $\mathscr{P}$-filtration which is given by $K(\mathscr{P})_{\leq v}=K(\mathscr{P}_{\leq v})\subset K(\mathscr{P})$. The ambient Euclidean space $\mathbb{R}^n$ also has a natural $\mathscr{P}$-filtration which is given by $\mathbb{R}^n_{\leq v}=|v|$. In this section we will construct a $\mathscr{P}$-filtered space $C(\mathscr{A})$, together with two $\mathscr{P}$-filtered homotopy equivalences,

$$\phi: C(\mathscr{A})\to\mathbb{R}^n$$
$$\pi: C(\mathscr{A})\to K(\mathscr{P})$$

each of which restricts to $\mathscr{P}_{\leq v}$-filtered homotopy equivalences,

$$\phi_{\leq v}: C(\mathscr{A})_{\leq v}\to|v|$$
$$\pi_{\leq v}: C(\mathscr{A})_{\leq v}\to K(\mathscr{P}_{\leq v})$$

and to $\mathscr{P}_{<v}$-filtered homotopy equivalences,

$$\phi_{<v}: C(\mathscr{A})_{<v}\to\bigcup_{w<v}|w|$$
$$\pi_{<v}: C(\mathscr{A})_{<v}\to K(\mathscr{P}_{<v}).$$

Definition. $C(\mathscr{A})$ is the subset of $\mathbb{R}^n\times K(\mathscr{P})$ consisting of the union,

$$C(\mathscr{A})=\bigcup_{v\in\mathscr{P}}|v|\times K(\mathscr{P}_{\geq v}).$$

The map $\phi: C(\mathscr{A}) \to \mathbb{R}^n$ and $\pi: C(\mathscr{A}) \to K(\mathscr{P})$ are the projections to the two factors. The filtration of $C(\mathscr{A})$ is given by

$$C(\mathscr{A})_{\leq v} = \pi^{-1}(K(\mathscr{P}_{\leq v})).$$

We also observe

$$C(\mathscr{A})_{<v} = \bigcup_{w<v} C(\mathscr{A})_{\leq w} = \pi^{-1}(K(\mathscr{P}_{<v})).$$

2.3. The Homotopy Equivalences

Proposition. *For any $v \in \mathscr{P}$, the map $\phi: C(\mathscr{A}) \to \mathbb{R}^n$ restricts to surjective maps*

$$\phi_{\leq v}: C(\mathscr{A})_{\leq v} \to |v| \quad \textit{and} \quad \phi_{<v}: C(\mathscr{A})_{<v} \to \bigcup_{w<v} |w|$$

with contractible fibres. The map $\pi: C(\mathscr{A}) \to K(\mathscr{P})$ restricts to maps

$$\pi_{\leq v}: C(\mathscr{A})_{\leq v} \to K(\mathscr{P}_{\leq v}) \quad \textit{and} \quad \pi_{<v}: C(\mathscr{A})_{<v} \to K(\mathscr{P}_{<v})$$

with contractible fibres.

Proof. It is easy to see that $\phi(C(\mathscr{A})_{\leq v}) \subset |v|$. Let us consider the preimage $\phi_{\leq v}^{-1}(p) = \phi^{-1}(p) \cap C(\mathscr{A})_{\leq v}$ of a point $p \in |v|$. Let $w \leq v$ denote the *smallest* element of $\mathscr{P}$ such that $p \in |w|$. Then $\phi^{-1}(p) = \{p\} \times K(\mathscr{P}_{\geq w})$, so

$$\phi^{-1}(p) \cap C(\mathscr{A})_{\leq v} = \{p\} \times K(\mathscr{P}_{[w,v]}).$$

Both of these sets are contractible because the posets $\mathscr{P}_{\geq w}$ and $\mathscr{P}_{[w,v]}$ have a minimal element. Similarly, $\phi_{<v}^{-1}(p) = \{p\} \times K(\mathscr{P}_{[w,v)})$, so it is contractible.

The fibres of π are also contractible: if $x \in K(\mathscr{P})$, then x lies in the *interior* of a unique simplex which corresponds to a chain of elements, $v_0 < v_1 < \ldots < v_r$ of $\mathscr{P}$. It is easy to see that $\pi^{-1}(x) = |v_0| \times \{x\}$, which is contractible. □

Corollary. *The homomorphism*

$$\pi_* \circ \phi_*^{-1}: H_i(\mathbb{R}^n, \bigcup \mathscr{A}) \to H_i(K(\mathscr{P}), K(\mathscr{P}_{<T}))$$

is an isomorphism for each i. □

2.4. Central Arrangements

If all the linear spaces in $\mathscr{A}$ pass through a single point (which we may take to be the origin in $\mathbb{R}^n$) then the arrangement is said to be *central.* In this case, reflections through the origin (i.e., multiplication by -1 in each coordinate axis) is an $\mathscr{P}$-filtration preserving involution $\tau: \mathbb{R}^n \to \mathbb{R}^n$. This gives rise to an involution $\tau: C(\mathscr{A}) \to C(\mathscr{A})$ by operation on the first factor.

Proposition. *If $\mathscr{A}$ is a central arrangement and τ denotes the reflection through the origin, then the map $\phi: C(\mathscr{A}) \to \mathbb{R}^n$ satisfies $\phi\tau = \tau\phi$, while the map $\pi: C(\mathscr{A}) \to K(\mathscr{P})$ satisfies $\pi\tau = \pi$.*

Proof. Obvious. □

2.5. Appendix: The Arrangement Maps to the Order Complex

The contents of this appendix are not needed for the development of our results, but give another, more geometric way to see the relationship between the order complex and the arrangement of affine spaces. However, this approach involves making a choice of control data on the stratification of $\mathbb{R}^n$, which introduces a few technical complications. We will later use the map defined here to give explicit cycle representatives of each of the homology classes in $M = \mathbb{R}^n - \bigcup A_i$.

Let X be a Whitney stratified space and let $\mathcal{S}$ denote the set of strata of X. This set is partially ordered by the closure relations: $V < W \Leftrightarrow V \subset \bar{W}$. If $K(\mathcal{S})$ denotes the order complex of $\mathcal{S}$, we have canonical $\mathcal{S}$-filtrations of both X and $\mathcal{S}$ which are given by:

$$X_{\leq V} = \bar{V}$$
$$(K(\mathcal{S}))_{\leq V} = K(\mathcal{S}_{\leq V}) \subset K(\mathcal{S}).$$

There is a standard construction in stratification theory which gives an $\mathcal{S}$-filtered map $X \to K(\mathcal{S})$ as follows: choose a system of control data (Part I, Sect. 1.5) on X. This consists of a tubular neighborhood $\pi_V: T_V \to V$ for each stratum V of X, and a tubular distance function $\rho_V: T_V \to [0, 1]$ with certain compatibility properties. These neighborhoods may be chosen so that:

(a) $T_V \cap T_W \neq \phi \Leftrightarrow$ either $V \subset \bar{W}$ or $W \subset \bar{V}$
(b) $\rho_V^{-1}(1) = \partial T_V$ is the boundary "sphere" bundle of T_V over V
(c) ρ extends to a continuous map $\bar{T}_V \to [0, 1]$ so that $\rho^{-1}(0) = \bar{V}$.

Definition. The multidistance map $G: X \to K(\mathcal{S})$ is given in barycentric coordinates by

$$\begin{aligned} G(x) = (1 - \rho_1(x))\, V_1 + (\rho_1(x)(1 - \rho_2(x))\, V_2 + \rho_1(x)\, \rho_2(x)(1 - \rho_3(x))\, V_3 + \dots \\ + \rho_1(x)\, \rho_2(x) \dots \rho_{k-1}(x)(1 - \rho_k(x))\, V_k \end{aligned}$$

whenever $V_1 < V_2 < \dots < V_k$ and $x \in T_{V_1} \cap T_{V_2} \cap \dots \cap T_{V_{k-1}} \cap V_k$ (and where we have written ρ_i instead of ρ_{V_i}).

Definition. Let $X = \mathbb{R}^n$ be stratified by the flats consisting of the intersection of a collection $\mathcal{A}$ of affine subspaces, and let $\mathcal{P}$ denote the corresponding partially ordered set. The map $G: \mathbb{R}^n \to K(\mathcal{P})$ is the canonical multidistance map defined above, with respect to some choice of control data on X.

Remarks. Changing the choice of control data will result in an $\mathcal{P}$-filtered homotopic map. In fact, it is possible to give a map $G': \mathbb{R}^n \to K(\mathcal{P})$ which is in the same $\mathcal{P}$-filtered homotopy class of G, without choosing a system of control data at all, by replacing the ρ_i with an appropriate multiple of the distance from the flat $|v_i|$.

If the arrangement was a central arrangement, then the map G can be chosen to be equivariant with respect to the involution which is given by multiplication by -1 (i.e., reflection through the origin).

Theorem. *The map G is an $\mathcal{P}$-filtered homotopy equivalence which, for each $v \in \mathcal{P}$ restricts to a $\mathcal{P}_{\leq v}$-filtered homotopy equivalence,*

$$G: |v| \to K(\mathcal{P}_{\leq v}).$$

Proof. This result is surprisingly difficult to verify, and since it is not needed in our development, we will not give the details. However, a homotopy inverse is given as follows: for each element $v \in \mathscr{P}$, choose a point $p(v)$ in the flat $|v|$. Map the vertices of $K(\mathscr{P})$ (which are in one to one correspondance with these elements $v \in \mathscr{P}$) to the corresponding points that were just chosen. Extend this map linearly over the simplices of $K(\mathscr{P})$. Since the flats are flat, this is an $\mathscr{P}$-filtration preserving map. $\square$

Chapter 3. Morse Theory of $\mathbb{R}^n$

3.1. The Morse Function

Throughout this chapter we will fix an arrangement $\mathscr{A}=\{A_1, A_2, \ldots, A_m\}$ of affine subspaces of $\mathbb{R}^n$, let $\mathscr{P}$ denote the corresponding partially ordered set of flats which are the intersections of the affine spaces, let T denote the unique maximal element of $\mathscr{P}$ corresponding to $\mathbb{R}^n$, and let $K(\mathscr{P})$ denote the order complex of $\mathscr{P}$. We denote by $M=\mathbb{R}^n-\bigcup\mathscr{A}$ the space of interest, i.e., the complement of the affine subspaces in $\mathscr{A}$.

The arrangement $\mathscr{A}$ gives rise to a Whitney stratification of $\mathbb{R}^n$, with one stratum

$$S(v)=|v|-\bigcup_{w<v}|w|$$

for each flat $v\in\mathscr{P}$. Thus, the closure $\overline{S(v)}=|v|$. The strata of $\mathbb{R}^n$ form a $\mathscr{P}$-decomposition of $\mathbb{R}^n$ (see Part I, Sect. 1.1), as well as a $\mathscr{P}$-filtration of $\mathbb{R}^n$.

For almost any point $q\in M\subset\mathbb{R}^n$, the function $f\colon\mathbb{R}^n\to\mathbb{R}$ which is given by

$$f(x)=\text{distance}^2(q, x)$$

is a Morse function on $\mathbb{R}^n$ (with respect to this particular Whitney stratification of $\mathbb{R}^n$) with distinct critical values (see Part I, Sect. 2.2).

3.2. Intuition Behind the Theorem

This Morse function is "perfect" for local reasons, i.e., for each isolated critical value v, the long exact sequence on homology for the pair $(M_{\leq v+\varepsilon}, M_{\leq v-\varepsilon})$ splits into short exact sequences. This is because each critical point is a minimum (so the tangential behavior is perfect) and every homology class in the (relative) homology of the normal Morse data comes from an absolute class in the homology of the link of the critical point: the normal Morse data is homotopy equivalent to the halflink mod its boundary (Part I, Sect. 3.9), $(\ell^+, \partial\ell)$ and since all the subspaces under consideration pass through the critical point p, multiplication by -1 gives an "antipodal" homeomorphism,

$$\tau\colon(\ell^+, \partial\ell)\to(\ell^-, \partial\ell).$$

Each relative cycle $\xi \in H_i(\ell^+, \partial \ell)$ can be glued to the "antipodal" cycle $\pm\tau(\xi)$ to give an absolute cycle,

$$\xi \pm \tau(\xi) \in H_i(L) = H_i(\ell^+ \cup_{\partial \ell} \ell^-).$$

To see that $\xi \pm \tau(\xi)$ is a cycle, we need to use the fact that $\partial_*(\xi) \in H_{i-1}(\partial \ell)$ is invariant (up to sign) under the antipodal map: this comes from the fact that every relative class $\xi \in H_i(\ell^+, \partial \ell)$ is the pullback under a map $\phi: \ell \to K(\mathscr{P})$, which restricts to an antipodally invariant map on the boundary $\partial \ell$.

3.3. Topology Near a Single Critical Point

We fix a single critical point p in some stratum $S(v) \subset |v|$ of the arrangement and set $\alpha = f(p)$. Let N be an affine subspace of $\mathbb{R}^n$ which is complementary to $|v|$, and which meets $|v|$ transversally at the point p. Choose N so that $df(p)(N) \neq 0$. Define $F: \mathbb{R}^n \to \mathbb{R}$ to be the affine function $F(x) = f(p) + df(p)(x - p)$. Choose $0 < \varepsilon \ll \delta \ll 1$ in accordance with Part I, Sect. 3.9, i.e., first choose $\delta > 0$ so small that the closed ball of radius δ, $B_\delta(p)$ intersects only those flats $|w|$ for which $w \geq v$, and so that the boundary $\partial B_\delta(p)$ is transverse to the flats $N \cap |w|$. Then choose $\varepsilon > 0$ so small that $F|(N \cap B_\delta(p))$ has no critical values in the interval $[\alpha - \varepsilon, \alpha + \varepsilon]$, except for the single critical value α. Recall that the *link* L of the point p is the space

$$L = M \cap N \cap \partial B_\delta(p).$$

We shall also be considering the $\mathscr{P}$-filtration of the sphere,

$$\bar{L} = N \cap \partial B_\delta(p)$$

which is given by the intersections with the flats. The upper halflink and its boundary are defined by:

$$(\ell^+, \partial \ell^+) = (B_\delta(p), \partial B_\delta(p)) \cap M \cap N \cap F^{-1}(\alpha + \varepsilon).$$

We shall also consider the closure of the upper halflink,

$$(\bar{\ell}^+, \partial \bar{\ell}^+) = (N \cap B_\delta(p) \cap F^{-1}(\alpha + \varepsilon), N \cap \partial B_\delta(p) \cap F^{-1}(\alpha + \varepsilon))$$

and the "embedded" halflink,

$$(L^+, \partial L^+) = (L \cap F^{-1}[\alpha, \alpha + \varepsilon], L \cap F^{-1}(\alpha))$$
$$(\bar{L}^+, \partial \bar{L}^+) = (\bar{L} \cap F^{-1}[\alpha, \alpha + \varepsilon], \bar{L} \cap F^{-1}(\alpha + \varepsilon)).$$

Recall that in Part I, Sect. 3.10 we found a stratification preserving homeomorphism between $(\bar{\ell}^+, \partial \bar{\ell})$ and $(\bar{L}^+, \partial \bar{L}^+)$ which restricted to a homeomorphism between $(\ell^+, \partial \ell)$ and $(L^+, \partial L^+)$. This is illustrated in the following two diagrams of $M = \mathbb{R}^2$ – three lines through the origin. The origin is treated as an isolated singularity and the upper halflink ℓ^+ is the solid line on the left hand diagram, while the embedded halflink L^+ (i.e., the upper half of the link) is the solid semicircle in the right hand diagram. F is the height function.

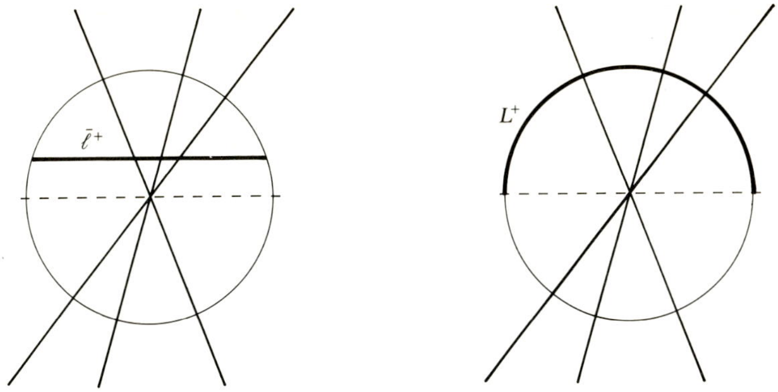

The upper halflink and the embedded halflink

Since N and F are linear, ℓ^+ is itself the complement (in the disk $\bar{\ell}^+$ of dimension $n-d(v)-1$) of an arrangement $\mathscr{B}=\{A_i \cap \bar{\ell}^+\}$ of affine spaces whose partially ordered set of flats is just $\mathscr{P}_{>v}$.

3.4. The Involution

Inversion through the point p is a distance preserving and $\mathscr{P}$-decomposition preserving involution on $F^{-1}(\alpha)$. It exchanges L^+ with L^- and it restricts to an involution $\tau: \partial \bar{L}^+ \to \partial \bar{L}^+$. This gives rise (by means of the above homeomorphism) to a $\mathscr{P}$-decomposition preserving involution $\tau: \partial \bar{\ell}^+ \to \partial \bar{\ell}^+$ which takes $\partial \ell^+$ to $\partial \ell^+$.

Proposition. *Let $\tau_*: H_i(\partial \ell^+) \to H_i(\partial \ell^+)$ denote the map induced on homology by the involution τ, and let $\partial_*: H_i(\ell^+, \partial \ell^+) \to H_{i-1}(\partial \ell^+)$ denote the connecting homomorphism. Then, $\tau_* \partial_* = (-1)^{n-d(v)-1} \partial_*$.*

Proof. First we will show that $\tau^* i^* = i^*$, where i^* is the cohomology restriction to the boundary,

$$i^*: H^{n-d(v)-i-1}(\bar{\ell}^+, \bigcup \mathscr{B}) \to H^{n-i-d(v)-1}(\partial \bar{\ell}^+, \bigcup \mathscr{B} \cap \partial \bar{\ell}^+).$$

Recall from Sect. 2 that there are two $\mathscr{P}_{>v}$-filtration preserving homotopy equivalences,

$$K(\mathscr{B}) \xleftarrow{\ \pi\ } C(\mathscr{B}) \xrightarrow{\ \phi\ } \bar{\ell}^+$$

where $K(\mathscr{B})$ is the order complex of $\mathscr{B}$. (We are making a slight abuse of notation, because $\bar{\ell}^+$ is a disk which is filtered by flat subdisks, rather than a Euclidean space which is filtered by affine subspaces.) Define $\partial C(\mathscr{B})$ to be $\phi^{-1}(\partial \bar{\ell}^+)$ and consider the following diagram:

$$\begin{array}{ccccc}
K(\mathscr{B}) & \xleftarrow{\ \pi\ } & C(\mathscr{B}) & \xrightarrow{\ \phi\ } & \bar{\ell}^+ \\
\Big\| & & \uparrow{\scriptstyle i} & & \uparrow{\scriptstyle i} \\
K(\mathscr{B}) & \xleftarrow{\ \pi_1\ } & \partial C(\mathscr{B}) & \xrightarrow{\ \phi\ } & \partial \bar{\ell}^+
\end{array}$$

Note that $\pi|\partial C(\mathscr{B})$ is a τ-equivariant map, because π collapses each flat to a single point, while τ preserves each flat. Thus, $\tau\pi_1=\pi_1$. For every cohomology class $\xi\in H^{n-i-d(v)-1}(\bar{\ell}^+, \bigcup\mathscr{B})$ there is a class $\xi'\in H^{n-i-d(v)-1}(K(\mathscr{B}), K(\mathscr{B}_{<T}))$ such that $\xi=\phi_*\pi^*(\xi')$, since ϕ and π are $\mathscr{P}$-filtration preserving homotopy equivalences (and where ϕ_* denotes $(\phi^*)^{-1}$). Therefore,

$$\tau^* i^*(\xi)=\tau^* i^* \phi_* \pi^*(\xi')=\phi_* \tau^* \pi_1^*(\xi')=\phi_* \pi_1^*(\xi')=i^* \phi_* \pi^*(\xi')=i^*(\xi).$$

The proposition now follows from Poincaré duality. The sign $(-1)^{n-d(v)-1}$ is important: it is the degree of the antipodal map on the sphere $\partial\bar{\ell}$. It enters in the commutativity of the second square in the following diagram, in which the vertical homomorphisms are the (relative) Poincaré duality isomorphism (i.e., cap product with the fundamental class):

$$\begin{array}{ccccc} H^*(\bar{\ell}^+, \bigcup\mathscr{B}) & \xrightarrow{i^*} & H^*(\partial\bar{\ell}, \bigcup\mathscr{B}) & \xrightarrow{\tau^*} & H^*(\partial\bar{\ell}, \bigcup\mathscr{B}) \\ \downarrow \cap[\bar{\ell}] & & \downarrow \cap[\partial\bar{\ell}] & & \downarrow \cap[\partial\bar{\ell}] \\ H_*(\ell^+, \partial\ell) & \xrightarrow{\partial_*} & H_*(\partial\ell) & \xrightarrow{\tau_*} & H_*(\partial\ell) \end{array}$$

To compute the sign we observe that for any $\xi\in H^*(\partial\bar{\ell}, \bigcup\mathscr{B})$ we have,

$$\tau_*(\xi\cap[\partial\bar{\ell}])=\tau_*(\tau^*\tau^*\xi\cap[\partial\bar{\ell}])=\tau^*\xi\cap\tau_*([\partial\bar{\ell}]=(-1)^{\dim(\partial\bar{\ell})+1}\tau^*(\xi)\cap[\partial\bar{\ell}].$$

3.5. The Morse Function is Perfect

As in Sect. 3.3, we fix a critical point p in some stratum $S(v)\subset|v|$ and let $\alpha=f(p)\in(a,b)\subset\mathbb{R}$ be the corresponding critical value. Suppose the closed interval $[a,b]$ contains no critical values of f other than α.

Theorem. *The homology of the Morse data at* p is given by*:*

$$H_i(M_{\le b}, M_{\le a})\cong H^{n-d(v)-i-1}(K(\mathscr{P}_{>v}), K(\mathscr{P}_{(v,T)})).$$

Furthermore, the long exact sequence for the homology groups of the pair $(M_{\le b}, M_{\le a})$ *breaks into a series of split short exact sequences,*

$$0\to H_i(M_{\le a})\to H_i(M_{\le b})\to H_i(M_{\le b}, M_{\le a})\to 0.$$

Proof. First note that p is a minimum for the function $f|S(v)$, because f is the distance from a point and $\overline{S(v)}=|v|$ is a flat. Thus, the Morse index λ of $f|S(v)$ is 0. By Part I, Sect. 10.8, the space $M_{\le b}$ can be obtained (up to homotopy) from the space $M_{\le a}$ by attaching the pair $(\ell^+, \partial\ell^+)$. (By Part I, Sect. 7.5, the upper halflink depends only on the differential $df(p)$ and is independent of the normal slice which is used in its definition, so it can be obtained from an F-level instead of from an f-level, as in Sect. 3.3 above.) By Poincaré duality and the homotopy equivalence of Sect. 2.3, we have

$$\begin{aligned} H_i(M_{\le b}, M_{\le a})&\cong H_i(\ell^+, \partial\ell^+)\cong H^{n-d(v)-1-i}(\bar{\ell}^+, \textstyle\bigcup\mathscr{B}) \\ &\cong H^{n-d(v)-1-i}(K(\mathscr{P}_{>v}), K(\mathscr{P}_{(v,T)})). \end{aligned}$$

It remains to show that the Morse function is "perfect", i.e., that the long exact sequence for the pair $(M_{\le b}, M_{\le a})$ splits into split short exact sequences. Consider the diagram which consists of the long exact sequences for the pairs $(M_{\le b}, M_{\le a})$ and $(L, \partial L^+)$:

$$
\begin{array}{ccccccccc}
 & & & & H_i(\ell^+, \partial\ell^+) & & & & \\
 & & & & \uparrow\cong & & & & \\
H_i(M_{\le a}) & \longrightarrow & H_i(M_{\le b}) & \xrightarrow{\beta} & H_i(M_{\le b}, M_{\le a}) & \xrightarrow{\partial_*} & H_{i-1}(M_{\le a}) & \longrightarrow & \\
\uparrow & & \uparrow j & & \uparrow \theta & & \uparrow & & \\
H_i(\partial L^+) & \longrightarrow & H_i(L) & \xrightarrow{\beta} & H_i(L, \partial L^+) & \xrightarrow{\partial_*} & H_{i-1}(\partial L^+) & \longrightarrow & \\
 & & & & \uparrow\cong & & & & \\
 & & & & H_i(\ell^+, \partial\ell^+)\oplus H_i(\ell^-, \partial\ell^-) & & & &
\end{array}
$$

The map θ is given by $\theta(x, y) = x$ and it restricts to an isomorphism $H_i(\ell^+, \partial\ell^+) \to H_i(M_{\le b}, M_{\le a})$. The involution τ acts on the bottom row and switches the two groups $H_i(\ell^+, \partial\ell^+)$ and $H_i(\ell^-, \partial\ell^-)$.

Let $\xi \in H_i(M_{\le b}, M_{\le a})$, say $\xi = \theta(\xi' \oplus 0)$. Then $\partial_*(\xi' \oplus (-1)^{n-d(v)}\tau(\xi')) = 0$ by Sect. 3.3 above, so there is a class $\eta \in H_i(L)$ such that $\beta(\eta) = \xi' \oplus (-1)^{n-d(v)}\tau(\xi')$. Consequently, $\beta j(\eta) = \theta\beta(\eta) = \theta(\xi' \oplus (-1)^{n-d(v)}\tau(\xi')) = \xi$, i.e., β is surjective. Careful examination of this argument shows that $\xi \mapsto \xi \oplus (-1)^{n-d(v)}\tau(\xi)$ is actually a splitting of β. □

3.6. Proof of Theorem A

The Morse function $f(x) = \text{distance}^2(x, q)$ has a single critical point on each stratum $S(|v|)$: it has a single critical point on each flat $|v|$ and the assumption that f is a Morse function implies that this critical point does not lie in any smaller flat $|w|$ (where $w < v$). The point q is also a critical point of f (which we associate to the top element $T \in \mathscr{P}$). By Theorem 3.5 (and induction), the homology group $H_i(M) = H_i(\mathbb{R}^n - \bigcup \mathscr{A})$ breaks into a sum (over the critical points of f) of the homology groups $H_i(M_{\le b}, M_{\le a})$ of the Morse data at each critical point. Except for the single critical point q, these groups are identified with certain cohomology groups of the order complex. In summary,

$$H_i(M; \mathbb{Z}) \cong \bigoplus_{v \in \mathscr{P}} H^{n-d(v)-i-1}(K(\mathscr{P}_{>v}), K(\mathscr{P}_{(v,T)}); \mathbb{Z})$$

where we have incorporated the contribution from the critical point q to the homology of M, by associating to the top vertex $T \in \mathscr{P}$ the group $\mathbb{Z} = H^{-1}(\phi, \phi)$ (i.e., the element T contributes a copy of $\mathbb{Z}$ in degree 0). □

3.7. Appendix: Geometric Cycle Representatives

It is possible to give a more direct proof that the Morse function is perfect, and to give geometric cycle representatives of the homology classes, by using the map $G: \mathbb{R}^n \to K(\mathscr{P})$ which was defined in Sect. 2.5.

Using the notation established in Sects. 3.4 and 3.5, we consider the long exact sequence for the pairs $(N^\delta_{\leq b}, N^\delta_{\leq a})$ and (L, L^-), where $N^\delta = N \cap M \cap B_\delta(p)$ is the normal slice through a single critical point p of the Morse function.

$$
\begin{array}{ccccccccc}
\longrightarrow & H_i(M_{\leq a}) & \longrightarrow & H_i(M_{\leq b}) & \longrightarrow & H_i(M_{\leq b}, M_{\leq a}) & \longrightarrow & H_{i-1}(M_{\leq a}) & \longrightarrow \\
& \uparrow & & \uparrow & & \uparrow \cong & & \uparrow & \\
\longrightarrow & H_i(L^-) & \longrightarrow & H_i(L) & \longrightarrow & H_i(L, L^-) & \longrightarrow & H_{i-1}(L^-) & \longrightarrow \\
& & & \cong \uparrow & & \uparrow \cong & & & \\
& & & H^{n-i-d(v)-1}(\bar{L}, \mathscr{B}) & & H_i(\ell^+, \partial \ell^+) & & & \\
& & & G^* \uparrow & & \uparrow \cong & & & \\
& & & \multicolumn{3}{c}{H^{n-i-d(v)-1}(K(\mathscr{B}), K(\mathscr{B}_{<T})) \cong H^{n-i-d(v)-1}(\bar{\ell}^+, \bigcup \mathscr{B})} & & &
\end{array}
$$

Since this diagram commutes, the homomorphism G^* is a splitting of the long exact sequence. Thus, the Morse function is perfect. Furthermore, since $K(\mathscr{B})$ is a simplicial complex, relative cohomology classes can be represented by simplicial "dual cells" ([G2]) in the first barycentric subdivision, $K'(\mathscr{B})$. But a relative cocycle ξ in $C^*(K'(\mathscr{B}), K'(\mathscr{B}_{<T}))$ is one whose support $|\xi|$ does not intersect the subspace $K(\mathscr{B}_{<T})$. Therefore, $G^{-1}(|\xi|)$ is the support of a geometric cycle in $\bar{L}$ which does not intersect $\bigcup \mathscr{B}$. We remark that since the map G is τ-equivariant, these cycles are τ-symmetric.

Chapter 4. Proofs of Theorems B, C, and D

4.1. Geometric Lattice

Recall that a partially ordered set $\mathcal{S}$ is a geometric lattice if

(a) it has a unique minimal elemcnt, $\mathbf{0}$

(b) for each $v \in \mathcal{S}$, all maximally ordered chains $\mathbf{0} = v_0 < v_1 < \ldots < v_p = v$ have the same number of elements, in which case we say the *rank* of v is p and we write $p = r(v)$.

(c) This function satisfies

$$r(v \wedge w) + r(v \vee w) \leq r(v) + r(w)$$

(d) every element is a join of elements of rank 1.

Now let $\mathcal{A} = \{A_1, A_2, \ldots, A_m\}$ be an arrangement of affine subspaces of $\mathbb{R}^n$ and let $\mathcal{P}$ denote the corresponding poset of flats (ordered by inclusion and with a unique maximal element T corresponding to $\mathbb{R}^n$). Suppose that there is a positive integer c such that each A_i has codimension c in $\mathbb{R}^n$ and that each flat $|v|$ has a codimension which is a multiple of c.

Proposition ([OS1]). *For any $v \in \mathcal{P}$, the poset $\mathcal{S} = (\mathcal{P}_{\geq v})^{op}$ is a geometric lattice, where "op" denotes the reverse partial ordering.*

Proof. This poset has a unique minimal element $\mathbf{0} = T$ and a unique maximal element $\mathbf{1} = v$. We set $r(w) = (n - \dim_{\mathbb{R}}(|w|))/c$. The equation

$$\dim(|w_1| + |w_2|) + \dim(|w_1| \cap |w_2|) = \dim(|w_1|) + \dim(|w_2|)$$

becomes

$$r(w_1 \wedge w_2) + r(w_1 \vee w_2) \geq r(w_1) + r(w_2)$$

since $|w_1| + |w_2|$ may not be a flat in the arrangement, but is contained in a smallest flat $|w_1 \wedge w_2|$. The elements of rank 1 are the atoms A_i which contain $|v|$, and every element is clearly a join of atoms. Finally, we observe that if $|w_1| \subset |w_2|$ are flats and if $|w_1| = |w_2| \cap A_j$ for some j, then

$$\operatorname{codim}(|w_1|) \geq \operatorname{codim}(|w_2|) \geq \operatorname{codim}(|w_1|) - c.$$

But we have assumed that both codimensions are multiples of c, so either $|w_2| \subset A_j$ or else

$$\operatorname{codim}(|w_1|) = \operatorname{codim}(|w_2|) + c.$$

Thus, property (b) is also satisfied. □

4.2. Proof of Theorem B

It is a theorem of Folkman [Fo] and Rota [Ro1] that for any geometric lattice $\mathcal{S}$, with unique maximal element **1** and minimal element **0**, and rank $r \geq 2$, the reduced homology of the order complex $K(\mathcal{S}')$ is given by

$$\dim \tilde{H}_p(K(\mathcal{S}')) = \begin{cases} 0 & \text{if } p \neq r-2 \\ |\mu(\mathbf{0}, \mathbf{1})| & \text{if } p = r-2 \end{cases}$$

where $\mathcal{S}' = \mathcal{S} - \{\mathbf{0}, \mathbf{1}\}$ and r is the rank of $\mathcal{S}$, and μ denotes the Möbius function. We apply this to $\mathcal{S} = (\mathcal{P}_{\leq v})^{\text{op}}$ and observe that $K(\mathcal{S}') = K((\mathcal{S}')^{\text{op}}) = K(\mathcal{P}_{(v, T)})$. Theorem B now follows from this identification and Theorem A (or, more directly from Remark 2 of Sect. 1.5 following Theorem A). □

4.3. Complements of Projective Spaces

Consider the situation of Sect. 1.7, i.e., $\mathcal{A} = \{A_1, A_2, \ldots, A_r\}$ denotes a collection of real linear (projective) subspaces of projective space $\mathbb{RP}^n$, $\mathcal{P}$ denotes the corresponding partially ordered set of flats, and to each $v \in \mathcal{P}$ we associate the dimension $d(v) = \dim_{\mathbb{R}}(|v|)$. Let m be the dimension of the largest flat (or flats). Fix a (partial) flag of subspaces

$$V^{n-m} \subset V^{n-m+1} \subset \ldots \subset V^{n+1} = \mathbb{R}^{n+1}$$

(with $\dim_{\mathbb{R}}(V^i) = i$) so that *each* element of the associated flag of projective spaces

$$\mathbb{P}(V^{n-m}) \subset \mathbb{P}(V^{n-m+1}) \subset \ldots \subset \mathbb{P}(\mathbb{R}^{n+1}) = \mathbb{RP}^n$$

is transverse to *each* of the flats in the arrangement. Let $\pi_i\colon V^i - \{0\} \to \mathbb{P}(V^i)$ denote the canonical projection and choose linear functions

$$f_i\colon V^i \to \mathbb{R}$$

(for $n - m + 1 \leq i \leq n+1$) so that $\ker(f_i) = V^{i-1}$. Define the following spaces:

$$\begin{aligned} M &= M_{n+1} = \mathbb{RP}^n - \bigcup \mathcal{A} \\ M_i &= M \cap \mathbb{P}(V^i) \\ \tilde{M}_i &= \pi_i^{-1}(M_i) \subset V_i. \end{aligned}$$

The flats in $\tilde{M}_i$ give a canonical stratification of V_i, which we refine (if necessary) by demanding that the origin be a zero-dimensional stratum. It follows that $f_i\colon V^i \to \mathbb{R}$ is a Morse function for this stratification, and it has a single critical point at the origin.

Lemma 1. *The projection $\pi_i\colon \tilde{M}_i \to M_i$ induces an isomorphism on homology,*

$$H_j((\tilde{M}_i)_{\leq \varepsilon}, (\tilde{M}_i)_{\leq -\varepsilon}) \cong H_j(\ell_i^+, \partial \ell_i^+) \cong H_j(M_i, M_{i-1})$$

where ℓ_i^+ denotes the upper halflink of the stratum $\{0\}$.

Proof of Lemma 1. Let $L_i = \tilde{M}_i \cap B_\delta(0)$ denote the link of the critical point 0. Thinking of f as a Morse function on $\tilde{M}_i \subset V^i$ (with a single critical point of index 0), by Part I, Sects 3.7 and 3.11 the local Morse data for f is equal to the normal Morse data for f, which is

$$(\ell_i^+, \partial\ell_i^+) \cong (L_i^+, \partial L_i^+)$$

where (see Sect. 3.3) $L_i^+ \subset L_i$ is the "embedded halflink",

$$(L_i^+, \partial L_i^+) = (L_i \cap f_i^{-1}[0, \infty), L_i \cap V^{i-1}).$$

However, the map π_i: $L_i \to M_i$ identifies antipodal points, takes ∂L_i^+ to M_{i-1}, and induces a homeomorphism of the quotient spaces, $L_i^+/\partial L_i^+$ and M_i/M_{i-1}. □

Lemma 2. *The long exact homology sequence (with $\mathbb{Z}/(2)$ coefficients) for the pair (M_i, M_{i-1}) splits into short exact sequences,*

$$0 \to H_j(M_{i-1}; \mathbb{Z}/(2)) \to H_j(M_i; \mathbb{Z}/(2)) \to H_j(M_i, M_{i-1}; \mathbb{Z}/(2)) \to 0.$$

Proof of Lemma 2. We wish to show that the boundary homomorphism

$$\partial_*: H_j(M_i, M_{i-1}; \mathbb{Z}/(2)) \to H_{j-1}(M_{i-1}; \mathbb{Z}/(2))$$

is zero. The projection π_i: $\tilde{M}_i \to M_i$ takes ∂L_i^+ to M_{i-1}. Thus, we have a commutative diagram

$$\begin{array}{ccccccccc}
\longrightarrow & H_j(L_i^+) & \longrightarrow & H_j(L_i^+, \partial L_i^+) & \xrightarrow{\partial_*} & H_{j-1}(\partial L_i^+) & \longrightarrow \\
& \downarrow{\scriptstyle \pi_*} & & \cong\downarrow{\scriptstyle \pi_*} & & \downarrow{\scriptstyle \pi_*} & \\
\longrightarrow & H_j(M_i) & \longrightarrow & H_j(M_i, M_{i-1}) & \xrightarrow{\partial_*} & H_{j-1}(M_{i-1}) & \longrightarrow
\end{array}$$

It suffices to show that $\pi_* \partial_*$: $H_j(L_i^+, \partial L_i^+; \mathbb{Z}/(2)) \to H_{j-1}(M_{i-1}; \mathbb{Z}/(2))$ is zero. If $\xi \in H_j(L_i^+, \partial L_i^+)$ then, by Proposition 3.4 we have

$$\tau_* \partial_*(\xi) = (-1)^{i-1} \partial_*(\xi).$$

Since $\pi_* \tau_* = \pi_*$, we have for i even,

$$\pi_* \partial_*(\xi) = \pi_* \tau_* \partial_*(\xi) = -\pi_* \partial_*(\xi)$$

so $\pi_* \partial_*(\xi) = 0$. On the other hand, for i odd, $\partial_*(\xi)$ lies in the $\mathbb{Z}/(2)$-invariant homology of $H_{j-1}(\partial L_i^+)$. It follows from standard arguments that $\pi_* \partial_*(\xi)$ is 0 (mod 2). (For example, using an equivariant triangulation of ∂L_i^+, it is possible to find a chain y so that $\partial\xi = y \pm \tau(y)$ and so that $\tau(\partial y) = -\partial y$. It follows that $\pi_*(\xi) = 2\pi_*(y)$.) □

4.4. Proof of Theorem C

For each i, the subspace $\mathbb{P}(V^i)$ is transverse to the flats in the arrangement $\mathscr{A}$. Therefore, the partially ordered set $\mathscr{P}_{(i)}$ of flats in the induced arrangement $\mathscr{A} \cap \mathbb{P}(V^i)$ is the same as the partially ordered set $\mathscr{P}$ of flats in the arrangement $\mathscr{A}$, except that all elements $v \in \mathscr{P}$ with $d(v) \le n-i$ have been omitted. This partially ordered set $\mathscr{P}_{(i)}$ also coincides with the partially ordered set of flats for the induced arrangement $\pi_i^{-1}(\mathscr{A}\cap V^i)\cap f_i^{-1}(\varepsilon)$. By Lemma 1 and Lemma 2 above, $H_j(M;\mathbb{Z}/(2))$ is a sum,

$$H_j(M;\mathbb{Z}/(2))=\bigoplus_{i=n-m}^{n+1} H_j(M_i, M_{i-1})=\bigoplus_{i=n-m+1}^{n+1} H_j(\ell_i^+,\partial\ell_i^+)\oplus H_j(\mathbb{RP}^{n-m-1})$$

since $M_{n-m}=\mathbb{RP}^{n-m-1}$ and $M_{n-m-1}=\phi$.

The remaining homology groups $H_j(\ell_i^+,\partial\ell_i^+)$ are identified by Poincaré duality with

$$\begin{aligned}H_j(\ell_i^+,\partial\ell_i^+)&=H^{i-j-1}(\bar{\ell}_i^+,\bigcup\mathscr{A}\cap\bar{\ell}_i^+)\\&=H^{i-j-1}(V_i\cap f_i^{-1}(\varepsilon), V_i\cap f_i^{-1}(\varepsilon)\cap\pi_i^{-1}(\bigcup\mathscr{A})).\end{aligned}$$

Which is identified (by the homotopy equivalence if Sect. 2.3) with

$$H^{i-j-1}(K(\mathscr{P}^{(i)}), K(\mathscr{P}^{(i)}_{<T})).$$

To simplify this formula, we define

$$\mathscr{P}^{(u)}=\{v\in\mathscr{P}\mid d(v)\geq u\}$$

and set $i=n-u+1$ to obtain

$$H_j(M)\cong\bigoplus_{u=0}^{m} H^{n-u-j}(K(\mathscr{P}^{(u)}), K(\mathscr{P}^{(u)}_{<T}))\oplus H_j(\mathbb{RP}^{n-m-1}).$$

4.5. Proof of Theorem D

This proof is parallel to Sects. 4.3 and 4.4, so we will only sketch it here. Choose generic *complex* subspaces $V^{n-m}\subset V^{n-m+1}\subset\ldots\subset V^{n+1}=\mathbb{C}^{n+1}$, with $\dim_{\mathbb{C}}(V^i)=i$, and let $\pi_i\colon V^i-\{0\}\to\mathbb{P}_{\mathbb{C}}(V^i)$ denote the projection. Let $M=M_{n+1}=\mathbb{CP}^n-\bigcup\mathscr{A}$, $M_i=M\cap\mathbb{P}_{\mathbb{C}}(V^i)$, and $\tilde{M}_i=\pi_i^{-1}(M_i)$. Choose *complex* linear functions $F_i\colon V^i\to\mathbb{C}$ so that $\ker(F_i)=V^{i-1}$, and let $f_i=\mathrm{Re}(F_i)\colon V^i\to\mathbb{R}$ denote the Morse function on $\tilde{M}_i$.

Lemma 1. *The projection $\pi_i\colon\tilde{M}_i\to M_i$ induces an isomorphism on homology,*

$$H_{j+1}((\tilde{M}_i)_{\leq\varepsilon},(\tilde{M}_i)_{\leq-\varepsilon})\cong H_j(\mathscr{L}_i,\partial\mathscr{L}_i)\cong H_j(M_i, M_{i-1})$$

where $\mathscr{L}_i$ denotes the complex link of the singular point $\{0\}$ in M^i.

Proof of Lemma 1. By Part II, Sect. 3.2 the local Morse data for f is

$$(\ell_i^+,\partial\ell_i^+)=(\mathscr{L}_i,\partial\mathscr{L}_i)\times(I,\partial I)$$

where I denotes the unit interval. Furthermore, by Part II, Sect. 2.3.5, the pair $(\mathscr{L}_i,\partial\mathscr{L}_i)$ can be realized as the pair

$$(\mathscr{L}_i,\partial\mathscr{L}_i)=\tilde{M}_i\cap\partial B_\delta(0)\cap(F_i^{-1}(\mathbb{R}^+), F_i^{-1}(0))$$

from which it follows that the quotient space $\mathscr{L}_i/\partial\mathscr{L}_i$ is homeomorphic to the quotient space M_i/M_{i-1}. $\square$

Lemma 2. *The long exact homology sequence (with $\mathbb{Z}$ coefficients) for the pair (M_i, M_{i-1}) splits into short exact sequences.*

Proof of Lemma 2. This follows from the same method as Lemma 2 of Sect. 4.3, but we replace L_i by $\mathscr{L}_i$ in the commutative diagram, observing that there is still an involution τ on $\partial\mathscr{L}_i$, that $\pi\tau=\pi$, and that $\dim_{\mathbb{R}}(\partial\mathscr{L}_i)$ is odd. □

Summary. By the method of Sect. 4.4, we have a direct sum decomposition

$$H_j(M)=\bigoplus_{i=n-m}^{n+1} H_j(M_i, M_{i-1})\cong \bigoplus_{i=n-m-1}^{n+1} H_j(\mathscr{L}_i, \partial\mathscr{L}_i)\oplus H_j(\mathbb{CP}^{n-m-1}).$$

By Poincaré duality and the homotopy equivalence of Sect. 2.3, we have

$$H_j(\mathscr{L}_i, \partial\mathscr{L}_i)=H^{2i-2-j}(\bar{\mathscr{L}}_i, \bar{\mathscr{L}}_i\cap\bigcup\mathscr{A})=H^{2i-2-j}(K(\mathscr{P}_{(i)}), K(\mathscr{P}_{(i)})_{<T})$$

where $\mathscr{P}_{(i)}=\{v\in\mathscr{P}\mid \dim_{\mathbb{C}}(|v|)>n-i\}$. As in Sect. 4.4, define

$$\mathscr{P}^{(u)}=\{v\in\mathscr{P}\mid \dim_{\mathbb{C}}(|v|)\geq u\}.$$

Then setting $i=n-u+1$ gives

$$H_j(M)=\bigoplus_{u=0}^{m} H^{2n-2u-j}(K(\mathscr{P}^{(u)}), K(\mathscr{P}^{(u)}_{<T}))\oplus H_j(\mathbb{CP}^{n-m-1}).\quad\square$$

Chapter 5. Examples

5.1. The Local Contribution May Occur in Several Dimensions

Consider the central arrangement $\mathscr{A}$ in $\mathbb{R}^3$ which consists of the coordinate hyperplanes and a skew line ℓ through the origin, as illustrated in the following picture.

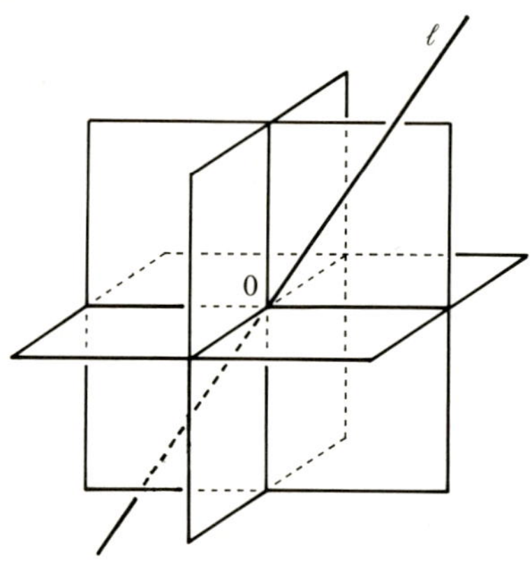

Let $M = \mathbb{R}^3 - \bigcup \mathscr{A}$. The partially ordered set of flats is the following:

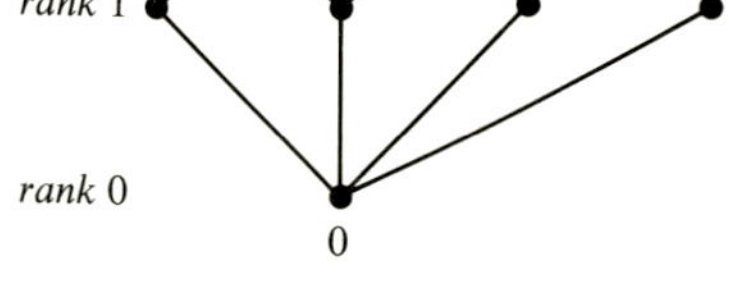

It consists of one vertex T (of rank 3) representing $\mathbb{R}^3$, three vertices (of rank 2) representing the coordinate planes, four vertices (of rank 1) representing the three coordinate axes and the skew line ℓ, and a single vertex of rank 0

representing the origin. Seven of the elements in this partially ordered set contribute a single copy of $\mathbb{Z}$ to the homology group $H_0(M)$ of degree 0. The flat corresponding to ℓ contributes a single copy of $\mathbb{Z}$ to the homology group $H_1(M)$. However, the flat corresponding to the origin contributes copies of $\mathbb{Z}$ to both $H_0(M)$ and $H_1(M)$.

5.2. On the Difference Between Real and Complex Arrangements

If $\mathscr{A}$ is an arrangement of complex subspaces of $\mathbb{C}^n$, then the homology of $\mathbb{C}^n - \bigcup \mathscr{A}$ may be computed by applying our formula for the real arrangement, which is obtained from $\mathscr{A}$ by forgetting the complex structure. The partially ordered set $\mathscr{P}$ of flats corresponding to this real arrangement will necessarily have a ranking function d which takes only even values. On the other hand, not every real arrangement of linear spaces with even-dimensional flats can be realized as the underlying set of a complex arrangement, as the following example shows: consider Pappus' configuration of complex lines in $\mathbb{C}^2$, i.e., choose 2 complex lines P and Q in $\mathbb{C}^2$. Then, choose three points (p_1, p_2, p_3) on P and three points (q_1, q_2, q_3) on Q. Join p_i with q_j for each $i \neq j$ and let I_1, I_2, and I_3 denote the points of intersection:

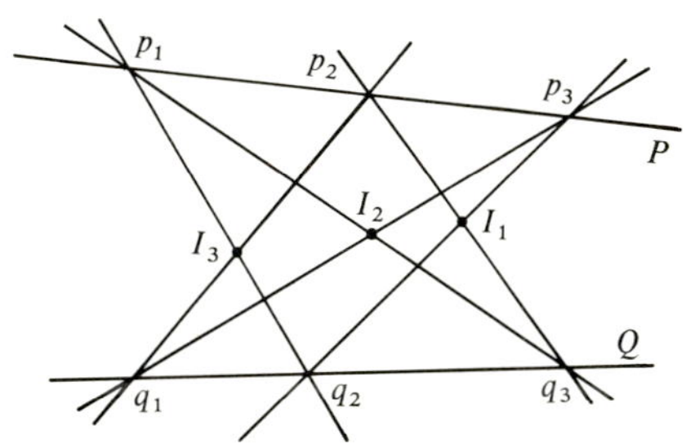

Recall that Pappus' theorem says that the intersection points I_1, I_2, and I_3 lie on a *complex* line, i.e., any complex line through I_1 and I_2 must also contain the point I_3. However, unless these points are (real)-collinear, there will exist many real two-dimensional planes which contain I_1 and I_2, but not I_3. So if we include such a plane in the configuration, we will have an arrangement of (real) even-dimensional flats which cannot be realized as the underlying set of an arrangement of complex spaces. (It is easy to guarantee that the intersection points I_1, I_2, and I_3 are not real-collinear. For example, take the two complex lines P and Q to be parallel, take the points p_i and q_i to correspond under parallel translation, but choose them so that p_1, p_2, and p_3 are not collinear within P.)

It would be interesting to find a characterization of those partially ordered sets with even valued ranking functions which arise from complex arrangements.

Bibliography

[Ab] Abraham R.: Transversality in manifolds of mappings. Bull. Amer. Math. Soc. *69* (1963), 470–474

[AR] Abraham R., Robbins H.: *Transversal Mappings and Flows.* W.A. Benjamin (N.Y.) 1967

[Al] Alexandrov P.S.: Diskrete Räume. Mat. Sb. (N.S.) *2* (1937), 501–518

[Ax] Alexander J.W.: A proof of the invariance of certain constants of analysis situs. Proc. Amer. Math. Soc. *16* (1915), 148–154

[AF1] Andreotti A., Frankel T.: The Lefschetz theorem on hyperplane sections. Annals of Mathematics *69* (1959), 713–717

[AF2] Andreotti A., Frankel T.: The second Lefschetz theorem on hyperplane sections. *Global Analysis*, papers in honor of K. Kodaira. D.C. Spencer and S. Iyanaga ed., Princeton University Press 1969

[AG] Andreotti A., Grauert H.: Théorème de finitude pour la cohomologie des espaces complexes. Bull. Soc. Math. France. *90* (1962), 193–259

[Ar] Arnold V.I.: The cohomology ring of the colored braid group. Mat. Zametki *5* (1969), 227–231; Math. Notes *5* (1969), 138–140

[Art] Artin M.: Théorème de finitude pour un morphisme propre. EGA4, expose XIV. Springer Lecture Notes in Mathematics, vol. 305, Springer-Verlag, N.Y. 1973

[Ba] Baclawski K.: Whitney numbers of geometric lattices. Advances in Math. *16* (1975), 125–138

[BaM] Baiada E., Morse M.: Homotopy and homology related to the Schoenflies problem. Ann. of Math. *58* (1953), 142–165

[Bar1] Barth W.: Transplanting cohomology classes in complex projective spaces. Amer. J. Math. *92* (1970), 951–967

[BL] Barth W., Larsen M.E.: On the homotopy types of complex projective manifolds. Math. Scand. *30* (1972), 88–94

[BFM] Baum P., Fulton W., MacPherson R: Riemann-Roch for singular varieties. Publ. Math. I.H.E.S. *45* (1976), 101–167

[BBD] Beilinson A., Bernstein J., Deligne P.: *Faisceaux Pervers.* Astérisque #100, Soc. Math. de France 1983

[Be] Benedetti R.: Density of Morse functions on a complex space. Math. Ann. *229* (1977), 135–139

[Ber] Bertini E.: Introduzione alla geometrica proiettiva degli iperspazi, Enrico Spoerri, Pisa 1977

[Bet] Betti: Sopra gli spazi di un numero qualunque di dimensioni. Annali di Mat. II, t. 4 (1871)

[Bi] Birkhoff G.: *Lattice Theory.* (3rd ed.) Amer. Math. Soc. Colloq. Publ. vol. 25. Amer. Math. Soc., Providence R.I. 1967

[Bir] Birkhoff G.D.: Dynamical systems with two degrees of freedom. Trans. Amer. Math. Soc. *18* (1917), 199–300

[BM] Blakers A., Massey W.S.: The homotopy groups of a triad, II. Ann. of Math. *55* (1952), 192–201

[Boa] Boardman J.: Singularities of differentiable maps. Publ. Math. I.H.E.S. *37* (1967), 21–57

[Bo1] Bott R.: On a theorem of Lefschetz. Mich. Math. J. *6* (1959), 211–216

[Bo2] Bott R.: Marston Morse and his mathematical works. Bull. Amer. Soc. *3* (1980), 907–745

[Bo3] Bott R.: An application of the Morse theory to the theory of Lie groups. Bull. Soc. Math. de France *84* (1956), 251–281

[Bo4] Bott R.: Lectures on Morse theory, old and new. Bull. Amer. Math. Soc. 7 (1982), 331–358

[Bo5] Bott R.: Nondegenerate critical manifolds. Ann. of Math. *60* (1954), 248–261

[Bo6] Bott R.: Periodicity theorem for the classical groups. Ann. of Math. *70* (1959), 179–203

[BS] Brasselet J.P., Schwartz M.H.: Sur les classes de Chern d'un ensemble analytique complèxe. Soc. Math. de France Astérisque *82–83* (1981), 93–147

[Br] Brieskorn E.: Sur les groupes de tresses (d'apres V.I. Arnold), Séminaire Bourbaki 1971/72, Springer Lecture Notes in Mathematics, vol. 317, Springer-Verlag, N.Y. 1973

[Bru] Bruce J.W.: Generic functions on semi algebraic sets. To appear in the Quarterly Journal of Mathematics, Oxford

[BC] Bruhat F., Cartan H.: Sur la structure des sous-ensembles analytiques réels. C.R. Acad. Sci. Paris *244* (1957) ser. A, 988–990

[BW] Bruhat F., Whitney H.: Quelques propriétés fondamentales des ensembles analytiques reals. Comm. Math. Helv. *33* (1959), 132–160

[Cai1] Cairns S.S.: Bull. Amer. Soc. *41* (1935), 529

[Cai2] Cairns S.S.: Ann. of Math. (2) *35* (1934), 579–587

[C1] Cartan H.: Variétés analytiques complèxes et cohomologie, Colloque sur fonctions de plusiers variables, Bruxelles (1953), 41–55

[C2] Cartan H.: Séminaire École Normale Sup., Paris 1951–52

[Ca] Cartier P.: Arrangements d'hyperplans: un chapitre de géometrie combinatoire. Séminaire Bourbaki (1980/81), exp. 561

[CGM] Cheeger J., Goresky M., MacPherson R.: L^2 cohomology and intersection homology of singular algebraic verieties. in *Seminar on differential geometry*, S.T. Yau (ed). Princeton University Press, Princeton N.J. 1982

[Ch] Cheniot D: Sur les sections transversales d'un ensemble stratifié. C. R. Acad. Sci. Paris. ser. A *275* (1972), A915–A916

[Ch2] Cheniot D: Une démonstration du théorème de Zariski sur les sections hyperplanes d'une hypersurface projective. Comp. Math. *27* (1973), 141–158

[Cl] Clemens H.: Degeneration of Kaehler manifolds. Duke Math. J. *44* (1977), 215–290

[Co] Conley C.: On a generalization of the Morse index. *Ordinary Differential Equations*. Academic Press N.Y. 1972, 27–33

[D1] Deligne P.: le groupe fondamental du complement d'une courbe plane. Seminaire Bourbaki #543 (Nov. 1979), Springer Lecture Notes in Mathematics, vol. 842. Springer-Verlag, N.Y. (1981)

[D2] Deligne P.: Les immeubles des groupes de tresses généralisés. Inv. Math. *17* (1972), 273–302

[D3] Deligne P.: Groupes de monodromie en géometrie algébrique. S.G.A. 7 (vol. 2). Lecture Notes in Mathematics, vol. 340, Springer-Verlag, N.Y., 82–164

[D4] Deligne P.: La conjecture de Weil II. Publ. Math. I.H.E.S., *52* (1980), 137–252

[DS] Denkowska Z., Stasica J.: Ensembles sous-analytiques à la Polonaises. To appear in Astérisque (Soc. Math. de France)

[DSW] Denkowska Z., Stasica J., Wachta K.: Stratification sous-analytiques avec les conditions de Whitney. Acta Math. Univ. Jagellonian, Poland (1984)

[Db] Dubson A.: Classes caractéristiques des variétés singulières. C. R. Acad. Sci. Paris ser. A. *287* (1978), 237–239

[Db2] Dubson A.: Calcul des invariants numériques des singularités et applications. Sonderforschungsbereich *40*, Univ. of Bonn 1981

[DuP] DuPlessis A.: On the genericity of topologically finitely-determined map-germs. Topology *21* (1982), 131–156

[Du] Durfee A.: Neighborhoods of algebraic sets. Trans. Amer. Math. Soc. *276* (1983), 517–530

[E] Eilenberg S.: Singular homology theory. Ann. of Math. (2) *45* (1944), 407–447

[Fa] Faltings G.: Formale Geometrie and homogene Räume. Inv. Math. *64* (1981), 123–165

[Fe1] Feldman E.A.: Geometry of immersions I. Trans. Amer. Math. Soc. *120* (1965), 185–224

[Fe2] Feldman E.A.: Geometry of immersions II. Trans. Amer. Math. Soc. *125* (1966), 181–215

[FK1] Fieseler K.H., Kaup L.: Singular Poincaré duality and intersection homology, in *Conference on Algebraic Geometry*. (J.B. Carrell, ed.) Proc. Conf. Can. Math. Soc. vol. 6, Amer. Math. Soc., Providence R.I. 1986

[FK2] Fieseler K.H., Kaup L.: Theorems of Lefschetz type in intersection homology, preprint

[FK3] Fieseler K.H., Kaup L.: Quasi-isomorphic perversities and obstruction theory for pseudomanifolds, preprint

[FK4] Fieseler K.H., Kaup L.: The vanishing invariants in intersection homology of complex spaces, preprint

[FK5] Fieseler K.H., Kaup L.: Vanishing theorems for the intersection homology of Stein spaces, to appear

[FK6] Fieseler K.H., Kaup L.: Intersection cohomology of q-complete spaces, preprint

[Fo] Folkman J.: The homology groups of a lattice. J. Math. and Mech. *15* (1966), 631–636

[Fox] Fox R.: Covering spaces with singularities. In: *Algebraic Geometry and Topology: a Symposium in Honor of S. Lefschetz.* Princeton University Press, Princeton N.J. 1957

[Fr1] Fritzsche K.: q-konvexe Restmengen in kompakten komplexen Mannigfaltigkeiten. Math. Ann. *221* (1976), 251–273

[Fr2] Fritzsche K.: Geometry and cohomology of certain domains in complex projective space. J. Reine Angew. Math. *323* (1981), 1–52

[Fu] Fulton W.: *Intersection Theory.* Springer-Verlag, N.Y. 1984

[FH] Fulton W., Hansen H.: A connectedness theorem for projective varieties, with applications to intersections and singularities of mappings. Ann. of Math. *110* (1979), 159–166

[FL1] Fulton W., Lazarsfeld R.: Connectivity and its applications in algebraic geometry. In: Proceedings of the first midwest algebraic geometry conference. Springer Lecture Notes in Mathematics, vol. 862, Springer-Verlag, N.Y. (1981), 26–92

[FL2] Fulton W., Lazarsfeld R.: On the connectedness of degeneracy loci and special divisors. Acta Math. *146* (1981), 271–283

[FM1] Fulton W., MacPherson R.: *A Categorical Framework for the Study of Singular Spaces.* Memoirs of the American Math. Soc. vol. 243. A.M.S., Providence R.I. 1981

[GK] Gerstner O., Kaup L.: Homotopiegruppen von Hyperflächenschnitten. Math. Ann. *204* (1973), 105–130

[Gi] Gibson, George C, Wirtmuller, Looijenga, duPlessis: Topological stability of smooth maps. Springer Lecture Notes in Mathematics, vol. 555, Springer-Verlag, N.Y. 1976

[Gin] Ginsburg V.: Characteristic varieties and vanishing cycles. Inv. Math. *84* (1986), 327–402

[Go] Goldstein N.: Ampleness and connectedness in complex G/P. Trans. Amer. Math. Soc. *274* (1982), 361–373

[GG] Golubitsky, Guillemin V.: *Stable Mappings and Their Singularities.* Graduate Texts in Mathematics, vol. 14, Springer-Verlag, N.Y. 1973

[G1] Goresky M.: Triangulation of stratified objects. Proc. Amer. Math. Soc. *72* (1978), 193–200

[G2] Goresky M.: Whitney stratified chains and cochains. Trans. Amer. Math. Soc. *261* (1981), 175–196

[GM1] Goresky M., MacPherson R.: Stratified Morse Theory. Proceedings of Symposia in Pure Mathematics *40*. Amer. Math. Soc., Providence R.I. 1983, 517–533

[GM2] Goresky M., MacPherson R.: *Stratified Morse Theory.* Springer-Verlag N.Y. 1987

[GM3] Goresky M., MacPherson R.: Morse theory and intersection homology. *Analyse et Topologie sur les Espaces Singuliers.* Astérisque *101*, Soc. Math. France (1983), 135–192

[GM4] Goresky M., MacPherson R.: Intersection homology theory. Topology *19* (1980), 135–162

[GM5] Goresky M., MacPherson R.: Intersection homology II. Inv. Math. *71* (1983), 77–129

[Gra] Grauert H., Riemenschneider O.: Kählersche Mannigfaltigkeiten mit hyper-q-konvexem Rand. In: *Problems in Analysis: A Symposium in Honor of Salomon Bochner.* Princeton University Press, Princeton N.J. 1970

[Gr1] Griffiths P.: Hermitian differential geometry, Chern classes, and positive vector bundles. In: Global Analysis: Papers in honor of K. Kodaira. Princeton University Press, Princeton, N.J. 1969, 185–251

[Gro] Grothendieck A.: Cohomologie locale des faisceaux cohérents et théorèmes de Lefschetz locaux et globaux. S.G.A.2, Masson (Paris) and North-Holland (Amsterdam) 1968

[GuR] Gunning R.C., Rossi H.: *Analytic Functions of Several Complex Variables.* Prentice-Hall, New Jersey 1965

[Hae1] Haefliger A.: Les singularités des applications differentiables. Sem. H. Cartan 1956/57 exp. #7

[H1] Hamm H.: Lokale topologische Eigenschaften komplexer Räume. Math. Ann. *191* (1971), 235–252

[H2] Hamm H.: On the vanishing of local homotopy groups for isolated singularities of complex spaccs. J. Reinc Angew. Matl. *323* (1981), 172–176

[H3] Hamm H.: Zum Homotopietyp Steinscher Räume. J. Reine Angew. Math. *338* (1983), 121–135

[H4] Hamm H.: Lefschetz theorems for singular varieties, *Singularities*. Proc. Symp. Pure Math. vol. 40, Amer. Math. Soc. (Providence R.I.) 1983

[H5] Hamm H.: Zum Homotopietyp q-vollständiger Räume. J. Reine Angew. Math. *364* (1986), 1–9

[H6] Hamm H.: Morse theory on singular spaces and Lefschetz theorems. To appear in Proceedings of Semester on Singularities, Warsaw 1985

[HL1] Hamm H., Lê D.T.: Un théorème de Zariski du type de Lefschetz. Ann. Sci. Ecole Norm. Sup. (3) *6* (1973), 317–355

[HL2] Hamm H., Lê D.T.: Lefschetz theorems on quasiprojective varieties. Bull. Soc. Math. de France *113* (1985), 123–142

[HL3] Hamm H., Lê D.T.: Local generalizations of Lefschetz Zariski theorems. Preprint, Univ. of Munster 1986

[Ha1] Hardt R.: Topological properties of subanalytic sets. Trans. Amer. Math. Soc. *211* (1975), 57–70

[Ha2] Hardt R.: Stratifications of real analytic mappings and images. Invent. Math. *28* (1975), 193–208

[Ha3] Hardt R.: Stratification via corank one projections. Proc. Symp. Pure Math. *40* (part 1). Amer. Math. Soc., Providence R.I. (1983), 559–566

[Hr] Hartshorne R.: Ample vectorbundles. Publ. Math. I.H.E.S. *29* (1966), 63–94

[HM] Henry H.P., Merle M.: Limites de normales, condition de Whitney et éclatement d'Hironaka. Proc. Symp. Pure Math. *40* (part 1). Amer. Math. Soc., Providence R.I. 1983, 575–584

[Hi1] Hironaka H.,: Subanalytic sets. *Number Theory, Algebraic Geometry and Commutative Algebra*, volume in honor of A. Akizuki, Kinokunya Tokyo 1973, 453–493

[Hi2] Hironaka H.: Stratification and flatness. *Real and Complex Singularities*, Nordic Summer School (Oslo, 1976). Sijthoff-Noordhoff, Groningen 1977

[Hi3] Hironaka H.: Introduction to real analytic sets and real analytic maps. Lecture Notes of Ist. Math. "Leonida Tonelli" Pisa 1973

[Hi4] Hironaka H.: Normal cones in analytic Whitney stratifications. Publ. Math. I.H.E.S. *36* (1969), 127–138

[Hö] Hörmander L.: Fourier integral operators I. Acta. Math. *127* (1971), 79–183

[J] Johnson F.: On the triangulation of stratified sets and singular varieties. Trans. Amer. Math Soc. *275* (1983), 333–343

[KT] Kambouchner A., Trotman D.: Whitney (a)-faults hard to detect. Ann. Scien. E. N. S. Paris *12* (1979), 465–471

[Kr1] Karchyauskas K.K.: A generalized Lefschetz theorem. Funct. Anal. Appl. *11* (1977), 312–313 (Funktional. Anal. i. Priložen. *11* (1977) no. 4, 80–81)

[Kr2] Karchyauskas K.K.: Homotopy properties of complex algebraic sets. Studies in Topology, Steklov Institute (Leningrad) 1979

[Kr3] Karchyauskas K.K.: Homotopy properties of complex algebraic sets. Thesis 1979

[K1] Kashiwara M.: Index theorem for constructible sheaves. In: *Systemes Differentiales et Singularities*. Astérisque *130*. Soc Math. de France, Paris (1985), 193–209

[K2] Kashiwara M.: Vanishing cycle sheaves and holonomic systems of differential equations. In: *Algebraic Geometry* (Tokyo/Kyoto, 1986) (M. Raynaud and T. Shioda, ed.). Springer Lecture Notes in Mathematics vol. 1016, pp. 134–142. Springer-Verlag, N.Y. 1986

[KS] Kashiwara M., Schapira P.: *Microlocal Study of Sheaves*. Astérisque *128*, Soc. Math. de France, Paris 1985

[KT1] Kato M.: Topology of k-regular spaces and algebraic sets. *Manifolds-Tokyo (1973)*. University of Tokyo press 1975, 153–159

[KT2] Kato M.: Partial Poincaré duality for k-regular spaces and complex algebraic sets. Topology *16* (1977), 33–50

[Kp1] Kaup L.: Zum Satz von Lefschetz über Hyperebenenschnitte. Verlag der Bayerischen Akademie der Wissenschaften, München 1967

[Kp2] Kaup L.: Poincaré-Dualität für Räume mit Normalisierung. Annalie della Scuola Normale Superiore di Pisa 26 (1972), 1–31

[Kp3] Kaup L.: Exakte Sequenzen für globale und lokale Poincaré-Homomorphismen. *Real and*

Complex Singularities, Oslo 1976. Proceedings of Nordic Summer School Symposium in Mathematics. Sijthoof and Noordhoff Publ. Oslo 1978, 267–296

[Kp4] Kaup L.: Zur Homologie projektiv algebraischer Varietäten. Ann. Sc. Norm. Sup. Pisa, XXVI, Fasc. I (1972)

[Kp5] Kaup L.: Eine topologische Eigenschaft Steinscher Räume. Nachrichten der Akademie der Wissenschaften, Göttingen 1966, Nr. 8

[KW] Kaup L., Weidner H.G.: Mayer-Vietoris-Sequenzen und Lefschetzsätze für mehrfache Hyperflächenschnitte in der Homotopie. Math. Zeit. *142* (1975), 243–269

[Ki1] King H.: Topological type of isolated critical points. Ann. Math. *107* (1978), 385–397

[Ki2] King H.: Topological type in families of germs. Inv. Math. *62* (1980), 1–13

[Kl] Kleiman S.: The transversality of a general translate. Comp. Math. *28* (1974), 287–297

[KB] Koopman, Brown: On the covering of analytic loci by complexes. Trans. Amer. Math. Soc. *34* (1932), 231–251

[Ku] Kuo T.C.: The ratio test for analytic Whitney stratifications. Proc. of Liverpool Singularities, Symp. I, Springer Lecture Notes in Mathematics vol. 192 (C.T.C. Wall, ed.), Springer-Verlag, N.Y. 1971, 141–149

[Lam] Lamotke K.: Questions about the proof of the hard Lefschetz theorem. Proc. Symp. Pure Math. *40* (1983) I, 663–667

[L] Larsen M.: On the topology of complex projective manifolds. Inv. Math. *19* (1973), 251–260

[La] Lazzeri F.: Morse theory on singular spaces. *Singularities a Cargese*. Astérisque #7–8, Soc. Math. France 1973

[Lê1] Lê D.T.: Sur les cycles évanouissants des espaces analytiques, C.R. Acad. Sci. Paris, Ser. A *288* (1979), 283–285

[Lê2] Lê D.T.: Le concept de singularité insolée de fonction analytique. Preprint, Ecole Polytechnique 1985 to appear in Advances in Mathematics

[Lê3] Lê D.T.: Vanishing cycles on analytic spaces. Kokyuroku series of Res. Inst. Math. Sc., Kyoto University. Proc. of symp. on algebraic analysis 1975

[Lê4] Lê D.T.: Some remarks on relative monodromy. In: *Real and Complex Singularities*, Nordic Summer School (Oslo). Sijthoof and Noordhoff, Groningen 1977

[LK] Lê D.T., Kato: Vanishing cycles on analytic sets. Proc. Symp. on Algebraic Analysis, R.I.M.S., Kyoto, July 1975

[LT1] Lê D.T., Teissier B.: Variétés polaires locales et classes de Chern des variétés singulières. Ann. of Math. *114* (1981), 457–491

[LT2] Lê D.T., Teissier B.: Cycles évanescents et conditions de Whitney II. Proc. Symp. Pure Math. *40* (part 2). Amer. Math. Soc., Providence R.I. 1983, 65–103

[Lef1] Lefschetz S.: A page of Mathematical biography. Bull. Amer. Math. Soc. *74*, 854–879

[Lef2] Lefschetz S.: *Analysis Situs et la Géometrie Algébrique*. Gauthier-Villars, Paris 1924

[Lef3] Lefschetz S.: *Topology*. Amer. Math. Soc. Colloq. Publ., N. Y. 1930

[Lef4] Lefschetz S.: *Algebraic Topology*. Amer. Math. Soc. Solloq. Publ. vol. 27, Amer. Math. Soc., N.Y. 1942

[LW] Lefschetz S., Whitehead J.H.C.: On analytical complexes. Trans. Amer. Math. Soc. *35* (1933), 510–517

[Lev] Levine H., Thom R.: *Singularities of Differentiable Maps*. Univ. of Bonn 1959 (see Math. Reviews *40* #896)

[Lo1] Lojasiewicz S.: Ensembles Semi-analytiqes. Preprint, I.H.E.S. 1972

[Lo2] Lojasiewicz S.: Triangulation of semi-analytic sets. Ann. Scu. Norm. di Pisa (1965), 449–474

[Lo3] Lojasiewicz S.: Sur la problème de la division. Studio Math. *18* (1959), 87–136

[Lo4] Lojasiewicz S.: Rozprawy Mathematyczne *22* (1961)

[LS] Lyusternik L., Schnirelmann L.: *Topological Methods in the Calculus of Variations*. Gosudarstv. Izdat. Techn.-Teor. Lit. Moscow 1930

[MP1] MacPherson R.: Chern classes for singular algebraic varieties. Ann. of Math. *100* (1974), 423–432

[MP2] MacPherson R.: Global questions in the topology of singular spaces. Proceedings of the International Congress of Mathematicians, Warsaw 1983

[MV] MacPherson R., Vilonen K.: Elementary construction or perverse sheaves. Inv. Math. *84* (1986), 403–436

[Ma1] Mather J.: *Notes on Topological Stability*. Mimeographed notes, Harvard University 1970

[Ma2] Mather J.: Stratifications and mappings. *Dynamical Systems* (M.M. Peixoto, ed.), Academic Press, N.Y. 1973

[Ma3] Mather J.: Stability of C^∞ mappings II: Infinitesimal stability implies stability. Ann. Math. *89* (1969), 254–291

[Ma4] Mather J.: Stability of C^∞ mappings III: Finitely determined map germs. Publ. Math. I.H.E.S. *35* (1969), 127–156

[Maz] Mazur B.: Morse theory. In: *Differential and Combinatorial Topology*. S.S. Cairns, ed. Princeton Univ. Press, Princeton, N.J. 1965, 145–165

[Mi1] Milnor J.: *Morse Theory*. Ann. of Math. studies #51, Princeton University Press, Princeton N.J. 1963

[Mi2] Milnor J.: Singular Points of Complex Hypersurfaces. Annals of Math. studies #61, Princeton University Press, Princeton N.J. 1968

[Mor] Morlet C.: Le lemme de Thom et Théorème de plongement de Whitney. Seminaire H. Cartan, 1961/62. exp. 4, 5, 6, 7, 8. Paris 1964

[Mo1] Morse M.: The analysis and analysis situs of a regular n-spread in $(n+r)$ space. Proc. Nat. Acad. Sci. U.S.A. *13* (1927), 813–817

[Mo2] Morse M.: Nondegenerate real-valued differentiable functions. Proc. Nat. Acad. Sci. U.S.A. *57* (1967), 32–38

[Mo3] Morse M.: The existence of nondegenerate functions on a compact differentiable n-manifold. Ann. Mat. Pura. Appl. *49* (1960), 117–128

[Mo4] Morse M.: Relations between the critical points of a real function of n independent variables. Trans. Amer. Math. Soc. *27* (1925), 345–396

[Mo5] Morse M.: The foundations of a theory of the calculus of variations in the large. Trans. Amer. Math. Soc. *30* (1928), 213–274

[Mo6] Morse M.: Mathematics and the arts, Bulletin of the Atomic Scientists *15* (1955), 55–59

[Nak] Nakai I.: On topological types of polynomial mappings. Topology *23* (1984), 45–66

[N] Narasimhan R.: On the homology of Stein spaces. Inv. Math. *2* (1967), 377–385

[NT] Navarro Aznar V., Trotman D.: Whitney regularity and generic wings. Ann. Inst. Fourier Grenoble *31* (1981), 87–111

[Ne] Newstead P.: A note on the homology groups of varieties defined by s equations. Bull. Lon. Math. Soc. *12* (1980), 183–186

[Og] Ogus A.: Local cohomological dimension of algebraic varieties, Ann. of Math. (2) *98* (1973), 327–366

[Ok] Oka M.: On the cohomology structure of projetive varieties. In: *Manifolds – Tokyo*. A. Hattori (ed.). University of Tokyo Press 1975

[Oko1] Okonek C.: Concavity, convexity and complements in complex spaces. Preprint, Univ. of Göttingen 1985

[Oko2] Okonek C.: A comparison theorem. Preprint, Univ. of Göttingen 1985

[Oko3] Okonek C.: Barth-Lefschetz theorems for singular spaces. Preprint, Univ. of Göttingen 1986

[OS1] Orlik P., Solomon L.: Combinatorics and topology of complements of hyperplanes. Inv. Math *56* (1980), 167–189

[OS2] Orlik P., Solomon L.: Unitary reflection groups and cohomology. Inv. Math. *59* (1980), 77–94

[OS3] Orlik P., Solomon L.: Complexes for reflection groups. Proc. Midwest algebraic geometry conference. Springer Lectures Notes in Mathematics, vol. 862, Springer-Verlag, N.Y. 1981, 193–207

[OS4] Orlik P., Solomon L.: Coxeter Arrangements. In: *Singularities*. Proc. of Symp. in Pure Math. vol. 40 (part 2). Amer. Math. Soc., Providence R.I. 1983, 269–291

[O] Orro P.: Conditions de regularité, espaces tangents et fonctions de Morse sur les espaces stratifies. Thesis, University of Paris, Orsay 1984

[OT1] Orro P., Trotman D.: Sur les fibres de Nash de surfaces a singularités isolées. C.R. Acad. Sci. Paris *299*, ser. I (1984), 397–399

[OT2] Orro P., Trotman D.: On regular stratifications and conormal structure of subanalytic sets. Bull. London Math. Soc. *18* (1986), 185–191

[Os] Osgood W.F.: *Lehrbuch der Funktionentheorie II*, Leipzig 1929

[PS] Picard E., Simart G.: *Theorie des Fonctions Algébriques de Deux Variables Independants*, tome 1, Gauthier-Villars, Paris 1897

[P1] Pignoni R.: Density and stability of Morse functions on a stratified space, Ann. Scuola Norm. Sup. Pisa (4) *4* (1979), 592–608

[P2] Pignoni P.: Morse functions and the strict Whitney conditions. Bol. Un. Mat. Ital A(5) *17* (1980), 1023–1028

[Po1] Poincaré H.: L'analysis situs. J. de l'Ecole Polytechnique (Paris) *1* (1895), 1–121. Reprinted in *Œuvres de H. Poincaré* t. VII. Gauthier-Villars (Paris) 1953

[Po2] Poincaré H.: Complement à l'analysis situs. Rendiconti del Circolo Matematico di Palermo *13* (1899), 285–343. Reprinted in *Œuvres de H. Poincaré* t. VII. Gauthier-Villars (Paris) 1953

[Po3] Poincare H.: Sur less courbes definies par les equations differentielles, part 3. Liouville J. (4) *1* (1885), 167–244. Reprinted in *Œuvres de H. Poincaré* tome 1, Gauthier-Villars, Paris 1953

[Ri] Riemann B.: Fragment. *Œuvres Mathématiques de B. Riemann*. A. Blanchard, Paris (1968), 44

[R] Rossi H.: Continuation of subvarieties of projective varieties. Amer. J. Math. *41* (1969), 565–575

[Ro1] Rota G.C.: On the foundations of combinatorial theory I. Theory of Mobius functions. Z. Wahrscheinlichkeitsrechnung *2* (1964), 340–368

[Ro2] Rota G.C.: On the combinatorics of the Euler characteristic. Studies in Pure Math. presented to Richard Rado (L. Mirsky, ed.) p. 221–233. Academic Press, London 1971

[Sab1] Sabbah C.: Géometrie des espaces conormaux. Soc. Math. de France Astérique *130* (1985), 161–192

[Sai] Saito M.: Modules de Hodge polarisables. Preprint, Institute for Advanced Study, Princeton N.J. 1986

[Sa1] Sard A.: The measure of the critical values of differentiable maps. Bull. Amer. Math. Soc. *48* (1942), 883–890

[Sa2] Sard A.: Images of critical sets. Ann. Math. *68* (1958), 247–259

[Sat] Sato K.: Local triangulation of real analytic varieties. Osaka Math. J. *15* (1963), 109–125

[Sc] Schwartz M.H.: Classes caractéristiques definées par une stratification d'une variété complexe. C.R. Acad. Sci. Paris *260* (1965), 3262–3264 and 3535–3537

[Se] Serre J-P.: Quelques problèmes globaux relatifs aux variétés de Stein. Colloque sur les fonctions de plusiers variables, Bruxelles (1953), 57–68 (reprinted in *Œuvres*, vol. 1, p. 259. Springer-Verlag, N.Y. 1986)

[S] Siersma D.: Singularities of functions on boundaries, corners, etc. Quart. J. Math. Oxford Ser. (2) *32* (1981), 119–127

[Sma1] Smale S.: Morse inequalities for a dynamical system. Bull. Amer. Math. Soc. *66* (1960), 43–49

[Sma2] Smale S.: The generalized Poincaré conjecture in higher dimensions. Bull. Amer. Math. Soc. *66* (1960), 373–375

[Sma3] Smale S.: A survey of some recent developments in differential topology. Bull. Amer. Math. Soc. *69* (1963), 131–145

[Sm1] Sommese A.: Concavity theorems. Math. Ann. *235* (1978), 37–53

[Sm2] Sommese A.: Theorems of Barth-Lefschetz type for complex subspaces of homogeneous complex manifolds. Proc. Nat. Acad. Sci. USA *74* (1977), 1332–1333

[Sm3] Sommese A.: Complex subspaces of homogeneous complex manifolds I, Transplanting theorems. Duke Math. J. *46* (1979), 527–548

[Sm4] Sommese A.: Complex subspaces of homogeneous complex manifolds II: Homotopy results. Nagoya Math. J. *86* (1982), 101–129

[Sm5] Sommese A.: A convexity theorem. Proc. Symp. Pure Math. *40* (part 2) 497–505. Amer. Math. Soc., Providence R.I. 1983

[Sm6] Sommese A.: Submanifolds of abelian varieties. Math. Ann. *237* (1978), 229–256

[SV] Sommese A., Van de Ven A.: Homotopy groups of pullbacks of varieties. To appear in Nagoya Math. Journal

[SoV] Soroni G., Villani V.: *q*-complete spaces and cohomology. Trans. Amer. Math. Soc. *125* (1966), 432–448

[St] Steenrod N.: *The Topology of Fibre Bundles.* Princeton University Press, Princeton N.J. 1951

[Ste1] Sternberg S.: *Lectures on Differentiual Geometry.* Prentice Hall (New Jersey) 1964

[Sw] Switzer A.: *Algebraic Topology: Homotopy and Homology.* Springer-Verlag, N.Y. 1975

[Tei] Teissier B.: Varietes polaires II: Multiplicités polaires, sections planes et conditions de Whitney, In: *Algebraic Geometry*, Proceedings, La Rabida 1981. Springer Lecture Notes in Mathematics, vol. 961, Springer-Verlag, N.Y. 1982

[T1] Thom R.: Les singularités des applications differentiables, Ann. Inst. Fourier (Grenoble) *6* (1955–56), 43–87

[T2] Thom R.: Singularities of differentiable Mappings, Proc. of Liverpool Singularities Symposium I, pp. 1–89. Springer Lecture Notes in Mathematics, vol. 192, Springer-Verlag, N.Y. 1971

[T3] Thom R.: Local topological properties of differentiable mappings, *Colloquium on Differential Analysis*, pp. 191–202. Tata Inst. Bombay 1964. Oxford Univ. Press, London and New York 1964

[T4] Thom R.: Propriétés differentiables locales des ensembles analytiques, Seminaire Bourbaki No. 281 (1964–1965)

[T5] Thom R.: Ensembles et Morphismes Stratifiés, Bull. Amer. Math. Soc. *75* (1969), 240–284

[T6] Thom R.: La stabilité topologique des applications polynomiales, Enseign. Math. (2) *8* (1962), 24–33

[T7] Thom R.: Un lemme sur les applications différentiables. Bol. Soc. Mat. Mexicana *1* (1956), 59–71

[T8] Thom R.: Sur une partition en cellules associée à une fonction sur une variété. C.R. Acad. Sci. Paris *228*, ser. A (1949), 973–975

[T9] Thom R.: Unpublished lecture at the Institute for Advanced Study, Princeton, N.J. 1957 (as cited in (AF2])

[T10] Thom R.: Sur l'homologie des variétés algébriques réelles. In: *Differential and Combinatorial Topology* (S. Cairns, ed.), Princeton University Press, Princeton N.J. 1965

[T11] Thom R.: Variétés differentiables cobordantes. C.R. Acad. Sci. Paris *236* (1953), 1733–1735

[T12] Thom R.: Quelques propriétés globales des variétés differentiables. Comm. Math. Helv. *28* (1954), 17–86

[Tr1] Trotman D.: Counterexamles in stratification theory: two discordant horns, Proceedings of Nordic Summer School, Oslo 1976. Sijthoff and Noordhoff, Groningen 1977

[Tr2] Trotman D.: Stability of transversality to a stratification implies Whitney (a)-regularity. Inv. Math. *50* (1979), 273–277

[Tr3] Trotman D.: Comparing regularity conditions on stratifications. Proc. Symp. Pure Math. *40* (part 2) 575–586. Amer. Math. Soc., Providence R.I. 1983

[Tr4] Trotman D.: A transversality property weaker than Whitney (a)-regularity. Bull. London Math. Soc. *8* (1976), 235–228

[VW] van der Waerden B.L.: Topologische Begründung der Kalküls der abzählenden Geometrie. Math. Ann. *102* (1929), 337–362

[Veb] Veblen O.: *Analysis Situs*, Amer. Math. Soc. Colloq. Publ., Amer. Math. Soc., Providence R.I. 1922

[V] Verdier J.L.: Stratifications de Whitney et théorème de Bertini-Sard. Inventiones Math. *36* (1976), 295–312

[Ve1] Verona A.: Integration on Whitney prestratifications. Rev. Roumaine Math. Pures Appl. *17* (1972), 1473–1480

[Ve2] Verona A.: Triangulation of stratified fibre bundles. Manuscripta Math. *30* (1979), 425–445

[Wall1] Wall C.T.C.: Geometric properties of generic differentiable manifolds. *Geometry and Topology.* Lecture Notes in Mathematics vol. 597. Springer-Verlag, N.Y. 1977, 707–774

[Wall2] Wall C.T.C.: Stratified sets: a survey. *Liverpool Singularities Symposium I.* Lecture Notes in Mathematics vol. 192. Springer-Verlag, N.Y. 1971

[Wa] Wallace A.H.: *Homology Theory of Algebraic Varieties.* Pergamon Press 1958

[W1] Whitney H.: Tangents to an analytic variety, Ann. of Math. *81* (1965), 496–549

[W2] Whitney H.: Local properties of analytic varieties. *Differentiable and Combinatorial Topology*, S. Cairns, ed., Princeton University Press, Princeton N.J. 1965, 205–244

[W3] Whitney H.: On singularities of mappings of Euclidean spaces I. Mappings of the plane into the plane, Ann. of Math. *62* (1955), 374–410

[W4] Whitney H.: Complexes of manifolds. Proc. Nat. Acad. Sci. U.S.A. *33* (1947), 10–11

[W5] Whitney H.: Singularities of mappings of Euclidean spaces. Symposium internacional de topología algebráica, Mexico City. UNESCO 1958, 285–301

[W6] Whitney H.: Elementary structure of real algebraic varieties. Ann. of Math. *66* (1957), 545–556

[W7] Whitney H.: Differentiable manifolds. Ann. of Math. *37* (1936), 645–680

[Z1] Zariski O.: On the Poincaré group of a projective hypersurface. Ann. Math. *38* (1937), 131–141

[Z2] Zariski O.: Studies in equisingularity I, II, III. Amer. J. Math. *87* (1965); *90* (1968)

[Z3] Zariski O.: Some open questions in the theory of equisingularity. Bull. Amer. Math. Soc. *77* (1971), 481–491

[Zas] Zaslavsky T.: Facing up to arrangements: face-count formulas for partitions of space by hyperplanes. Mem. Amer. Math. Soc. *154* (1975)

Index

Ergebnisse der Mathematik und ihrer Grenzgebiete, 3. Folge

A Series of Modern Surveys in Mathematics

Volume 1: **A. Fröhlich**

Galois Module Structure of Algebraic Integers

1983. X, 262 pages. ISBN 3-540-11920-5

Volume 2: **W. Fulton**

Intersection Theory

1984. XI, 470 pages. ISBN 3-540-12176-5

Volume 3: **J. C. Jantzen**

Einhüllende Algebren halbeinfacher Lie-Algebren

1983. V, 298 Seiten. ISBN 3-540-12178-1

Volume 4: **W. Barth, C. Peters, A. Van de Ven**

Compact Complex Surfaces

1984. X, 304 pages. ISBN 3-540-12172-2

Volume 5: **K. Strebel**

Quadratic Differentials

1984. 74 figures. XII, 184 pages.
ISBN 3-540-13035-7

Volume 6: **M. J. Beeson**

Foundations of Constructive Mathematics

Metamathematical Studies

1985. XXIII, 466 pages. ISBN 3-540-12173-0

Volume 7: **A. Pinkus**

n-Widths in Approximation Theory

1985. X, 291 pages. ISBN 3-540-13638-X

Volume 8: **R. Mañé**

Ergodic Theory and Differentiable Dynamics

Translated from the Portuguese by Silvio Levy
1987. 32 figures. Approx. 330 pages.
ISBN 3-540-15278-4

Contents: Measure Theory. - Measure-Preserving Maps. - Ergodicity. - Expanding Maps and Anosov Diffeomorphisms. - Entropy. - Bibliography. - List of Notations. - Subject Index.

Volume 9: **M. Gromov**

Partial Differential Relations

1986. IX, 363 pages. ISBN 3-540-12177-3

Volume 10: **A. L. Besse**

Einstein Manifolds

1986. 22 figures. XII, 510 pages.
ISBN 3-540-15279-2

Volume 11: **M. D. Fried, M. Jarden**

Field Arithmetic

1986. XVII, 458 pages. ISBN 3-540-16640-8

Volume 12: **J. Bochnak, M. Coste, M.-F. Roy**

Géométrie algébrique réelle

1987. 44 figures. X, 373 pages.
ISBN 3-540-16951-2

Table des matières: Introduction. - Corps ordonnés, corps réels clos. - Ensembles semi-algébriques. - Variétes algébriques réelles: définitions et exemples. - Algébre réelle. - Le principe de Tarski-Seidenberg comme outil de transfert. - Le 17e problème de Hilbert. - Spectre réel. - Fonctions de Nash. - Stratification. - Places réelles. - Topologie des ensembles algébriques sur un corps réel clos. - Fibrés vectoriels algébriques. - Fonctions polynômiales ou régulières à valeurs dans les sphères. - Modèles algébriques des variétes C^{∞}. - Anneaux de Witt en géométrie algébrique réelle. - Bibliographie. - Index des notations. - Index.

Springer-Verlag
Berlin Heidelberg New York
London Paris Tokyo

Springer